INTERNATIONAL CENTRE FOR MECHANICAL SCIENCES

COURSES AND LECTURES - No. 311

NUMERICAL METHODS
AND
CONSTITUTIVE MODELLING
IN GEOMECHANICS

EDITED BY

C.S. DESAI

UNIVERSITY OF ARIZONA

G. GIODA

UNIVERSITY OF UDINE
AND
POLITECNICO DI MILANO

Springer-Verlag   Wien GmbH

Le spese di stampa di questo volume sono in parte coperte da
contributi del Consiglio Nazionale delle Ricerche.

This volume contains 234 illustrations.

In order to make this volume available as economically and as
rapidly as possible the authors' typescripts have been
reproduced in their original forms. This method unfortunately
has its typographical limitations but it is hoped that they in no
way distract the reader.

ISBN 978-3-211-82215-9   ISBN 978-3-7091-2832-9 (eBook)
DOI 10.1007/978-3-7091-2832-9

PREFACE

For many problems in geomechanics that cannot be solved by using conventional procedures, the modern numerical methods provide a powerful tool for analysis and design. These methods, characterized by the discretization of continua and by the use of algorithms suited for programmable computers, allow solution of complex problems involving nonlinear and time dependent material behaviour, arbitrary geometries, initial or in situ conditions, multi-phase media, different loadings (static, quasi-static, cyclic and dynamic), and such environmetal factors as temperature, fluids and pollutant movement. As a result, numerical methods and associated important topics, such as constitutive modelling for mechanical response of soils, rocks, interfaces and joints, mathematical aspects related to convergence, accuracy, stability of time integration schemes, and computer aided engineering and graphics, have received significant attention of researchers and practicing engineers.

The present volume is based on the lectures presented during the advanced course on Numerical Methods in Geomechanics Including Constitutive Modelling organized by the International Centre for Mechanical Sciences, Udine, Italy during 10-14 July 1989. It presents a review of the current state and of up-to-date developments of a wide range of topics including theoretical aspects, laboratory testing, field verification and practical applications of numerical methods and constitutive models.

In particular, the problems discussed include various numerical methods, theory, testing and verification of constitutive models; implementation of constitutive models in nonlinear computational procedures; verification with respect to laboratory and field observation of boundary value problems; mathematical programming; unconfined seepage flows; dynamic and consolidation analysis of saturated and partially saturated non linear porous media; interpretation of field measurements; time integration for nonlinear and dynamic problems. The material in this volume would be useful to students, researchers and practicing engineers.

We would like to express our appreciation to the authors for presenting their lectures and for preparing the manuscripts for publication, to Professor Giovanni Bianchi, Secretary General of CISM, and to the highly capable staff of CISM for assistance during the course and in the publication of this volume.

Chandrakant S. Desai, Tucson   *Giancarlo Gioda, Udine*

CONTENTS

Page

Preface

CONTENTS

MODELLING AND TESTING:
IMPLEMENTATION OF NUMERICAL MODELS
AND
THEIR APPLICATION IN PRACTICE

C.S. Desai

University of Arizona, Tucson, Arizona, USA

PART 1.   INTRODUCTION AND SCOPE

INTRODUCTION

The complexity of many geomechanical systems necessitates the use of modern numerical methods such as the finite element, boundary element and finite difference procedures. With the almost unlimited power of the current and future computers, these methods can provide extremely powerful tools for analysis and design of engineering systems with complex factors that was not possible or very difficult with the use of the conventional methods, often based on closed-form analytical solutions.

The computational power available will be of limited use unless the mechanical response of materials constituting the engineering systems was properly defined and implemented. There are other factors that also play an important role toward robust, consistent and reliable solutions from computer methods. Constitutive modelling of geologic materials, which are usually more complex than many other engineering materials, assumes a highly vital role in this endeavor.

Appropriate field and/or laboratory tests that simulate various factors such as nonlinearity, nonhomogeneity, insitu stresses, states of stress and strain, stress paths, discontinuities, types of loading and existence of fluid and thermal effects that influence the mechanical behavior are essential in the development of constitutive models.

The scope of these notes is:

1.   A very brief review of finite element methods,

2.   Mathematical, testing, and verification aspects for constitutive modelling of geologic materials and discontinuities (joints and interfaces) with special emphasis on the hierarchical single surface approach developed by the author and co-workers, and

3.   Implementation of the models in nonlinear finite element procedures for solution of practical and realistic boundary value problems.

Finite Element Analysis [19]

The literature on finite element method is very wide and it is assumed that the readers are familiar with the details of this method. Hence, only a very brief review is presented here with statements of factors that are important for reliable and efficient numerical solutions.

The element equations are given by

$$[m] \{\ddot{q}\} + [c] \{\dot{q}\} + [k] \{q\} = \{Q(t)\} \tag{1-1}$$

where [m], [c] and [k] are mass, damping and stiffness matrices, respectively, {q} is the vector of nodal displacements, {Q(t)} is the vector (time dependent) of nodal forces, and the overdot denotes time derivative.

For static analysis, the first two terms in Eq. (1-1) do not appear. Although geometric nonlinearity involving large strains and deformations is important for some situations, here attention is given to material nonlinearity which is mainly embedded in the stiffness matrix [k]. Details of the hierarchical approach for constitutive modelling developed by the author and co-workers are included.

Material Nonlinearity: Solids and Discontinuities (Joints and Interfaces)

Soils, rocks and concrete assumed to be continuous solids contain discontinuities like joints, fissures and cracks, while interfaces exist between structural and geological materials. Development of appropriate models that can define various special modes of deformation, no slip, slip, debonding, rebonding, interpenetration, require considerations different from those for surrounding solid materials. Analytical modelling and testing for joints and interfaces are included herein in the general framework used for the solid materials; modelling for the latter is described in details.

Other Factors

In addition to material nonlinearity, other factors such as efficient and convergent time integration schemes for field and dynamic problems, and use of up-to-date developments in computer aided engineering and graphics also play an important role in obtaining reliable and efficient numerical solutions.

Brief statements of some new and efficient time integration schemes developed by the author and co-workers are included here.

Field and Coupled Problems

Behavior of fluid in the pores of geologic materials can have significant influence with reference to flow (seepage) of the fluid assuming the material to be rigid, and resulting hydraulic forces, and coupled deformation and fluid response. The author has developed a method called residual flow procedure (RFP) for transient and steady state free surface flow. Also, the author and co-workers have developed finite element procedure for dynamics of coupled nonlinear response of soils. Brief descriptions of these developments are included herein.

Applications

A number of applications of the finite element method, in which the proposed constitutive models have been implemented for static and dynamic problems in geomechanics are also described. They involve verifications of the numerical results with laboratory and field observations.

PART 2.   HIERARCHICAL SINGLE SURFACE CONSTITUTIVE MODELS

2.1   INTRODUCTION AND GENERAL DESCRIPTION

The authors and co-workers have developed a hierarchical single surface (HISS) approach for constitutive modelling that allows for progressive development of models of higher grades corresponding to different levels of complexities. One of the basic ingredients of this concept is the single surface yield and potential functions. A brief historical review of multi-surface and single surface models is given in Appendix 2-1.

In the HISS concept, the model for initially isotropic material, hardening isotropically with associative plasticity, is treated as the basic, δ_0, that involves zero deviation from normality (δ) of the plastic strain increment to the yield surface, F. Models of higher grades, isotropic hardening with nonassociative response due to friction (δ_1), nonassociative response due to factors such as friction and kinematic hardening (δ_2) are obtained by superimposing modifications or corrections to the basic δ_0 model. Figure 2-1 shows schematic of various models.

The above HISS approach is applied to characterize behavior of both "solids" and "discontinuities;" the latter is based on the specializations of the models for solids. The research philosophy in developing the models integrates empirical data of laboratory tests using advanced and new testing devices and verification of models with respect to laboratory and (field) observations of boundary value problems. Intuitive considerations also play a role in this integrated approach.

Table 2-1 shows a brief overview of models for solids and discontinuities and references to relevant publications (App. 2-1). Explanations for the associated material constants are given subsequently in Table 2-2.

BRIEF DESCRIPTIONS

In view of the length limitations, and because the details of the models are given in various references (Appendix 2-1 and Table 2-1), only brief statements of various models are presented herein. However, emphasis is given to those (δ_0, δ_1) models for which details of the determination of constants, laboratory testing and verification are stated in these notes, see Part 5.

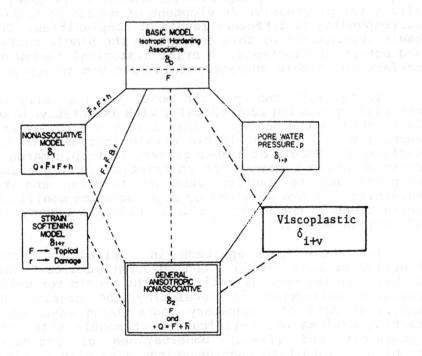

Figure 2-1.   Hierarchical Single Surface Approach
for Solids: Interfaces/Joints

General Polynomial Function

The basic polynomial form proposed for the modelling, which is a special case of the concept proposed in [13] is general, and can be used to develop models in the context of various theories such as hypoelasticity, plasticity and viscoplasticity. However, at this time, the models are developed in the context of the theory of plasticity.

Model Description

Over the last decade or so, the author and co-workers have conducted research to evolve a unified modelling concept that can eliminate the need for multisurface yield functions. The following single surface yield and plastic potential functions are one of the main accomplishments of this research. Further details of the historical developments are given in Appendix 2-1.

Yield Function F

$$F \equiv \left(\frac{J_{2D}}{p_a^2}\right) - F_b F_s = 0 \tag{2-1}$$

where J_{2D} = second invariant of the deviatoric stress tensor,

$$F_b = -\alpha \left(\frac{J_1}{p_a}\right)^n + \gamma \left(\frac{J_1}{p_a}\right)^2 = \text{basic function, describing the shape}$$

in the $J_1 - \sqrt{J_{2D}}$ space. $F_s = (1 - \beta S_r)^m$ = shape function, describing the shape in octahedral planes, S_r = stress ratio,

here, $S_r = \frac{\sqrt{27}}{2} J_{3D}/J_{2D}^{-\frac{3}{2}}$, γ, β, m = material response functions (assumed constants for some versions) associated with the ultimate behavior, and m = -0.5 as a reference parameter, α = hardening function, J_1 = first invariant of the stress tensor, p_a = atmospheric pressure with stress units, and n = phase change or transition (from contraction to dilation) parameter. Typical plot of F in various stress spaces are shown in Figs. 2-2 and 2-3 for a soil and rock, respectively.

Potential Function Q

For the associative model (δ_0), the potential function Q is equivalent to F (F $\equiv$ Q), whereas for the nonassociative model (δ_1), Q is defined as a modification of F as [4, 24, 44, 45, 57]

$$Q \equiv F + h (J_i, \xi) \quad i = 1, 2 \text{ and } 3 \tag{2-2}$$

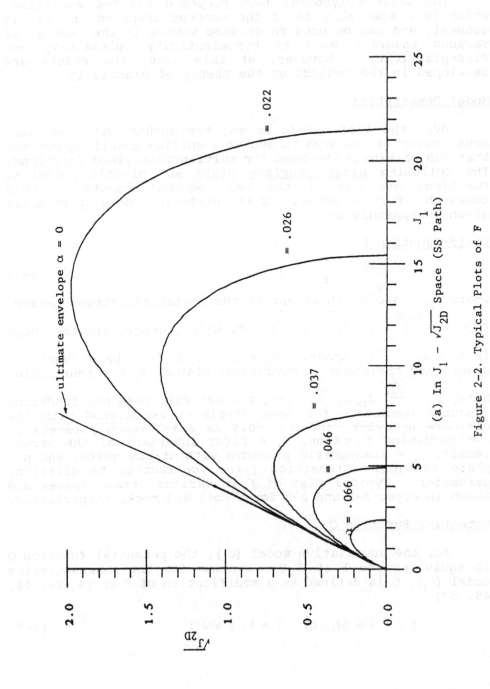

(a) In $J_1 - \sqrt{J_{2D}}$ Space (SS Path)

Figure 2-2. Typical Plots of F

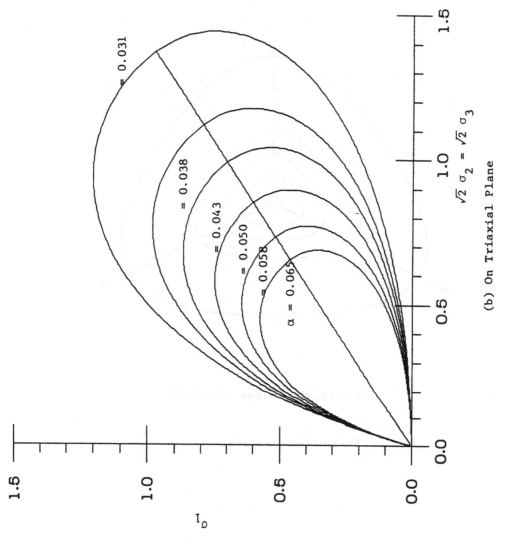

(b) On Triaxial Plane

Figure 2-2.   (cont'd.)

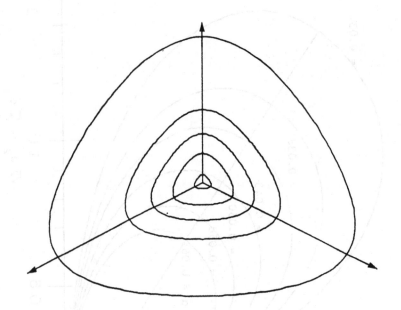

(c) Octahedral Plane (β = 0.46)

Figure 2-2. (cont'd.)

where h = correction function, $\xi = \int (d\varepsilon_{ij}^P d\varepsilon_{ij}^P)^{1/2}$, = trajectory of the plastic strains, and J_i = (i = 1, 2 and 3) = first, second and third invariants of the stress tensor.

Here Q is defined as

$$Q = (\frac{J_{2D}}{P_a^2}) - [\alpha_Q (\frac{J_1}{P_a})^n + \gamma (\frac{J_1}{P_a})^2] F_s$$ (2-3)

where $\alpha_2 = \alpha + \kappa (\alpha_o - \alpha) (1 - r_v)$, $r_v = \frac{\xi_v}{\xi}$, ξ_v = volumetric part of ξ, $\alpha_o = \alpha$ at beginning of shear loading, and κ = nonassociative parameter calculated from the ultimate slope of $\varepsilon_1 - \varepsilon_v$ plot; in general, is a function of factors such as (initial) density and stress path.

For some geologic materials, the shape of the ultimate surface changes with the mean pressure (J1). This behavior can be captured by defining [39, 40, 46]:

$$F_s = [\exp (\beta_1 \frac{J_1}{P_a} - \beta S_r]^m$$ (2-4)

or by defining [39, 40]

$$\beta = \beta_o \exp (-\beta_1 \frac{J_1}{P_a})$$ (2-5)

This allows for curved ultimate yield surface in $J_1 - J_{2D}$ space and variation of shape (in octahedral plane) e.g., from nearly triangular at low pressures, to circular at very high pressures, as shown subsequently, Fig. 2-3(c). It was found for some soils, the nonassociative response depends on the stress path [106]. Then κ can be expressed as a function of stress path as

$$\kappa = \kappa_1 + \kappa_2 S_r$$ (2-6)

The above discussion illustrates the versatility of the HISS approach in constitutive modelling where the basic model is modified to capture increasingly complex factors that affect behavior of (geologic) materials.

Hardening or Growth Functions

The response function   can be expressed as

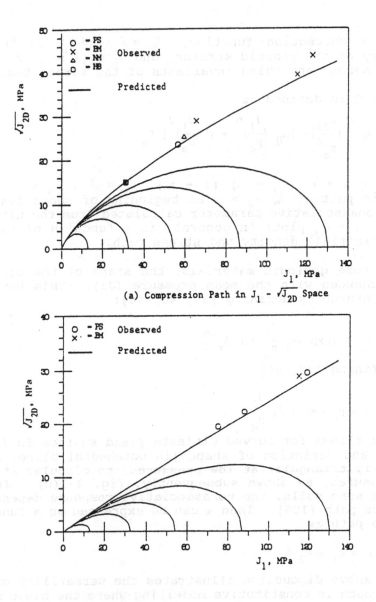

(a) Compression Path in J_1 - $\sqrt{J_{2D}}$ Space

(b) Extension Path
    in J_1 - $\sqrt{J_{2D}}$ Space

PS - Present Study
BM - Brian Mound
          (Wawersik et al. 1980)
NM - New Mexico
          (Wawersik and Hannum,
          1980)
HB - Hackberry
          (Wawersik et al. 1979)

Figure 2-3. Plots of F for Rock Salt

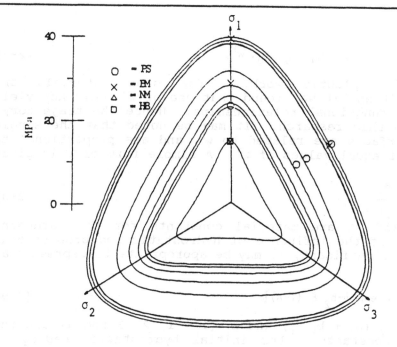

(c) Octahedral

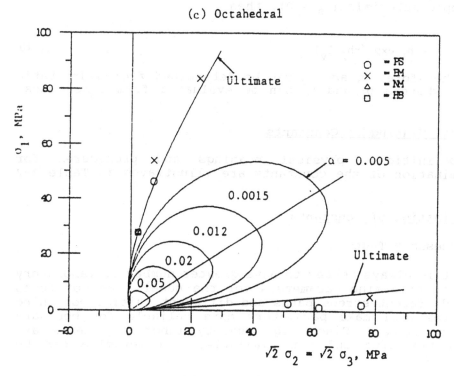

(d) Triaxial

Figure 2-3.   (cont'd.)

$$\alpha = \alpha \ (\xi, \ \xi_v, \ \xi_D, \ r_v, \ r_D, \ w^P) \tag{2-7}$$

where w^P = plastic work. For some materials and discontinuities, it was found that use of w^P would not yield consistent functional relationships. Hence the trajectory is used in this research. It may be noted that the single (yield) surfaces are related to w^P, and are proportional to contours of equal values of w^P. A simple form of α is given by

$$\alpha = \frac{a_1}{\eta_1 \xi_1} \tag{2-8}$$

where a_1 and η_1 are material constants for the hardening behavior. If the influence of hydrostatic and proportional loading is significant, it may be appropriate to express α as [30, 60]

$$\alpha = b \ exp \ [-b_2 \ \xi \ (1-A)] \tag{2-9}$$

where $A = \xi_D/(b_3 + b_4 \ \xi_D)$ and b_i $(i = 1, 2, 3$ and $4)$ are the hardening constants. For initial hydrostatic loading of isotropic materials $(\xi_D = 0)$, then

$$\alpha = b_1 \ exp \ (-b_2 \ \xi_v) \tag{2-10}$$

and, therefore, b_1 and b_2 can be determined from hydrostatic tests, whereas b_3 and b_4 can be evaluated from other shear tests.

Physical Meaning of Constants

Definitions, physical meanings and procedures for determination of the constants are illustrated in Table 2-2 [47].

Determination of Constants

(a) Necessary Tests

It is always better to have greater number of laboratory tests to determine parameters for the model. Use of tests under different stress paths and different initial confining pressures will make parameters more reliable for boundary value problems. Since high quality laboratory tests are always expensive and not available, a compromise can be developed.

It is desirable to have a few tests in compression,
extension and simple shear stress paths. It is advisable to
use at least one extension path with conventional triaxial
compression test, even though an approximate set of constants
can be obtained from only <u>one CTC test</u> [106].

Summary

Absolute minimum	One CTC (standard triaxial or multiaxial) with $\phi_c = \phi_E$
Recommended minimum	1 CTC, 1 (TE or CTE)

Desirable   (a)   3 CTC   1 (TE)   1 SS
                     (b)   >3 CTC   >1 TE   > 1 SS +
                                                          1TC

Common stress paths used in the laboratory testing of
geologic material are illustrated in Fig. 2-4. Typical
stress-strain and volume change curves for a sand are shown
in Fig. 2-5. Here $\tau_{oct} = \frac{1}{3} [(\sigma_1 - \sigma_2)^2 + (\sigma_2 - \sigma_3)^2 + (\sigma_3 - \sigma_1)^2]^{1/2}$.

Methods for Determination of Constants

Elastic Constants (E, ν)

Elastic moduli are calculated from unloading/reloading
slopes of shear tests, Fig. 2-6. Average values from all the
tests are used. This is done in the program HISS-DOD1,
described later in Part 5 [18] for finding constants, based
on laboratory test data.

Relations between E and ν and unloading slope (S) for
various stress paths are given in Table 2-3 where S_i = slope
of the unloading/reloading curve τ_{oct} vs. ε_i.

Based on given unloading/reloading slopes (Fig. 2-6),
average (representative) E and ν values are evaluated. If
unloading/reloading responses are not available, E and ν
values must be provided from other consideration such as
initial slopes of observed curves. A subroutine in the program
HISS-DOD1 calculates corresponding unloading/reloading slopes
for each stress path to be used later, for evaluation of
plastic strains from total strains.

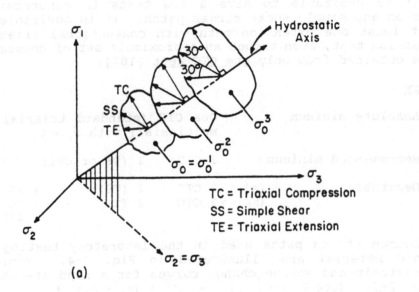

(a)

TC = Triaxial Compression
SS = Simple Shear
TE = Triaxial Extension

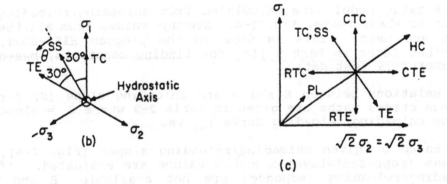

(b)

(c)

Figure 2-4. Various stress-paths followed in laboratory tests

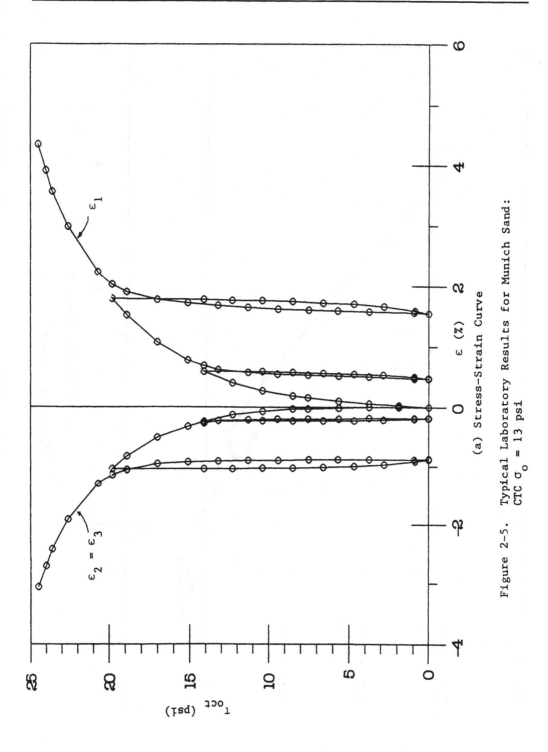

Figure 2-5.   Typical Laboratory Results for Munich Sand:
CTC σ_o = 13 psi

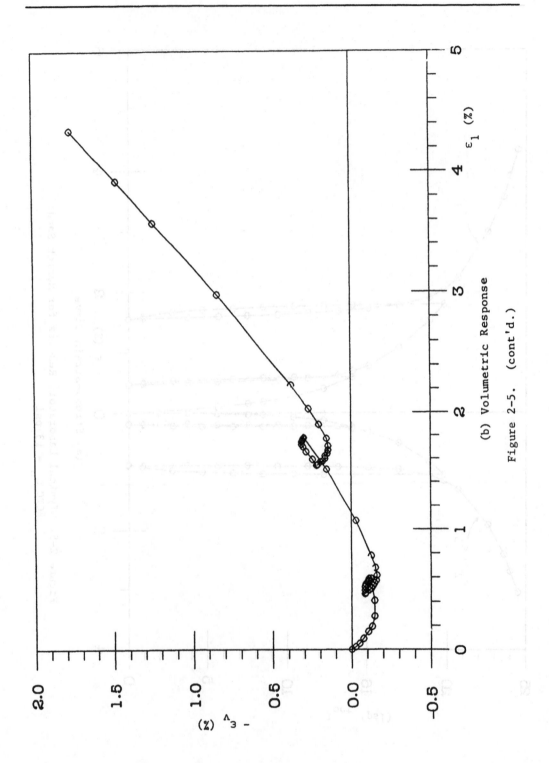

(b) Volumetric Response

Figure 2-5. (cont'd.)

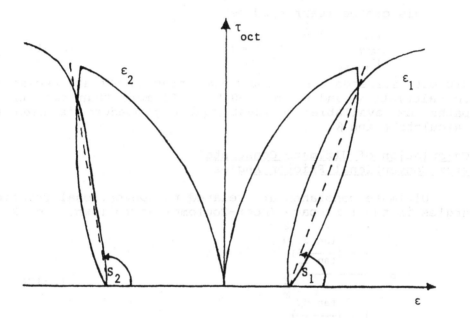

Figure 2-6.   Unloading/Reloading Slopes

Ultimate Constants (γ, β and $m = -0.5$)

These constants are evaluated in the subroutine FAIL in the program HIPAR (Part 5). The value of m for many geologic materials is found to be equal to -0.50.

At the ultimate, the hardening parameter α is zero. Therefore, F becomes

$$J_{2D} - (\gamma J_1^2)(1 + \beta S_r)^m = 0 \tag{2-11a}$$

This can be rearranged as

$$\left(\frac{J_{2D}}{J_1^2}\right)_{ult}^{\frac{1}{m}} \gamma^{-\frac{1}{m}} - S_r \, \beta = 1 \tag{2-11b}$$

Ultimate stresses for at least two stress paths are necessary to calculate γ and β (m = -0.5). If more than two stress paths are available, a least square procedure is used to calculate γ and β.

Calculation of Ultimate Constants from Conventional Fricton Angles

Ultimate constants are related to conventional friction angles in the τ-σ space (Mohr-Coulomb) as follows, Fig. 2-7.

$$\beta = \frac{1 - \left(\dfrac{\tan \theta_C}{\tan \theta_E}\right)^{\frac{2}{m}}}{1 + \left(\dfrac{\tan \theta_C}{\tan \theta_E}\right)^{\frac{2}{m}}} \tag{2-12a}$$

where $\tan \theta_C = \dfrac{2}{\sqrt{3}} \dfrac{\sin \phi_C}{3 - \sin \phi_C}$, $\tan \theta_E + \dfrac{2}{3} \dfrac{\sin \phi_E}{3 + \sin \phi_E}$, and θ_C, θ_S, θ_E, ϕ_C, ϕ_S and ϕ_E are as shown in Fig. 2-7.

$$\sqrt{\gamma} = \frac{\tan \theta_C}{(1-\beta)^{\frac{m}{2}}} = \frac{\tan \theta_E}{(1-\beta)^{\frac{m}{2}}} \tag{2-12b}$$

If only one CTC test is available (for a cohesionless material, i.e. c = 0), ϕ_C can be calculated, and ϕ_E can be adopted as equal to ϕ_C [106]. If a relation between the

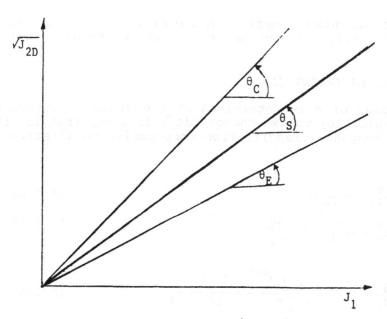

(a) Ultimate Envelopes in $\sqrt{J_{2D}}$ - J_1 Space

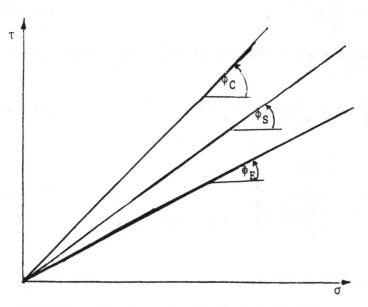

(b) Ultimate Envelopes in Mohr-Coulomb (τ-σ) Space

Figure 2-7.   Ultimate Envelopes in Different Stress Spaces

ultimate stress under compression and extension paths with J_1 is available, [39, 40], the value of ϕ_E can be found from such relations.

Phase Change Parameter (n)

The value of n is determined at the state of stress at which the (plastic) volume change ($d\varepsilon_v^P$) is zero; that is, the volumetric response transits from compressive to dilative.

Flow Rule

$$d\varepsilon_{ij}^P = \lambda \frac{\partial Q}{\partial \sigma_{ij}} \tag{2-13a}$$

$$d\varepsilon_v^P = d\varepsilon_{ii}^P = 3\lambda \frac{\partial Q}{\partial J_1} \tag{2-13b}$$

Since $\lambda \neq 0$

$$\frac{\partial Q}{\partial J_1} = 0 - \left(- \frac{\alpha_Q \, n \, J^{n-1}}{P_a^n} + \frac{2\gamma \, J_1}{P_a^2}\right) F_s \tag{2-14a}$$

Since $J_1 \, F_s = 0$

$$\frac{J_1}{P_a} = \left(\frac{2\gamma}{\alpha_Q \, n}\right)^{\frac{1}{n-2}} \tag{2-14b}$$

From the consistency condition (i.e., stress point is on the yield surface),

$$\frac{J_{2D}}{J_1^2} - \left[-\alpha \left(\frac{2}{\alpha_Q \, n}\right) + \gamma\right] F_s = 0 \tag{2-15a}$$

Therefore,

$$n = \frac{2 \left(\frac{\alpha}{\alpha_Q}\right)}{1 - \left(\frac{J_{2D}}{J_1^2}\right) \frac{1}{F_s} \gamma}\Bigg|_{\text{at zero volume change}} \tag{2-15b}$$

Since $\alpha \cong \alpha_Q$ at this point,

$$n = \cfrac{2}{1 - (\cfrac{J_{2D}}{J_1^2}) \cfrac{1}{F_s} \gamma} \Bigg|_{\text{at zero volume change}}$$ (2-15c)

In general, n is a function of factors such as initial density and confining stress.· At this time, a constant average value of n is used. n is calculated in a subroutine HISS-DOD1 in the program [Part 5, Ref. 18]. There is an option to back predict the points where plastic volume change is zero using the average n value to check the validity of the averaging.

Hardening Parameters

a_1 and n_1 for the Simple Form, Eq. (2-8)

From the consistency condition at yield $F = 0$. Therefore,

$$\alpha = [\gamma - (\frac{J_{2D}}{J_1^2}) \frac{1}{F_s}] (\frac{P_a}{J_1})^{n-2}$$ (2-16a)

Also, from Eq. (2-7)

$$[1 \; \ell n \; \xi] \left\{ \begin{matrix} \ell n \; a_1 \\ n_1 \end{matrix} \right\} = \ell n \; \alpha$$ (2-16b)

α and ξ can be calculated for each point on the (observed) stress-strain curve. When all the points are considered, this results in a set of simultaneous equations in the form

$$[A] \; \{x\} = \{b\}$$ (2-17a)

where

$$[A] = \begin{bmatrix} 1 & \ell n \; \xi_1 \\ 1 & \ell n \; \alpha_2 \\ & \vdots \\ 1 & \ell n \; \xi_n \end{bmatrix}, \; \{b\} = \left\{ \begin{matrix} \ell n \; \alpha_1 \\ \ell n \; \alpha_2 \\ \vdots \\ \ell n \; \alpha_n \end{matrix} \right\}, \; \{x\} = \left\{ \begin{matrix} \ell n \; a_1 \\ n_1 \end{matrix} \right\}$$ (2-17b)

This set of equations is solved for $\ln a_1$ and η_1 using a least square procedure. Hydrostatic compression test can also be used to calculate a_1 and η_1. In this case, Eq. (2-16a) simplifies to

$$\alpha = \gamma \left[\frac{p_a}{J_1}\right]^{n-2} \tag{2-18}$$

The rest of the procedure is similar to that for shear tests.

b_1, b_2, b_3 and b_4 for the Split Form, Eq. (2-9)

(a) b_1 and b_2 from Hydrostatic Tests with
     Special Form of Eq. (2-9)

$$\alpha = b_1 \exp(-b_2 \xi_v) \tag{2-19a}$$

$$(1 - \xi_v) \left\{\frac{\ln b_1}{b_2}\right\} = (\ln \alpha) \tag{2-19b}$$

A set of simultaneous equations is obtained by considering all the points on the hydrostatic compression stress-strain curve.

$$[A] = \begin{bmatrix} 1 & -\xi_v(1) \\ & \\ & \\ 1 & -\xi_v(n) \end{bmatrix}, \quad \{b\} = \left\{\begin{array}{c} \ln \alpha_1 \\ \vdots \\ \vdots \\ \ln \alpha_n \end{array}\right\}, \quad \{x\} = \left\{\begin{array}{c} \ln b_1 \\ \\ b_2 \end{array}\right\} \tag{2-19c}$$

This can be solved for $\ln b_1$ and b_2 using the least square procedure.

b_3 and b_4 from Shear Tests

Now, from Eq. (2-9)

$$\alpha = b_1 \exp(-b_2 \xi (1-a)) \tag{2-20a}$$

$$\ln \alpha = \ln b_1 - b_2 \xi (1-a) \tag{2-20b}$$

$$a = \frac{(\ln \alpha - \ln b_1)}{b_2 \, \xi} + 1 \qquad (2\text{-}20c)$$

but
$$a = \frac{\xi_D}{b_3 + b_4 \, \xi_D} \qquad (2\text{-}20d)$$

$$[1 \quad \xi_D] \begin{Bmatrix} b_3 \\ b_4 \end{Bmatrix} = (\frac{\xi_D}{a}) \qquad (2\text{-}20e)$$

A set of simultaneous equations can be obtained by considering all the points on the stress-strain curve in a shear test:

$$[A] \ \{x\} = \{b\} \qquad (2\text{-}20f)$$

$$[A] = \begin{bmatrix} 1 & \xi_D \ (1) \\ & \\ & \\ 1 & \xi_D \ (n) \end{bmatrix} \quad \{b\} = \begin{vmatrix} (\frac{\xi_D}{a}) \ (1) \\ \\ (\frac{\xi_D}{a}) \ (n) \end{vmatrix} , \ \{x\} = \begin{Bmatrix} b_3 \\ b_4 \end{Bmatrix}$$

This can be solved for b_3 and b_4 using the least square procedure.

Nonassociative Parameter

As a simplification, parameter κ in the plastic potential Q is assumed as constant and is determined from the conditions near the ultimate in a subroutine in the program HISS-DOD1 (Part 5). Basic steps in evaluating κ are given below.

From the Flow Rule

$$d\varepsilon_{11}^P = \lambda \frac{\partial Q}{\partial \sigma_{11}} \qquad (2\text{-}21a)$$

and

$$d\varepsilon_v^P = \lambda \, [\frac{\partial Q}{\partial J_1} \frac{\partial J_1}{\partial \sigma_{11}} + \frac{\partial Q}{\partial J_1} \frac{\partial J_1}{\partial \sigma_{22}} + \frac{\partial Q}{\partial J_1} \frac{\partial J_1}{\partial \sigma_{33}}] \qquad (2\text{-}21b)$$

$$= 3\lambda \frac{\partial Q}{\partial J_1} \tag{2-21c}$$

$$(\frac{d\varepsilon_v^P}{d\varepsilon_{11}^P}) = (3\frac{\partial Q}{\partial J_1})/(\frac{\partial Q}{\partial \sigma_{11}}) = \nu^P \tag{2-21d}$$

Further details are available elsewhere [45, 47, 60].

2.2   ANISOTROPIC HARDENING MODELS, δ_2, δ_{2+p}

The model is developed as part of a hierarchical single surface approach described previously. The approach permits evolution of models of progressively higher grades from a basic model representing isotropic hardening associative behavior. Hence, the first stage in the development of the general anisotropic representation is to define the basic isotropic hardening associative model (δ_o) model that can be considered characteristic of an assumed fundamental state of the material. Observed behavioral trends would usually deviate from those of the basic model. Then material attributes such as friction, inherent and induced anisotropy, and the deformation history can be considered to cause deviations or corrections from the basic model. This is achieved by introducing a second order tensor whose evolution is governed by the deformation history and the level of induced anisotropy is defined. The validity of the model and the proposed concepts are investigated with respect to special (multiaxial) stress path tests on specimens of three sands, described later.

A number of investigators [9, 10, 80, 88] have proposed models for isotropic and kinematic hardening in geologic materials. These models have been widely recognized and used. The proposed concept is based on similar theories of elastoplasticity. However, it contains various new aspects such as (i) the hierarchical concept in which induced anisotropy is introduced as a correction (defined through deviation from normality) to the basic model, (ii) introduction of a new additional measure of induced anisotropy, (iii) a formulation that continuously carries the growth of induced anisotropy, and (iv) motion of a single potential function that translates in the "fixed" field of yield surfaces. As a consequence, the number of constants is relatively smaller than that for comparable capability of other models, and their physical meanings can be defined.

Details of the proposed anisotropic model are given in
[45, 99, 100]. Here only a brief description of the model is
presented for the sake of completeness, while the main
attention is given to the verification of the new concept and
definition for induced anisotropy, laboratory test results and
verifications for three sands.

Figure 2-8 illustrates schematically details of the
proposed model in the J_1 - $\sqrt{J_{2D}}$ stress space. The surface, Q,
translates in a fixed field of infinite number of isotropic
yield surfaces F, in the stress space; F ranges from an
initial surface F_0 that demarcates the initial elastic region
to the ultimate envelope, which is defined as the locus of
asymptotic stress values from observed stress-strain behavior.
Surface Q can intersect the fixed surfaces at the stress point
during loading, unloading and reloading.

Surfaces F serve to (i) define virgin loading and
initiation of unloading and reloading, (ii) govern magnitude
of plastic strain increments and (iii) retain memory of
maximum prestress. Surface Q acts as the plastic potential
during virgin and nonvirgin loading, and governs the
directions of plastic strain increments. During nonvirgin
loading (unloading and reloading), Q also serves as a loading
surface allowing for elastic behavior within its domain, and
plastic behavior beyond its boundaries.

In general, Q may be of any shape or size and may expand,
translate or rotate in the stress space, and is a function of
σ_{ij}, a_{ij} and α_m. Tensor, a_{ij}, is the state variable that
reflects the development and evolution of induced anisotropy.

With the hierarchical concept, Q can now be expressed in
the same form as that of F as

$$Q = \bar{J}_{2D} - (\sigma_0 \bar{J}_1^n + \gamma \bar{J}_1^2) 1-\beta \bar{S}_r)^m \qquad (2-22)$$

Here the overbar denotes quantities associated with a modified
stress tensor $\bar{\sigma}_{ij}$ given by

$$\bar{\sigma}_{ij} = \sigma_{ij} - a_{ij} \qquad (2-23)$$

For convenience, in this study, Q is assumed to remain
unchanged in shape, size and orientation, with its size the
same as that of the initial yield surface F_0, with constant
value of hardening parameter α_0, and its location in the
stress space is defined by a_{ij}.

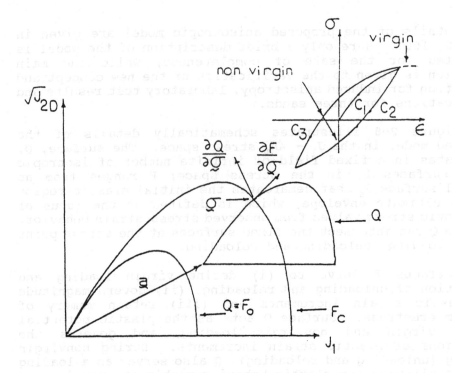

Figure 2-8. Schematic of Proposed Model

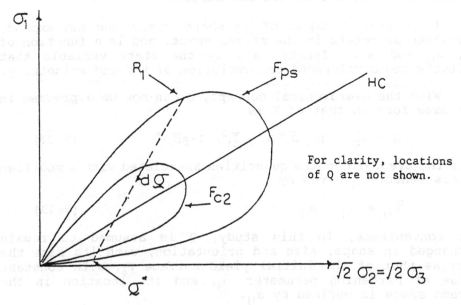

For clarity, locations
of Q are not shown.

Figure 2-9. Nonvirgin Loading

Flow Rule

For every stress point in the stress space, there exists a corresponding yield surface F_c (Fig. 2-8). The material is said to be in the virgin state if all the yield surfaces corresponding to its previous stress history are completely contained within the current yield surface F_c, and the stress increment is directed outwards of F_c. If the stress increment is not directed outward of F_c, then F_c is retained as the memory of the maximum prestress, F_{ps}, until the stress point leaves its bounds once again.

The plastic flow rule may be written as

$$d\varepsilon_{ij}^P = \lambda \frac{\partial Q/\partial \sigma_{ij}}{||\partial Q/\partial \sigma_{rs}||} \qquad (2\text{-}24)$$

where $d\varepsilon_{ij}^P$ is the incremental plastic strain and λ is a scalar of proportionality whose value is λ_v and λ_{nv}, depending on whether the loading is virgin or nonvirgin, respectively, and $||\ ||$ denotes norm. Equation (2-24) may be expressed as

$$d\varepsilon_{ij}^P = \bar{g}_1 \delta_{ij} + \bar{g}_2 (\sigma_{ij} - a_{ij})$$
$$+ \bar{g}_3 (\sigma_{ik} - a_{ik})(\sigma_{kj} - a_{kj}) \qquad (2\text{-}25)$$

where $\bar{g}_i = \bar{g}_i (J_r, \lambda, \alpha, \beta, \lambda, m, n)$, $r = 1, 2, 3$ and δ_{ij} is Kronecker delta. It can be seen that the tensor, a_{ij}, governs the direction of $d\varepsilon_{ij}^P$.

During __virgin loading__, using the consistency condition $dF_c = 0$, λ_v is derived as

$$\lambda = \lambda_v = - \frac{-\dfrac{\partial F_c}{\partial \sigma_{ij}} d\sigma_{ij}}{\dfrac{\partial F}{\partial \xi}} \qquad (2\text{-}26)$$

During the __nonvirgin loading__, an interpolation function is postulated for λ_{nv}. In defining the interpolation function, it has been ensured that (a) the degree of maximum prestress is incorporated in the formulation through surface F_{ps} and

(b) a smooth transition occurs from the nonvirgin to virgin loading, and from elastic to plastic behavior. The effect of prestress is introduced by identifying a stress level σ_{ij}^*, on F_{ps}, which represents the intersection of the extension of $d\sigma_{ij}$ from σ_{ij} with F_{ps}, Fig. 2-9. With these stipulations, and on the basis of a study of observed unloading and reloading behavior of sands under various stress paths and various levels of loading, the following expression is postulated for λ_{nv}:

$$\lambda_{nv} = [\lambda_v]_{\sigma_{rs}^*} [1+\theta \bar{R}^{h_1}]^{-1} = \frac{(\frac{\partial F_{ps}}{\partial \sigma_{ij}})_{\sigma_{rs}^*} d\sigma_{ij}}{\frac{\partial F_{ps}}{\partial \xi} [1+\theta \bar{R}^{h_1}]} \qquad (2-27)$$

where h_1 is a material constant and θ is given by

$$\theta = h_2 (\frac{\Delta_{ps}}{J_1^*})^{h_3} \qquad (2-28)$$

Here h_2, h_3 are material constants, $\Delta_{ps} = (\gamma/\alpha_{ps})^{1/n-2}$ denotes J_1 intercept of F_{ps} and J_1 is the first invariant of σ_{ps}, and $\bar{R}$ is the ratio that depends on the relative position of the current stress point within the prestress surface and is expressed in terms of the hardening parameter, α, associated with the current surface, prestress surface and the surface from which the last stress reversal took place. Thus $\bar{R}$ is expressed as

$$\bar{R} = [\frac{(S_r)_* /\alpha_{ps} - (S_r)_c /\alpha_{ps}}{(S_r)_c /\alpha_c - (S_r)_R /\alpha_R}]^{(1-\beta S_r^*)^{-m/2}} \qquad (2-29)$$

where subscripts *, c and R refer to stress points $\ddot{\sigma}_{ij}$, σ_{ij} and the level of stress reversal σ_{ij}^R, respectively.

Comments

During the nonvirgin case, if neutral loading occurs, the numerator in Eq. (2-28) may not be zero, which can cause discontinuity in λ_{nv}. It is possible to use schemes used in other models such that σ_{ij}^* is defined as the conjugate point of σ_u, in which case the numerator will be automatically zero. However, for most cases studied, the procedure used herein gives satisfactory results. Hence, at this time, it is used, and the numerator is set to zero if neutral (nonvirgin) loading occurs, which may be rare.

For special loading cases such as proportional cyclic loading, the stress point σ_{ij}^* coincides with the origin. This would create singularity. In this study, proportional loading tests are not considered. However, for such cases, the procedure would need modifications.

It may be mentioned that the procedures used here and in other models to generate λ_{nv} are essentially geometrical schemes, and there is no a priori or closed-form proof that such schemes represent the general behavior. The proof is obtained only in terms of prediction of observed behavior for specific classes of materials. In this sense, the proposed procedure yields satisfactory results for the three sands considered; however, it may need modifications for general application [99, 100].

Translation Rule

When an initially isotropic material is subjected to non-hydrostatic loading, anisotropy is induced due to changes in the composition and fabric of the material [3, 4]. The effect of induced anisotropy on the material response is introduced through a correction to the position of $(a_{ij})_{iso}$ corresponding to Q for the isotropic-associative (δ_o) model, (Fig. 2-10):

$$a_{ij} = (a_{ij})_{iso} + d_{ij} \qquad (2\text{-}30)$$

where d_{ij} is a correction tensor reflecting the level of induced anisotropy. In terms of the deviatoric and hydrostatic components, d_{ij} may be written as, Fig. 2-10:

$$d_{ij} = \Pi\ \delta_{ij} + \mu_{ij} \qquad (2\text{-}31)$$

Equation (2-10) can now be written in incremental form as

$$da_{ij} = d\ (a_{ij})_{iso} + d\ (d_{ij}) \qquad (2\text{-}32)$$

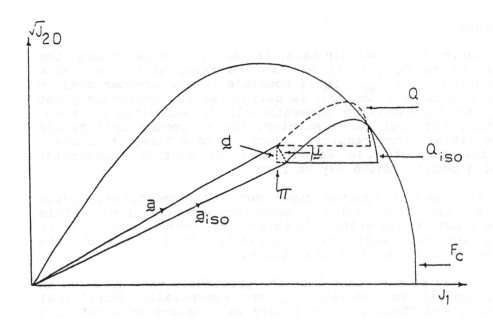

Figure 2-10 Location of Q and Q_{iso}

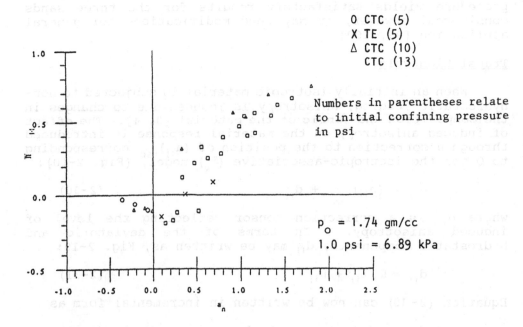

O CTC (5)
X TE (5)
△ CTC (10)
CTC (13)

Numbers in parentheses relate
to initial confining pressure
in psi

P_o = 1.74 gm/cc
1.0 psi = 6.89 kPa

Figure 2-11 Plot of $\overline{\Pi}$ vs. a_n for Leighton Buzzard Sand

where

$$d\ (d_{ij}) = d\Pi\ \delta_{ij} + d\ \mu_{ij} \tag{2-33}$$

Quantity $d\ (a_{ij})$ gives the incremental translation rule for Q for the associative behavior. During the virgin loading, the associative behavior requires that the (unit) normals to F_c and Q coincide. Then it can be shown that

$$d\ (a_{ij})_{iso} = p_1\ d\sigma_{ij} + p_2\ \sigma_{ij} \tag{2-34}$$

where $p_1 = 1 - (\alpha_c/\alpha_o)^{1/n-2}$ and $p_2 = \dfrac{1}{2-n}\ (\dfrac{\alpha_c}{\alpha_o}^{3-n})^{1/n-2}$.

During the nonvirgin loading, the position of Q for the associative behavior is postulated as

$$d\ (a_{ij})_{iso} = [(a_{ij}^*)_{iso} - (a_{ij})_{iso}\ ]\ \frac{||d\sigma_{rs}||}{||\sigma_{rs}^* - \sigma_{rs}||} \tag{2-35}$$

where $(a_{ij})_{iso}$ is the known associative position and $(a_{ij})_{iso}^*$ is the isotropic-associative position relevant to σ_{ij}^*, Fig. 2-9.

Correction Tensor d_{ij}

Increments $d\Pi$ and $d\mu_{ij}$ in Eq. (2-33) are defined such that the growth and demise of deviation from normality due to induced anisotropy are allowed for.

A new scalar measure, a_n, representing the level of anisotropy, is defined and is associated with $d\Pi$. The quantity $d\Pi$ corrects the volumetric response predicted by the isotropic-associative model. Because plastic strains predominate in soils, it is assumed that induced anisotropy occurs mainly due to plastic strains. The level of anisotropy is also dependent on the state of stress; the latter is expressed through elastic strains.

An additional measure of anisotropy, a_n, is chosen by assuming that a sufficient condition for compliance with normality is obtained if the orientation of plastic strain tensor (in the strain space) coincides with that of the elastic strain tensor. This allows correlation between deviation from normality with the deviation of total plastic strain direction from the total elastic strain directions. With these assumptions, Π is postulated as

$$\Pi = \Pi \ (R_e - R_p) = \Pi \ (a_n) \qquad (2\text{-}36a)$$

where

$$a_n = \frac{I_1^e}{\sqrt{I_2^e}} - \frac{I_1^p}{\sqrt{I_2^p}} = R_e - R_p \qquad (2\text{-}36b)$$

Here I_1 and I_2 = first and second invariants of the strain tensor ε_{ij}, the superscripts e and p refer to elastic and plastic strains, respectively, and the ratio I_1/I_2 denotes a measure of orientation of the state of strain in the strain space. When the material is in the elastic state, R_p is indeterminate and is assumed to be equal to R_e; thus $a_n = 0$, and no induced anisotropy occurs.

Based on comprehensive series of laboratory tests on three sands (Munich, Leighton Buzzard and Fuji River) described later, a linear relation is obtained for Π as

$$\Pi = h_4 \ a_n \qquad (2\text{-}37a)$$

and

$$d\Pi = h_4 \ da_n \qquad (2\text{-}37b)$$

where h_4 is a material parameter and represents a constant of proportionality with units of stress. Figure 2-11 shows typical variation of a_n vs. normalized value of $\Pi = [-3\Pi/(\gamma/\alpha_o)^{1/n-2}]$, for the Leighton Buzzard sand, under several different virgin and nonvirgin stress paths.

The second term in Eq. (2-33) is postulated as

$$d\mu_{ij} = b \ dE_{ij}^{\,p} \qquad (2\text{-}38)$$

where $dE_{ij}^{\,p}$ is the deviatoric part of the incremental plastic strain and b is a scalar. Since $d\Pi \ \delta_{ij}$ and $d\mu_{ij}$ are coupled, they are subjected to the constraint that during translation and plastic flow, the stress point σ_{ij} remains on the surface of Q; that is,

$$dQ = 0 \qquad (2\text{-}39a)$$

Equation (2-39a) yields the value of b as

$$
b = \frac{\frac{\partial Q}{\partial \overline{\sigma}_{ij}} [d\overline{\sigma}_{ij} - d (a_{ij})_{iso} - h_4 \, da_n \, \delta_{ij}]}{\frac{\partial Q}{\partial \overline{\sigma}_{ij}} dE_{ij}^P}
$$

(2-39b)

Equations (2-32) to (2-38) are now combined to give the general anisotropic translation rule for Q as

$$
da_{ij} = d (a_{ij})_{iso} + h_4 \, d (a_n) \, \delta_{ij}
$$
$$
+ b \, dE_{ij}^P
$$

(2-40)

where $d (a_{ij})_{iso}$ is given by either Eq. (2-34) or (2-35), depending on whether the loading is virgin or nonvirgin.

Comments

   The above formulations satisfy a number of observed phenomena and intuitive expectations. During hydrostatic loading of an initially isotropic material, both elastic and plastic strains are purely volumetric, $R_e = R_p$, $a_n = 0$. As a result, $\mu_{ij} = d_{ij} = 0$ and no deviation from normality is predicted. If, however, hydrostatic loading is preceded by a shear load-unload cycle from original hydrostatic state of stress, the state of elastic strains is purely volumetric whereas the total plastic strains are not. Therefore, $R_e \neq R_p$ and a_n, μ_{ij} and d_{ij} are all non zero. The model then predicts deviation from normality.

   During shear loading of an initially isotropic material from a hydrostatic state of stress, the directions of incremental elastic and plastic strains do not coincide and, therefore, the values of R_e and R_p change at different rates causing the measure of anisotropy to change from zero. This phenomenon has been experimentally observed and reported by a number of researchers [3, 76, 100]. If the loading is such that the orientation of total plastic strain R_p approaches the orientation of total elastic strains R_e, the magnitude of a_n decreases and a gradual demise of induced anisotropy, and hence of deviation from normality, occurs. This phenomenon may be observed in the case of hydrostatic loading following a shear load-unload cycle. Experimental evidence showing this fading process of induced anisotropy is reported in the literature [3, 76, 100].

When an initially isotropic material is loaded from a stress-free state with no change in the principal directions of stress, from Eqs. (2-25), (2-30), (2-32) and (2-38), it is seen that the tensors σ_{ij}, a_{ij} and $d\varepsilon_{ij}^P$ are all coaxial; that is, they have the same principal directions. If the principal directions of stress were to change, σ_{ij} and a_{oj} would no longer have coincident principal directions. From Eq. (2-25), $d\varepsilon_{ij}^P$ would have the same principal directions as the tensor ($\sigma_{ij} - a_{ij}$), which is, in general, neither coaxial with σ_{ij} nor a_{ij}. Thus, the model is capable of capturing non-coaxiality during rotations of principal stress directions.

Elastic Behavior and Intersection of F and Q. Whenever elastic behavior initiates with the stress increment directed inwards of Q, it is assumed that Q reverts to its isotropic position given by $(a_{ij})_{iso}$. Q remains in this position until the state of stress reaches its boundary again. Once σ_{ij} reaches of boundary of Q, plastic behavior is resumed, Eq. (2-31) applies once more and Q is reset accordingly. This is done to avoid any inconsistencies or ambuiquities regarding the type of loading that might arise due to the fact that F_c and Q intersect each other. Such a possibility is illustrated at point C in Fig. 2-12(a), where a nonvirgin loading sequence ABCD occurs. Figure 2-12(b) illustrates the resetting process that eliminates the ambuiguity. The constant accumulation and evolution of induced anisotropy is not interrupted by this process. Any changes in induced anisotropy and d_{ij} that occurred during the elastic behavior would be reflected when Q is reset according to eq. (2-31). The net effect of this procedure is to change the range of elastic behavior to a small extent [100].

The nonassociative model may violate the stability conditions for certain loading paths. This does not necessarily imply unstable material response because even under such loading paths, the materials tested herein exhibited stable response. The proposed rate independent model based on the test data from quasi static or slow cyclic tests can be applied for many materials such as cohesionless soils subjected to dynamic loadings with low frequencies. For materials involving rate effects, it will be more appropriate to develop and use rate dependent models.

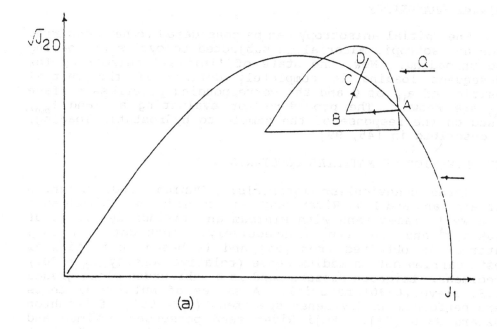

(a)

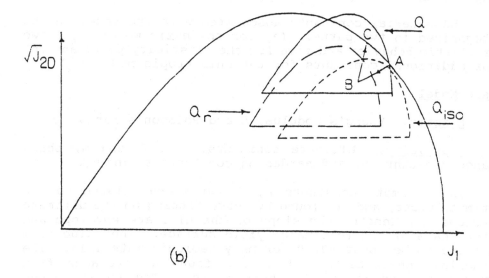

(b)

Figure 2-12 Resetting of Q During Elastic Behavior

Initial Anisotropy

The initial anisotropy can be considered to have occurred when an isotropic material is subjected to cycles of loading and unloading. Now, the state of 'initial' anisotropy for subsequent loading is completely defined if the initial position of a_{ij}^{o} of Q and the corresponding prestress surface F_{pso} are known. The procedure for evaluating a_{ij}^{o} and F_{pso}, based on the response of the sample to hydrostatic loading, is described in [45, 99].

DETERMINATION OF MATERIAL CONSTANTS

Three cohesionless materials: Munich sand, Leighton Buzzard sand and Fuji River sand, are considered. Munich sand is a well graded sand with minimum and maximum densities of 1.69 t/m^3 and 2.07 t/m^3, respectively. Test data for this material is obtained from [94] and is based on multiaxial tests carried out on medium dense (relative density, D_r = 70%) specimens. Leighton Buzzard sand is a subrounded close graded (U.S. sieve 20-30) material. A series of multiaxial tests were performed on dry dense specimens (D_r = 95%) of Leighton Buzzard sand [99]. Fuji River sand possesses maximum and minimum void ratios of 1.08 and 0.53, respectively. The test results used here are obtained from drained conventional triaxial tests on loose (initial void ratio = 0.74) saturated samples, and are reported in [102].

The material constants associated with the model can be categorized in two parts: (a) for the basic model (δ_0), two for elastic behavior and five for the plasticity part; and (b) four additional constants for the anisotropic model (δ_2):

Basic Model

Elastic. Young's modulus, E and Poisson's ratio, ν.

Plasticity. Ultimate constants: γ, β and m; phase change constant, n, and hardening constants a_1 and η_1.

The ultimate constants γ, β and m are related to the ultimate state, and are found by curve fitting on the ultimate envelope. γ denotes the slope of the ultimate envelope and the shape of the (ultimate) yield surface. The value of m is found to be about -0.50 for many geologic materials. The hardening constants a_1 and η_1 are found as averages from compression and extension shear tests. The phase change parameter is related to the point of transition from contractive to dilative behavior. A minimum of two compression and one extension tests are required to find the constants. In view of length limitations, details of the determination of these constants are not given herein; they are available in [24, 27, 45, 99]. Details of the additional constants for the anisotropic behavior are given below.

Anisotropic. The three constants h_i (i = 1, 2, 3) are found from unloading-reloading loops of (cyclic) compression and extension tests.

The interpolation constants h_1, h_2 and h govern the magnitude of plastic strains during nonvirgin loading. Equation (2-27) may be expressed as

$$\ell n \ (A) = \ell n \ (\theta) + h_1 \ \ell n \ (\overline{R}) \qquad\qquad (2\text{-}41)$$

where

$$A = \frac{- [\frac{\partial F_{ps}}{\partial \sigma_{ij}}]_{\sigma^*} \ d\sigma_{ij}}{\frac{\partial F}{\partial \xi} \ ||d\varepsilon^P_{ij}||} - 1.0.$$

By plotting ℓn (A) vs. ℓn ($\overline{R}$) from the available stress-strain data during unloading and reloading, the values of and h_1 are obtained as the intercept and slope, respectively, Fig. 2-13(a). Equation (2-30) may be written as

$$\ell n \ (\theta) = \ell n \ (h_2) + h_3 \ \ell n \ (\frac{\Delta ps}{J_1^*}) \qquad\qquad (2\text{-}42)$$

A plot, Fig. 2-13(b), of ℓn (θ) vs. ℓn ($\frac{\Delta ps}{J_1^*}$) from the test mentioned above gives the values of h_2 and h_3.

The translation parameter h_4 is found from stress-strain data during virgin and/or nonvirgin loading as the slope of the line of best fit through the points plotted in Fig. 2-11.

The material constants for the three sands are shown in Table 2-4.

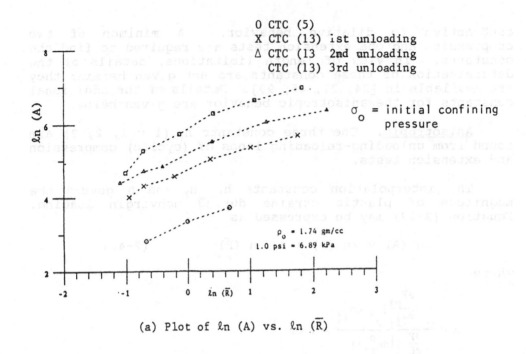

(a) Plot of ℓn (A) vs. ℓn (R̄)

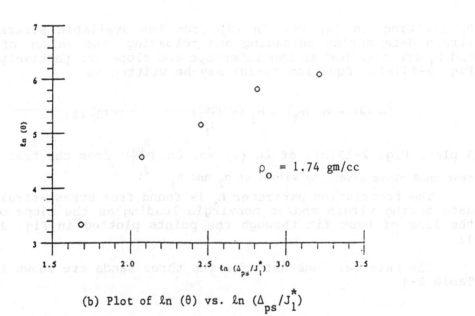

(b) Plot of ℓn (θ) vs. ℓn (Δ_{ps}/J_1^*)

Figure 2-13 Evaluation of Interpolation Constants
for Leighton Buzzard Sand

VERIFICATION OF NONASSOCIATIVE CONCEPT

Measure of Anisotropy a_n and Deviation from Normality

Verification 1: The suitability of using a_n as the measure of anisotropy and its correlation to deviation from normality δ are verified for the Munich sand. For this purpose, contours of equal values of a_n and equal values of deviation from normality at different stages during various compression (CTC, TC) and extension (TE, CTE) stress paths for the Munich sand were plotted. Typical results for the compression tests are shown in Fig. 2-14. A comparison of Figs. 2-14(a) and 2-14(b) indicates that deviation from normality and the proposed measure of induced anisotropy (a_n) show similar trends. Similar results are also found for the extension tests [99].

Verification 2: Five special series of tests were performed on the Leighton Buzzard sand with initial relative density D_r = 95% [99, 100]. In the first series, a hydrostatic compression HC test was performed by applying loading from $_0$ = 0 to 20 psi (0 to 140 kPa). The paths followed in the remaining four series are depicted in Figs. 2-15(a) to 2-15(d). First, a HC load (δ_0) was applied; e.g., 0-1 in Fig. 2-15. Then the specimen was subjected to compression, Fig. 2-15(a), extension, Fig. 2-15(b) and compression/extension, Figs. 2-15(c) and 2-15(d), loading and unloading cycles; the maximum prestress during these cycles is denoted by $R = \sigma_1/\sigma_0$. Each cycle started loading from a given level of σ_0 and returned to that σ_0 at the end of unloading; the sequences of the paths followed are indicated by number 1, 2, 3....

The postulated measure, a_n, Eq. (2-36b), is evaluated at various stages during the cycles, specifically at the ends of the unloadings. This measure represents the development and accumulation of induced anisotropy. It is affected by the previous history of the loading-unloading cycles. Typical values of a_n at σ_0 = 15 psi (103 kPa) for the five series are shown in Table 2-5, as starting values. Table 2-5 also shows values of a_n at the end of the HC loading from σ_0 = 15 to 20 psi (103 to 138 kPa).

The observed (laboratory) values of ε_1, ε_2 and ε_3 during the HC loadings are noted. By assuming radial isotropy ($\varepsilon_2 = \varepsilon_3$), which was observed in the laboratory test results, the physical measure of deviation from isotropic behavior is defined from the change of $\gamma = \varepsilon_1 - \varepsilon_3$. Plots of γ vs. σ_0 for

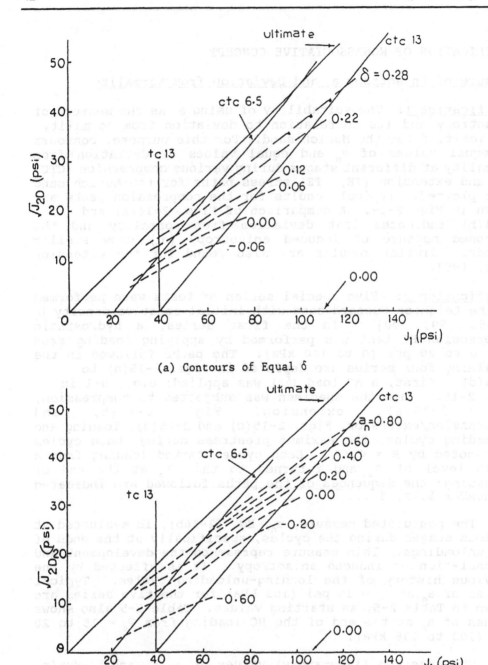

(a) Contours of Equal δ

(b) Contours of Equal a_n

Figure 2-14 Comparison of a_n and δ for Munich Sand –
Compression Paths

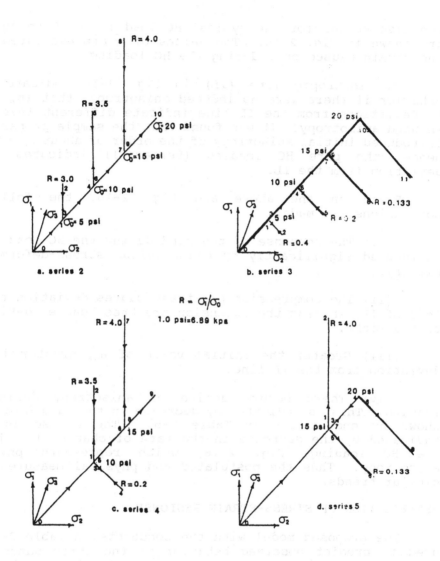

Figure 2-15 Stress Paths for Special Test Series

the five series for the typical HC load $(\sigma_0$ = 15 to 20 psi) are shown in Fig. 2-16. The values of γ are evaluated from the strain caused only during the HC loading.

The isotropic line (IL) in Fig. 2-16 indicates the behavior if there were no induced anisotropy; that is, a_n = 0. Deviations from the IL line indicate different levels of induced anisotropy. It was found that the sample preparation introduced initial anisotropy of the order of about a_n = 0.05. Hence, the pure HC loading (Series 1) indicates small deviation from the IL.

Based on the above and Fig. 2-16, the following conclusions are made:

(i) The response of the sand during the HC loading is influenced significantly by the previous stress-deformation history.

(ii) The compression prestress causes deviation to the left of IL, whereas the extension prestress causes deviation to the right.

(iii) Greater the initial value of a_n, greater is the deviation from the IL line.

(iv) There occurs demise of anisotropy during HC loading. This is indicated by decrease in the values of a_n as shown by end values in Table 2-5. The demise is also indicated by the decrease in the rate of change of during the HC loading, Fig. 2-16, which represents physical anisotropy. Thus the postulated and physical measures show similar trends.

VERIFICATION OF STRESS-STRAIN RESPONSE

The proposed model with the constants in Table 2-5 was used to predict observed behavior of the three sands [99, 100]. The predictions were obtained by integrating the following incremental constitutive equations, starting from initial (hydrostatic) states:

$$\{d\sigma\} = [C^{ep}] \{d\varepsilon\} \qquad\qquad (2-43)$$

where $\{d\sigma\}$ and $\{d\varepsilon\}$ = vectors of incremental stresses and strains, respectively, and $[C^{ep}]$ = tangent constitutive matrix.

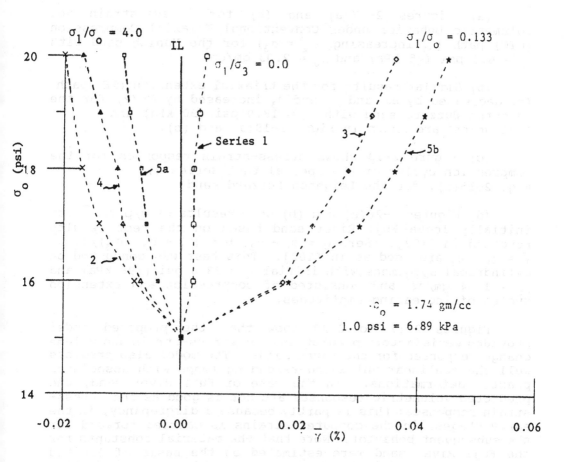

Figure 2-16 Effect of Preloading on HC Response (σ_o = 15 to 20 psi);
See Table 2 from Starting and End Values of a_n

In view of the length limitations, only a few typical comparisons between predictions and observations are described below.

(a) Figures 2-17(a) and (b) for stress-strain and volumetric behavior under Conventional Triaxial Compression (CTC) path (σ_1 increasing, $\sigma_2 = \sigma_3$) for the Munich sand with σ_0 = 6.5 psi (45 kPa) and ρ_0 = 2.03 gm/cm^3.

(b) Similar results for the triaxial extension (TE) path (σ_1 decreased by $\Delta\sigma$ and σ_2 and σ_3 increased by $\Delta\sigma/2$) for the Leighton Buzzard sand with σ_0 = 13.0 psi (90 kPa) and ρ_0 = 1.71 gm/cm^3 are shown in Figs. 2-18(a) and (b).

(c) Figure 2-19 shows stress-strain responses for the compression cycles of the special test Series 2, Fig. 2-15(a), for the Leighton Buzzard sand.

(d) Figures 2-20(a) and (b) show results of q/p vs. $\overline{\gamma}$ for initially loose Fuji River sand based on the test results reported in [102]. Here q = $\sigma_1 - \sigma_3$, p = ($\sigma_1 + \sigma_2 + \sigma_3$)/3, $\overline{\gamma} = \varepsilon_1 - \varepsilon_3$ are used as in [102]. This test was conducted on cylindrical specimens with initial σ_0 = 28.4 psi (196 kPa) and ρ_0 = 1.54 gm/cm^3 and consisted of compression and extension cycles of increasing amplitudes.

Figures 2-17 to 2-20 show that the proposed model provides satisfactory predictions for stress-strain and volume change responses for the three sands. The model also predicts well the nonlinear unloading-reloading loops with associated plastic deformations. In the case of Fuji River sand, the predicted volumetric responses are not as good as the stress-strain responses; this is partly because a discrepancy, in the early stages, in the computed strains is carried forward for the subsequent behavior. Note that the material constants for the Fuji River sand were estimated on the basis of limited test data reported in [102]; in view of this, the correlation is considered to be satisfactory.

Pore Water Pressure (δ_{2+p}-Model)

The model δ_2 is modified to include water pressure p by expressing effective stress tensor $\overline{\sigma}_{ij}$ as

$$\overline{\sigma}_{ij} = (\sigma_{ij}^T - p\,\delta_{ij}) - a_{ij} \qquad (2-44)$$

where σ_{ij}^T = total stress tensor and a_{ij} = tensor denoting location of Q in effective stress space.

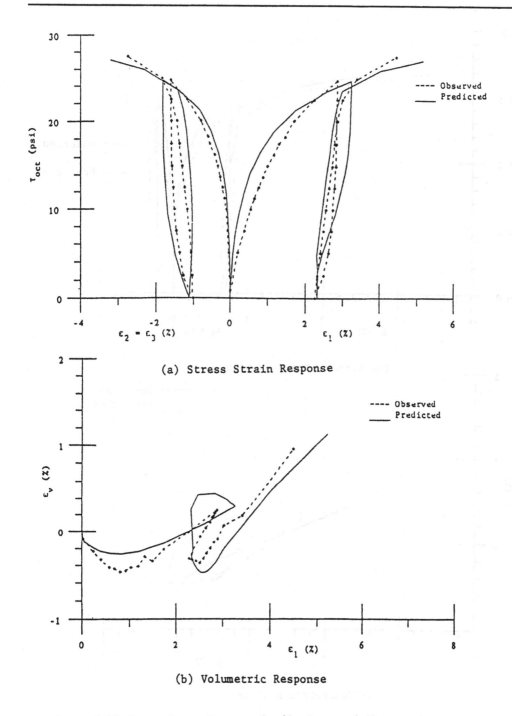

(a) Stress Strain Response

(b) Volumetric Response

Figure 2-17 Comparisons Between Predictions and Observations
for Munich Sand Under CTC (σ_o = 6.5 psi) Test

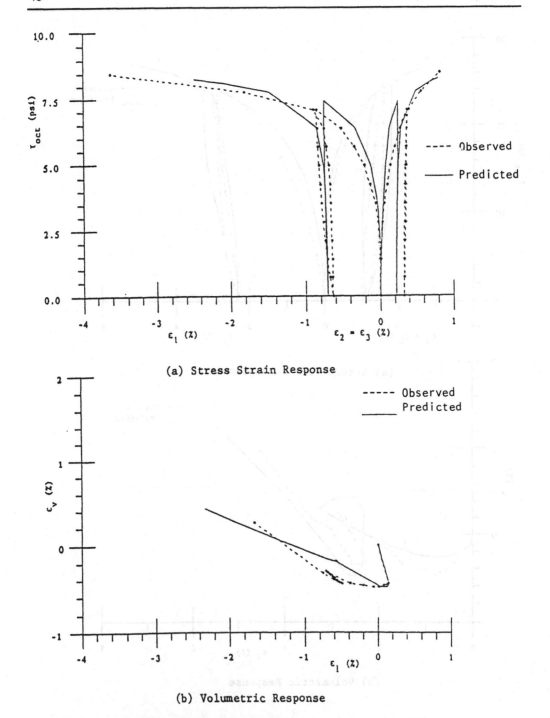

(a) Stress Strain Response

(b) Volumetric Response

Figure 2-18 Comparisons Between Predictions and Observations
for Leighton Buzzard Sand Under TE (σ_o = 13 psi) Test

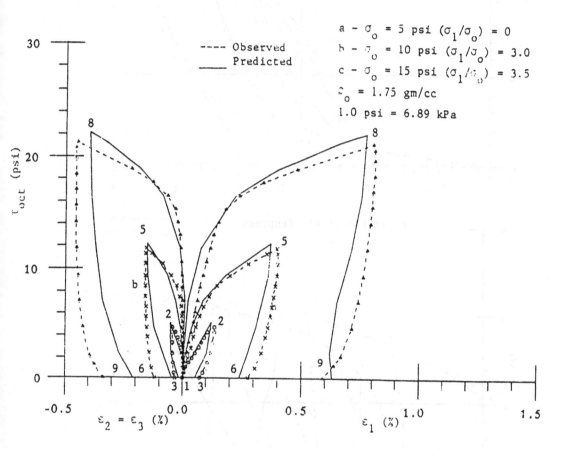

Figure 2-19 Comparison of Stress-Strain Responses for the CTC Segments
for Leighton Buzzard Sand - Test Series 2, Fig. 2-15

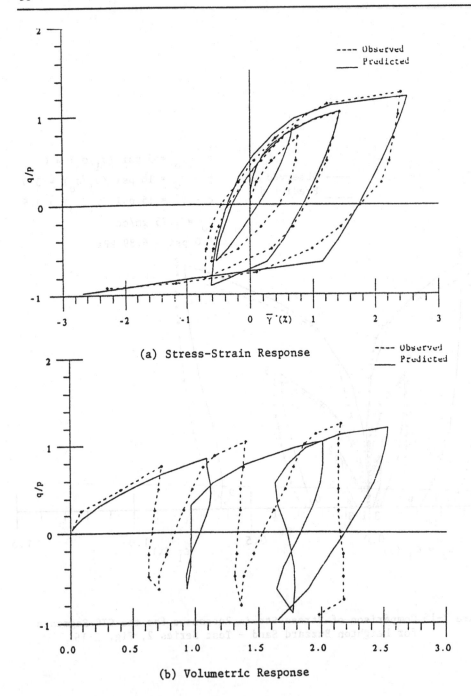

(a) Stress-Strain Response

(b) Volumetric Response

Figure 2-20 Comparisons Between Predicted and Observed Responses
for Fuji River Sand - Cyclic Triaxial Test (σ_o =
28.4 psi)

Details of implementation of δ_{2+p} model in the context of dynamics of porous media are given elsewhere [28, 29]. Modification of the model for cohesive soils is given in [107].

Conclusions

A general anisotropic nonassociative model that accounts for effects of initial and induced anisotropy, stress history and stress path is presented. The model is capable of simulating the behavior of soils during monotonic loading, load reversals, complex stress paths and cyclic loading. A hierarchical approach is adopted in developing the model. It permits evolution of models of progressively higher grades from a basic isotropic hardening associative model.

The mechanism of the proposed model is relatively simple. The parameters have been kept to a minimum, and they are easily determined based on commonly available laboratory test results, are related to specific states during deformation, and possess physical meanings. The new concepts used in the development of this model were verified on the basis of a series of special stress path tests on the sand specimens. The predictive capability of the model is verified with respect to laboratory multiaxial test data under various stress paths of virgin and nonvirgin loading for three different sands of initial dense and loose states.

2.3 MODELLING OF CONTACTS: INTERFACES AND JOINTS

Over the last two decades or so, the authors and co-workers have considered the problem of modelling interfaces and joints. Here various models have been used starting from the hyperbolic simulation of shear stress (τ) vs. relative displacement (u_r) curves for shear (stiffness) response, while arbitrarily high values are used for the normal (stiffness) response [15, 48]. In the context of cyclic behavior of interfaces, models have been proposed and used by using mathematical functions to simulate loading, unloading and reloading behavior [22,38, 52, 54, 55, 111]. Here, a new cyclic multi degree-of-freedom (CYMDOF) shear device has been used for testing of interfaces and joints [14].

The new idea of thin-layer element, Fig. 2-21, is proposed and used to incorporate the shear and normal response in the context of the finite element method [23, 35, 41, 48, 11]. This element has been found to be successful in

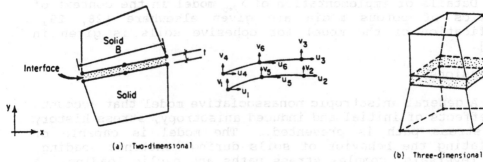

(a) Two-dimensional

B = (average) contact
dimension

(b) Three-dimensional

Thin-layer interface element

Figure 2-21

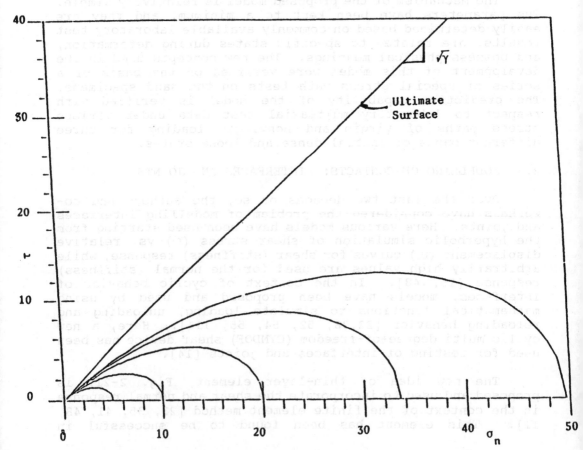

Figure 2-22   Typical Schematic of Yield Function for Joint

providing improved solutions for a number of static and dynamic problems [15, 22, 23, 35, 41, 48, 111]. The basic idea here is to replace the interface zone by a thin layer that has different constitutive properties but its finite element formulation is similar to that of the solid elements.

Important Comments

The problem of contacts which is dependent upon factors such as shear and normal stresses and deformations, asperities at contact, damage and degradation, stiffening and existence of fluids is highly complex. It is observed that such a behavior may not be simulated satisfactorily by using simulation of given $\tau - u_r$ curves, and by adopting empirical values of normal stiffness. In particular, it is believed that the normal response of contact is highly important for realistic simulation of interfaces and joints, and it is not appropriate to assign arbitrarily high values of normal stiffness during contact (compressive normal stress), and very low values when contact breaks (tensile normal stress).

The modelling and testing approach followed by the author and co-workers takes into account both shear and normal responses including laboratory measurements of shear and normal stresses and deformations, damage of asperities and degradation, stiffening and fluid pressures.

The current trend of our research is to use the hierarchical single surface (HISS) concept as a specialization for joints and interfaces [25, 26, 54, 78, 81]. Brief description of this procedure for joints and interfaces is given below.

Hierarchical Single Surface Plasticity Models

Based on the analogies between three-dimensional and planar joints or interfaces, Table 2-6, Eq. (2-1) is specialized as

$$F = \tau^2 + \alpha \ \sigma_n^n - \gamma \ \sigma_n^2 = 0 \qquad (2-45)$$

Since Eq. (2-45) is of the same form as Eq. (2-1), properties of Eq. (2-1) are retained, including $\alpha = 0$ on the ultimate surface, convexity and orthogonality of the intersection with the σ_n axis for the practical ranges of behavior considered herein.   Figure 2-22 presents a plot of surfaces F in the   versus σ_n space.

With Eq. (2-45), the joint does not possess tensile strength.   If it has tensile strength, the same function can be modified so that it is extended in the negative region of σ_n.   This can be achieved by defining modified shear stress $\tau' = \tau + R$, where R is proportional to the tensile strength [39].

Nonassociative Behavior

Shear behavior of many joints exhibit nonassociative response; that is, the plastic strain increments are not orthogonal to the yield surface F.   To account for this response, the correction function approach is used; here the plastic potential function, Q, is defined as

$$Q = F + h\ (\sigma_n,\ \xi) \qquad\qquad (2-46)$$

where h = correction function.   As a simplification, the correction is introduced through the modification of the growth function α to α_Q as

$$\alpha_Q = \alpha + \kappa\ (\alpha_I - \alpha)\ (1 - r_v) \qquad\qquad (2-47a)$$

where κ = material parameter, α_I = value of α at the end of initial normal loading or at the initiation of nonassociative response, and r_v = ratio ξ_v/ξ.   Thus the correction function h is given by

$$h\ (\sigma_n,\ \xi) = \kappa\ (\alpha_I - \alpha)\ (1 - r_v)\ \sigma_n^n \qquad (2-47b)$$

Constants

The elastic shear and normal stiffnesses k_s and k_n, respectively, are determined from unloading and reloading slopes of the (laboratory) shear and normal responses.

The parameter γ is determined from the ultimate envelope representing the asymptotic stress from τ-u curves under different normal stresses, σ_n.   The phase change parameter n is found based on the point at which the normal deformation v_r transits from contractive to dilative.

The growth or hardening function is given by

$$\alpha = a_{\xi}^{\ b} \qquad\qquad (2-48)$$

where a and b are constants, and ξ is defined in Table 2-6. These constants are found from measured shear responses. The nonassociative parameter k is found based on the slope of u_r-v_r curve near the ultimate.

LABORATORY TESTS ON JOINTS

A series of quasi-static direct shear tests was performed on simulated joint surfaces using the translation portion of the CYMDOF device [25, 26, 54, 55].

Description of Samples

Simulated joint surfaces were prepared by casting concrete against a structured metal surface. The structured surface had a sawtooth shaped pattern intended to simulate asperities on the joint surface; Fig. 2-23 depicts the sawtooth shaped asperity pattern. Idealized (rock) joints were constructed by casting top and bottom samples such that when placed together, the asperities from the top and bottom produced a closely mated joint. Different joint geometries were simulated by varying the angle of the sawtooth asperity side with respect to the horizontal. Samples were cast with asperity angles i = 0, 5, 7, 9 and 15 degrees; the first denotes flat geometry. Note that the actual surface characteristics in each case were essentially the same, corresponding to the metal forms against which the concrete was cast.

Measurements

The normal and shear loads were measured by using STRAINSERT Load Cells (Nos. 1 and 2, Fig. 2-24), with capacity of 80 KN (18,000 lbs) under static loading, and ± 53.4 KN (± 12,000 lbs) under cyclic loading. The relative (shear) displacements were measured by installing Linear Variable Differential Transformers (LVDT's) marked No. 3 in Fig. 4.

In the case of the normal displacements, however, locating the LVDT's on the top of the upper sample holder or outside and away from the joint (Nos. 7 and 8), which is commonly done, can involve difficulties due to various reasons such as possible rotation of the box and inaccuracy due to the distance from the joint surface. In order to investigate this problem, additional measurement devices were installed, Fig. 2-24; this included two LVDT's mounted on brackets at the joint (Nos. 5 and 6), and special inductance coils (No. 4),

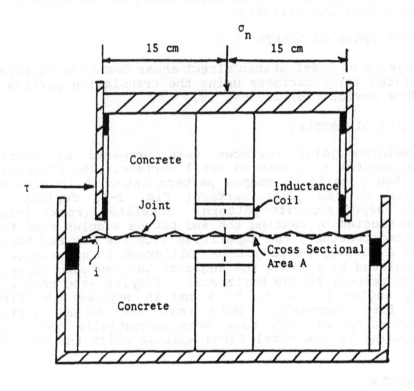

Figure 2-23 Schematic of Test Box and Joint Specimen

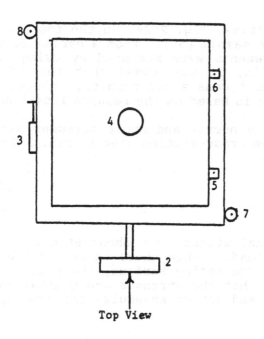

Top View

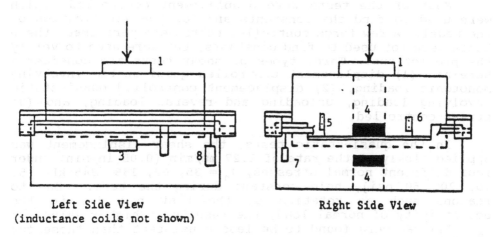

Left Side View
(inductance coils not shown)

Right Side View

Key

1. Vertical load cell
2. Horizontal load cell
3. Horizontal displacement LVDT
4. Inductance coils

5. Inside front displacement sensor
6. Inside back displacement sensor
7. Outside front vertical LVDT
8. Outside back vertical LVDT

Figure 2-24 Placement of Various Displacement Measuring Devices

mounted in cavities, Fig. 2-24, in the top and bottom halves
of the concrete sample [55]. From a series of tests in which
normal displacements were measured by using the LVDT's and
inductance coils, it was found that the inductance coils
provided the most consistent results. Hence, the test data
reported herein is based on the results from inductance coils.

The average normal and shear stresses were computed on
the basis of the cross-section area A, Fig. 2-23, of the joint
surface. Hence,

$$\sigma_n = \frac{V}{A} \tag{2-49a}$$

$$\tau = \frac{F}{A} \tag{2-49b}$$

where σ_n = normal stress, τ = shear stress, V = normal load
and F = shear load. Since there exists only a finite number
of contacts at the mating surface, these definitions involve
the assumption that the stresses are nominal and are uniform
over the joint and act on an equivalent area A, Fig. 2-23.

Tests

Most of the tests were displacement controlled, which
were used to find the constants and for the verification of
the model. A few force controlled tests were performed; these
tests were not used to find constants, but were used to verify
the predictions. Three types of shear tests are considered
herein: (1) displacement controlled quasi-static involving
monotonic loading, (2) displacement controlled quasi-static
involving loading, unloading and reverse loading, and (3)
stress controlled.

In the quasi-static tests, the shear displacement was
applied slowly at the rate of 1.27 mm/min (0.05 in/min) under
four different normal stresses, σ_n = 35, 69, 138, 345 kPa [5,
10, 20, 50 psi], held constant during the test. Due to
reasons such as rotation of the test box due to the
eccentricity of normal load, the results at higher values of
σ_n = 345 kPa were found to be less consistent than those for
the lower values of σ_n.

Samples were initially mated prior to shear, and the
tests were carried up to the maximum shear displacement of
about 0.635 cm (0.25 in) starting from the datum (0) and
following the path (0-1) up the asperity for the forward
loading; Fig. 2-25 depicts the path in the $\tau - \sigma_n$ space with
respect to yield surfaces, F. For tests involving loading,

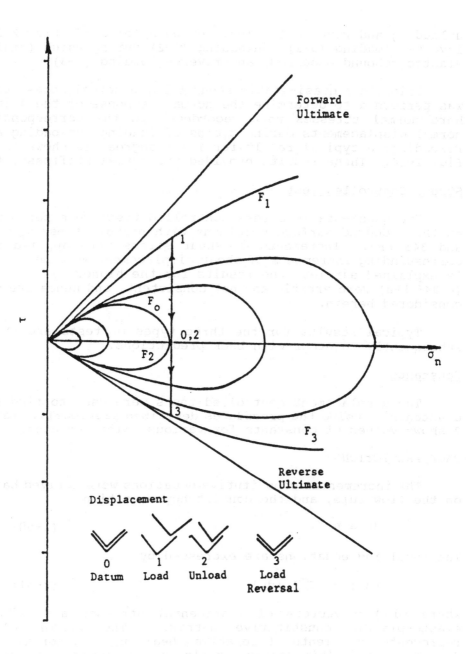

Figure 2-25 Yield Surfaces During Loading,
Unloading and Reverse Loading

unloading and reverse loading, the path followed (Fig. 2-25) involves loading (0-1), unloading (1-2) during which (small) elastic rebound occurred, and reverse loading (2-3).

Prior to a quasi-static shear test, a normal stress test was performed to determine the normal response of the joint. Here normal stresses were recorded with the corresponding normal displacements during cycles of loading, unloading and reloading; a typical result for i = 9 degrees is shown in Fig. 2-26. These results provided the normal stiffness, k_n.

Stress Controlled Tests

Two quasi-static stress controlled tests were performed on the 9-degree surface specimens with normal stress σ_n = 35 and 345 kPa. Increments of shear stress were applied and corresponding increments of shear displacement were measured. As explained earlier, the results for the higher σ_n (= 345 kPa) were erratic and not consistent, and hence are not considered herein.

Typical results for the three types of tests are shown subsequently together with back predictions.

Constants

The displacement controlled tests were used to find the constants by using the procedures described previously. Table 2 shows values of constants for various joint surfaces.

BACK PREDICTIONS

The incremental constitutive equations were derived based on the flow rule, and the consistency condition

$$dF = 0 \qquad\qquad\qquad (2-50)$$

The resulting equations are expressed by

$$\{d\sigma\} = [C^{ep}] \, \{d\varepsilon\} \qquad\qquad\qquad (2-51)$$

where $\{d\sigma\}$ = vectors of incremental stresses and $[C^{ep}]$ = elasto-plastic constitutive matrix. The vector $\{d\varepsilon\}$ represents increments of relative shear, u_r, and normal, v_r, displacements. This vector can also be expressed in terms of strains by using approximate definition as $\varepsilon = u_r/t$, $\varepsilon_n = v_r/t$ where t = (small) finite thickness of the joint [48].

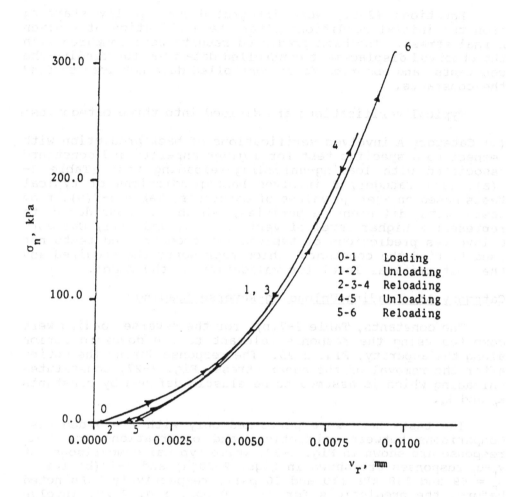

Figure 2-26   Normal Stress versus Vertical Displacement
For a Typical Normal Stress Test for 9° Surface

Equations (2-51) were integrated numerically starting from the initial conditions after the application of a given normal stress. The back predicted results were compared with the observed displacement controlled data used for finding the constants, and observed force controlled data not used to find the constants.

Typical verifications are divided into three categories:

(i) Category A involves verifications of back prediction with respect to a specific test for a given asperity and constants associated with loading-unloading-reloading test (Table 2-7(a), (ii) Category B involves back predictions of typical tests based on average values of constants, Table 2-7(a), from tests with different asperities, which is considered to represent a higher level of verification, and (iii) Category C involves predictions of typical force controlled tests not used to find the constants, which represents the required and the most critical level for validation of the model.

Category A: Loading-Unloading-Reverse Loading

The constants, Table 2-7(b), for the reverse loading were computed using the response relevant to the downward motion along the asperity, Fig. 2-25. The response during the motion after the removal of the shear stress, Fig. 2-27, constitutes unloading which is assumed to be elastic defined by constants k_n and k_s.

The tests here were performed only with i = 9 degrees. Comparisons between predictions and observations for τ -u_r response are shown in Fig. 2-27, while typical comparisons for v_r-u_r responses are shown in Figs. 2-28(a) and 2-28(b) for σ_n = 69 and 138 kPa (10 and 20 psi), respectively. As noted before, the predictions for σ_n = 50 psi, Fig. 2-27, involve significant discrepancy. The observed v_r-u_r responses during loading and reverse loading show little difference, Fig. 2-28. Although the predictions from the nonassociative model show differences with observations, the predictions are considered to be satisfactory.

Category B: Average Constants

Constants n, a, b and   showed relatively lower levels of variations, and hence were averaged as shown in the last column of Table 2-7(a). Since the constant γ shows consistent variation with asperity angle, it was considered to be a

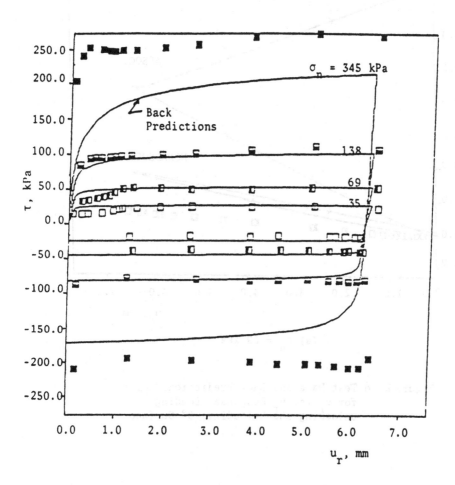

Figure 2-27 Test Data and Back Predictions for
τ vs. u_r Response; Loading, Unloading
and Reverse Loading, 9° Surface

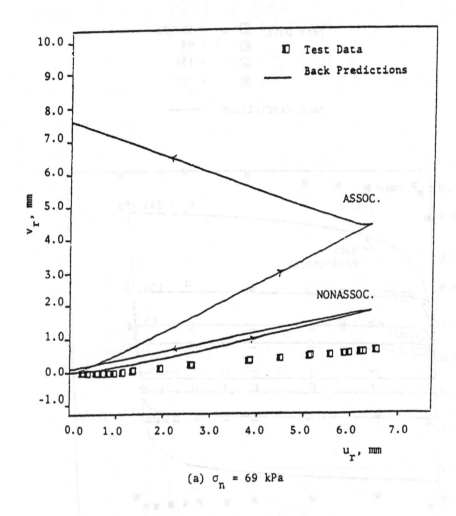

(a) σ_n = 69 kPa

Figure 2-28 Test Data and Back Predictions for
          for v_r vs. u_r Response, Loading,
          Unloading and Reloading, 9° Surface

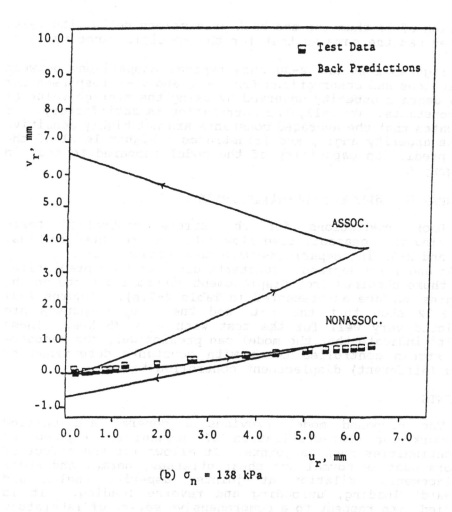

(b) $\sigma_n = 138$ kPa

Figure 2-28 (Continued)

function of i; thus for the back predictions below, its value adopted was the same as that for the specific asperity.

Figures 2-29 and 2-30 show typical comparisons between predictions and observations for $\tau - u_r$ and $v_r - u_r$ responses for i = 9 degrees asperity observed by using the average value of the constants. Overall, the correlation is satisfactory. It indicates that the averaged constants are not highly sensitive to the asperity angle, and illustrates a higher level of the back prediction capability of the model compared to that in Category A.

Category C: Stress Controlled Tests

Back predictions for the stress controlled tests utilizing the nonassociative flow rule are presented in Figs. 2-31 and 2-32 in comparisons with observations for $\sigma_n =$ 35 kPa and i = 9 degrees. Constants used for back predictions were those obtained from displacement controlled tests on the 9-degree surface as presented in Table 2-7(b). Figures 2-31 and 2-32 show that the $\tau - u_r$ and the $v_r - u_r$ responses are predicted very well for the test with σ_n = 35 kPa. These results indicate that the model can predict well the response from stress controlled tests with constants determined by using (different) displacement controlled tests.

COMMENTS

The proposed model provides a general and unified framework for characterization of mechanical response of discontinuities such as joints. It allows for the effect of factors such as normal and shear stresses, normal and shear displacements, dilation at joints, asperity angles and (forward) loading, unloading and reverse loading. It is verified with respect to a comprehensive series of laboratory tests. The model is derived as a specialization of the general hierarchical approach for solids, it is relatively simple, possesses a number of advantages compared to other available models of similar capabilities, and involves constants that can be determined from standard laboratory shear tests. One of the advantages of the model is that it provides for the use of the same plasticity framework for both the solid (parent) rock and joints. As a consequence, it can be implemented easily in solution procedures such as the finite element method for boundary value problems involving discontinuous media with solid and joint components.

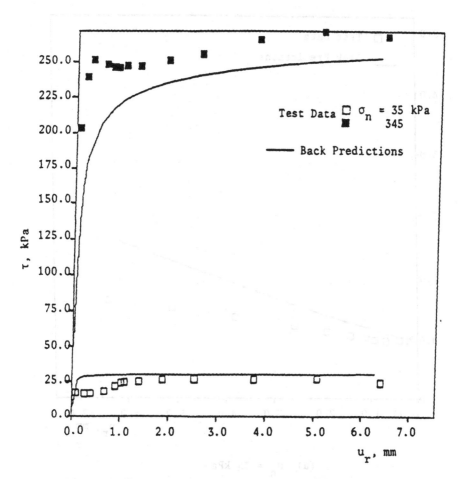

Figure 2-29 Test Data and Back Predictions for
τ vs. u_r Response with Average Constants,
and Nonassociative Model, 9° Surface

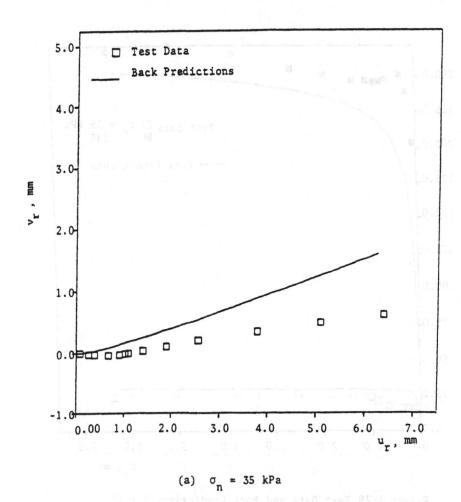

(a) σ_n = 35 kPa

Figure 2-30 Test Data and Back Predictions for
v_r vs. u_r Response with Average
Constants and Nonassociative Model,
9° Surface

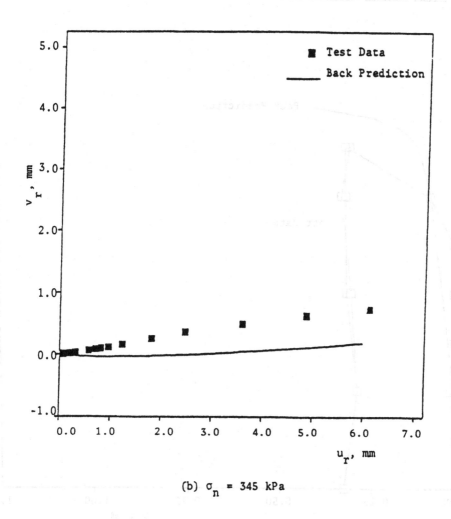

(b) σ_n = 345 kPa

Figure 2-30 (Continued)

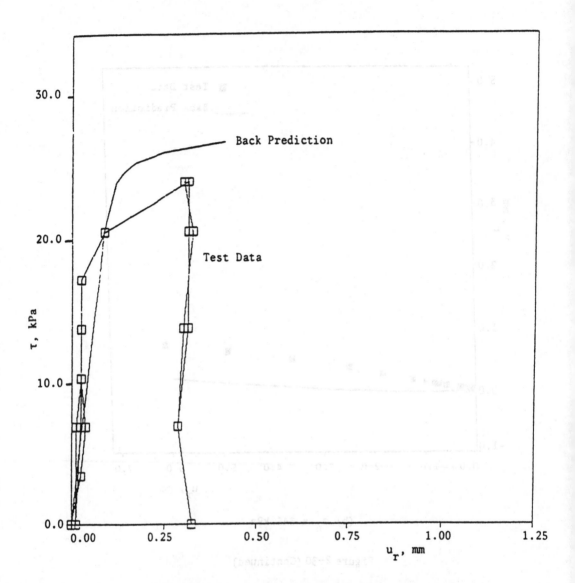

Figure 2-31 Test Data and Back Predictions for
τ vs. u_r, Stress Controlled Response,
9° Surface, σ_n = 35 kPa

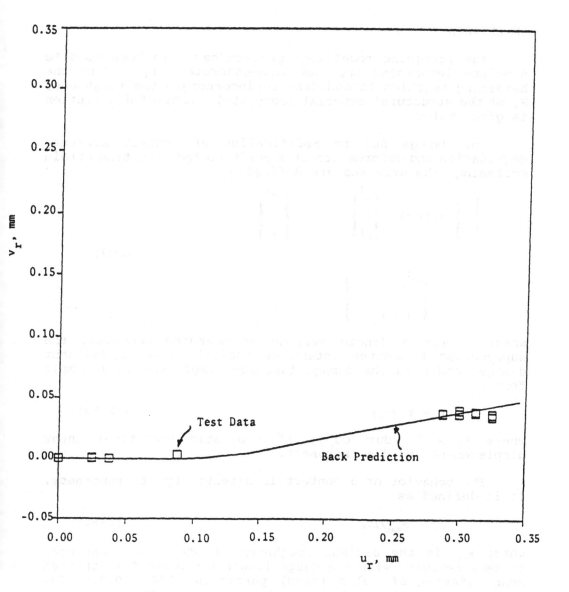

Figure 2-32 Test Data and Back Predictions for Stress Controlled v_r vs. u_r Response, 9° Surface, σ_n = 35 kPa

INTERFACES

The foregoing modelling approach has also been used to simulate interfaces such as sand-concrete [81]. Here the hardening function is modified to incorporate the roughness, R, of the structural material (concrete). A brief description is given below.

For damage due to modification of contact surface, degradation and deformation of asperities and resulting strain softening, the stresses are defined as

$$\left\{ \begin{array}{c} \sigma \\ \tau \end{array} \right\} = (1-r) \left\{ \begin{array}{c} \sigma^t \\ \tau^t \end{array} \right\} + r \left\{ \begin{array}{c} \sigma^t \\ 0 \end{array} \right\}$$

(2-52)

$$= \left\{ \begin{array}{c} \sigma^t \\ (1-r)\,\tau^t \end{array} \right\}$$

where σ and τ denote average or measured response, the superscript t denotes intact or topical response (without damage) and r is the damage function, expressed in a simple form as

$$r = r\,(\xi_D)$$

(2-53)

where $\xi_D = \int (du_r^p \, du_r^p)$, u_r^p = plastic (relative) shear displacement and d = increment.

The behavior of a contact is affected by its roughness. It is defined as

$$R_n = R_{max}/D_{50}$$

(2-54)

where R_{max} is the maximum roughness of steel (or concrete) surface measured within a gauge length L = D and D is the 50% mean diameter of solid (sand) particles [104, 105]. The hardening parameter is now defined as

$$\alpha = \alpha\,(\xi_v,\, \xi_D,\, R_n)$$

(2-55)

where $\xi_v = \int (dv_r^p \, dv_r^p)^{1/2}$. Explicit expressions for α, α_0, r and r_u are given by

$$\alpha = \gamma \, e^{-a\xi_v} \, (\xi_D)^b$$

$$\alpha_Q = \alpha + \gamma \, (1 - \tfrac{2}{r}) \, [1 - (k_1 + k_2 R) \, (1 - \tfrac{\gamma}{r_u})]$$

$$r = r_u \, (1 - e^{-A \, (\xi_D)^2}) \qquad\qquad (2-56)$$

$$r_u = \frac{\tau_{max} - \tau}{\tau_{max}}$$

where $R = R_n/R_{n \, critical}$, τ_{max} = maximum shear stress, τ_r = residual shear stress, and a, b, r, r_u, k_1, k_2 are material constants.

Application

The plasticity model for interfaces is used to predict laboratory simple shear behavior of interfaces between Toyoura Sand and steel conducted by [104, 105]. The tests were performed under various initial densities for sand, roughness for steel and normal stress σ_n. Measurements included shear and normal shear and normal deformations, and shear and normal stresses.

Figures 2-33(a) and (b) show typical comparisons of predictions and laboratory observations for τ/σ vs. relative shear displacements, u_r, and vertical displacements v_r vs. u_r for four different roughnesses and $\sigma = 98$ kPa, and relative sand density = 90%.

2.4 GENERALIZED "DISTURBED STATE" CONCEPT

A generalized "disturbed state" concept is being developed by the author and co-workers [2, 16, 66, 108] in the context of the HISS modelling approach. In this approach, special characteristics such as frictional, anisotropy and physical damage (microcracking, etc.) are referred to as disturbance (damage, changes, or perturbations) in the material with respect to basic or reference state(s). Thus damage that may constitute any change in the deforming material is considered as the disturbance from a reference state. One of the reference states is often adopted as the behavior of the initially isotropic material, hardening isotropically and following associative behavior, that is, the δ_o-model.

Figures 2-34 and 2-35 show schematics of the concept. Here the response of the densest state of the material is adopted as the top reference boundary, and that under the critical state as the other (lower) reference state. Then the observed response may lie between the two reference states. The plasticity model is then developed in terms of the two reference responses and the parameter D that defines the

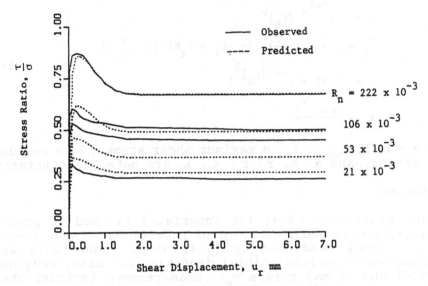

(a) Stress Ratio, τ/σ vs. Shear Displacement, u_r

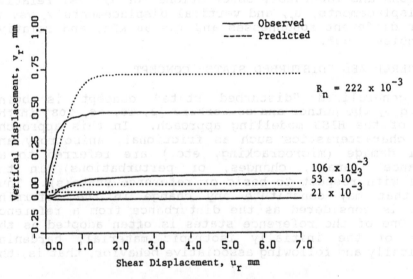

(b) Vertical Displacement, v_r, vs. Shear Displacement, u_r

Figure 2-33 Comparisons of Predictions and Observations for
Sand/Concrete Interfaces

CONCEPT OF $(\delta_r{}^*)$

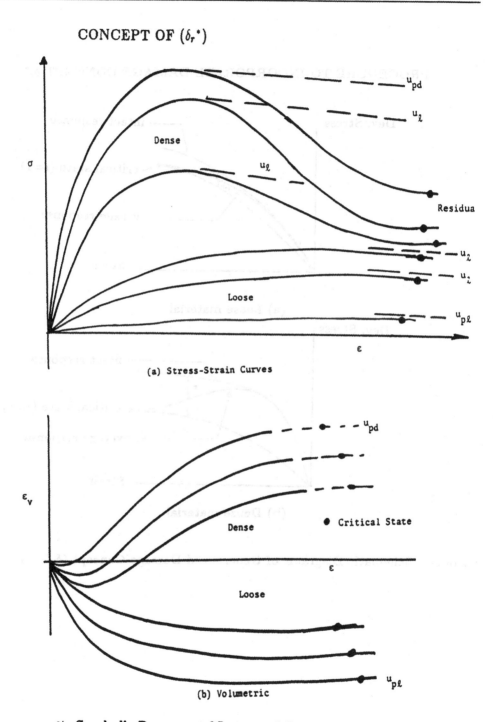

(a) Stress-Strain Curves

(b) Volumetric

Figure 2-34 Symbolic Response of Loose and Dense Materials

PROCEDURE TO INCORPORATE DAMAGE CONCEPT-$\delta_r{}^*$

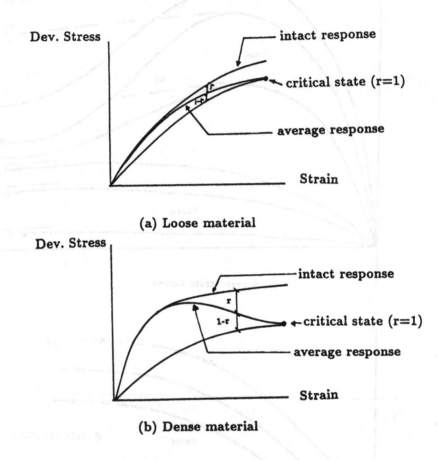

(a) Loose material

(b) Dense material

Figure 2-35 Schematic Diagrams of Generalized Damage Concept ($\delta_r{}^*$)

relative position of the observed response with respect to the densest and the critical responses.

The above approach is currently developed for entire (loose to dense) range of the material behavior under static, quasi-static and cyclic loading.

Appendix 2-1

SINGLE SURFACE YIELD AND POTENTIAL FUNCTION PLASTICITY MODELS:
A REVIEW [17]

INTRODUCTION

Recently, Kim and Lade [67] and Lade and Kim [74, 75] have reported the use of single surface yield (and potential) functions for describing hardening or continuous yielding of geologic materials. The objective of this contribution is to present a limited review of the single surface functions in the context of the theory of plasticity, and indicate that the models and the concept proposed by Lade and Kim [67, 74, 75] are essentially the same as the hierarchical single surface (HISS) concept proposed by Desai and co-workers [4, 13, 24, 26, 30, 39, 40, 45, 46, 49, 56, 57, 96, 100].

Definition

A schematic of the yield function, F, for a typical material, developed by the author and co-workers [4, 13, 24, 26, 30, 40, 45, 46, 49, 56, 57, 96, 100] in the context of the hierarchical single surface (HISS) approach is shown in Fig. 2-36. A single surface plasticity model involves a single mathematical function for describing yielding of a (geologic) material. Thus, the same function defines the entire process of yielding from the start of loading up to failure peak or ultimate condition, and includes the portions above and below the phase change curve, Fig. 2-36(a), that defines the state where volume change transits from contraction to dilation. The final or ultimate state is defined through the final yield surface (F_u) corresponding to the asymptotic states of stress from the observed stress-strain behavior. It provides unique ultimate state towards which the deformation process approaches, and the conventionally defined failure, peak or critical states lie below the ultimate state or coincide with it.

REVIEW

The following review is provided in relation to the single surface approach. It is not intended to provide a detailed review of the large body of literature, and only some of the works directly relevant to the subject of associative and nonassociative plasticity are reviewed. Table 2-8 shows

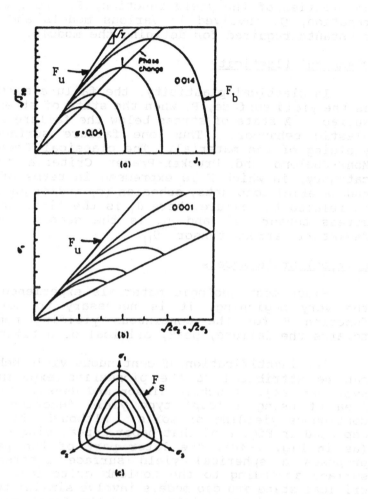

Figure 2-36 Plot of F in Various Stress Spaces:
(a) In $J_1 - \sqrt{J_{2D}}$) Space; (b) In
Triaxial Space; (c) In Deviatoric Space

an overview of the yield function, F, and plastic potential function, Q, involved in various models and the number of constants required for defining the models.

Classical Plasticity

In classical plasticity, the failure surface is adopted as the yield surface, F, when the state of stress reaches that surface. A state of stress below the failure surface denotes elastic behavior. Thus one failure surface defines the yielding of the material. The classical Tresca, von Mises, Mohr-Coulomb and Drucker-Prager Criteria fall into this category, in which F is expressed in terms of J_{2D} and/or J_1, and material constants, cohesion (c) and/or angle of friction, ϕ, related to failure, here J_1 is the first invariant of the stress tensor, σ_{ij} and J_{2D} is the second invariant of the deviatoric stress tensor, S_{ij}.

Critical State Models

Since most geologic materials experience yielding from the very beginning, it is necessary to define the yield function F for the continuous yielding behavior leading towards the failure, peak, critical or ultimate.

The identification of continuous yield behavior of soils can be attributed to the pioneering experimental work by Hvorslev [64]. Drucker, Gibson and Henkel [51] proposed the idea of using a 'cap' type yield function to define the continuous yielding of soils. Although the yield surfaces depicted in Fig. 8 of that paper appear single and continuous (as in Fig. 2-40), the later part of the paper, Fig. 11, proposes a spherical yield surface intersecting failure surface according to the Coulomb criterion. The following critical state and cap models involve similar two surfaces for defining continuous yielding and the final critical state or failure behavior.

In the critical state concept proposed by Roscoe and co-workers [86, 90, 91], a yield function F_y defines the continuous yielding, while the final yield is defined by the critical state at which no volume change occurs during further shearing. The locus of the critical state points, F_c, defines the critical state line (CSL). The surface F_y represents a single surface; however, the model involves two surfaces, a part of F_y below F_c, and F_c for description of the entire behavior. Poorooshasb [85] and Tatsuoka and Ishihara [103] recommended a procedure for determination of yield functions in the $p' - q$ space, where $p' = (\sigma_1' + 2\sigma_2')/3$ and $q = \sigma_1' - \sigma_3'$ based on cylindrical triaxial test configuration ($\sigma_2' = \sigma_3'$). Nova and Wood [83] presented an alternative procedure to develop experimentally the family of yield loci based on simple drained tests with standard triaxial device. Analytical expressions were developed for plastic potential and yield functions by considering various experimental loading paths, and it was shown that they lead to a family of continuous yield functions, similar to those shown in Fig. 2-40. The procedure was applied to behavior of two soils leading to the yield surfaces bounded by the critical state line, thus involving two surfaces for the behavior of the soils considered.

The cap model, proposed by DiMaggio and Sandler [50] is similar to the critical state model, except that a composite failure yield surface F_1 replaces the CSL. Again, the yield surfaces F_y allows for continuous yielding, and at failure, F_1 is used as separate yield surface.

The critical state and cap models can suffer from various limitations: (a) for states of stress below the CSL, only compressive volumetric strains are predicted, and dilative strains are predicted only if F_c is used as yield surface. Since many materials exhibit volume increase (dilation) before the critical state or peak stress is reached, the model cannot predict such behavior; (b) the yield surface is circular in the principal stress space; hence, the model cannot allow for path dependent failure or ultimate response, as it occurs for many geologic materials; (c) the hardening or yielding is defined through total volumetric plastic strains (or void ratio), thus the definition of hardening does not include the effect of deviatoric plastic strains; (d) the yield surface, F_y, intersects the p-axis orthogonally. Hence, the coupled effect of the mean pressure on shear strains is not included; and (e) if F_c (or F_1) is used in computations, the model involves singularity at the intersection of F_y and F_c (or F_1), which may cause computational difficulties.

Matsuoka and Nakai Model

Based on a comprehensive series of laboratory tests on soils, Matsuoka and Nakai [79], identified the following failure criterion:

$$F = \frac{J_1 \cdot J_2}{J_3} - \text{constant} = 0 \qquad (2\text{-}57)$$

where J_i (i = 1, 2, 3) = first, second and third invariants of the stress tensor, σ_{ij}. F in Eq. (1) can represent an open failure surface in the stress space, Fig. 2-37.

Lade's Model

Based on comprehensive laboratory tests (on sands) Lade and associates [72, 73] proposed the function F, Fig. 2-41 as

$$F = \frac{J_1^3}{J_3} - \text{constant} = 0 \qquad (2-58)$$

which can expand in the stress space and approach the failure surface F_1. Subsequently, a spherical yield cap, Fig. 2-41, was added to allow for continuous yielding of the material. This two surface model involves a singularity at the intersection of the cap and F_1. Also, as shown in Table 2-8, the model involves 13 elasticity and plasticity constants.

Hierarchical Single Surface (HISS) Models

Around 1975, Desai and co-workers considered the foregoing models, and implemented some of them in nonlinear finite element procedures for static, dynamic and consolidation analysis of problems in geomechanics. However, in view of the above limitations, systematic research that integrates theoretical, experimental and verification phases was initiated so as to develop a unified approach for modelling of (geologic) materials. One of the ingredients of this hierarchical approach has been development of a single mathematical function for yield and potential functions [4, 13, 24, 26, 30, 40, 45, 46, 49, 56, 57, 96, 100]. In the HISS approach, the yield function F (and plastic potential, Q) is expressed as a complete polynomial in J_1, $J_2^{1/2}$ and $J_3^{1/3}$ [13]. A simplified truncated form of F was developed and is given by [24, 45, 57]:

$$F = \frac{J_{2D}}{p_a^2} - F_b \cdot F_s \qquad (2\text{-}59a)$$

where

$$F_b = -\alpha \left(\frac{J_1}{p_a}\right)^n + \left(\frac{J_1}{p_a}\right)^2$$

$$F_s = (1 - \beta S_r)^m$$

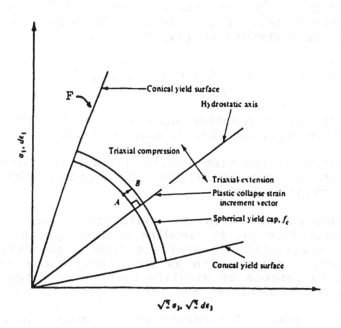

Figure 2-37 Surfaces in Lade, and Matsuoka and Nakai Models

S_r = stress ratio $\frac{\sqrt{27}}{2} J_{3D}/J_{2D}^{1.5}$, p_a = atmospheric pressure and α, n, γ, β and m are response functions or parameters. The basic function F_b is related to F in J_1 - $\sqrt{J_{2D}}$ space, Fig. 2-40(a), and the shape function F_s is related to F in the principal stress space, Fig. 2-36(c). The plastic potential function, Q, is expressed as [24, 45, 57]:

$$Q = F + h (J_i, \xi) = \frac{J_{2D}}{p_a^2} - F_b \cdot F_s + h (J_i, \xi) \qquad (2-59b)$$

where h = correction function and $\xi = \int (d\varepsilon_{ij}^P d\varepsilon_{ij}^P)^{1/2}$, and $d\varepsilon_{ij}^P$ = total incremental plastic strain tensor.

In addition to removing the foregoing limitations of the critical state and cap models, the HISS approach includes the following considerations:

(1) Theoretical basis in conjunction with the form invariance principle [4, 45] leading to development of models for special attributes of materials from a generalized and unified concept, referred to as the hierarchical approach. The concept is capable of handling both solid materials and discontinuities (interfaces and joints) [26, 78, 81].

(2) Development of nonassociative models based on the correction of F to lead to the potential function, Q, Eq. (2-59b).

(3) Development of the hardening function in terms of the total trajectory of plastic strains, ξ, including deviatoric and volumetric components; in the original critical state and cap models, hardening is defined only in terms of the volumetric plastic strains. Here it was found that use of the trajectory of total plastic strains provides more consistent relationships than the use of plastic work, W^P. It may be noted that the yield surfaces F are related to and are functions of plastic work contours. In other words, it can be shown that, in general, the plastic work contours have the same form and follow the yield surfaces.

(4) The final yield surface corresponds to the unique ultimate state; thus ambiguities due to the use of various definitions such as peak and failure are avoided.

(5) The concept allows for easy incorporation of the change in the shape and size of F, and tension and cohesion of materials.

(6) Development of as simple a model as possible with optimal number of material constants with appropriate physical meanings. As a result, the number of constants involved in the associative (δ_o) and nonassociative (δ_1) models is smaller or equal to other models of comparable capabilities, Table 2-8.

(7) Extensive cylindrical triaxial and multiaxial laboratory tests on soils, rocks and concrete have been performed to define the constants in models together with development of guidelines for minimum number of tests required for evaluation of the constants.

(8) Verification of the models with respect to laboratory tests used to find the constants and those not used to find the constants, and implementation of the models in nonlinear finite element procedures for two- and three-dimensional static and dynamic analysis including verification of the numerical procedures with respect to laboratory and field observations for various boundary value problems. Here, one of the advantages is that the single surface avoids the singularity that exists in the two and multisurface models.

(9) The HISS approach allows for the development of models of progressively increased complexities such as nonassociative behavior (δ_1) [24, 45, 57], anisotropic hardening with fluid pressure (δ_{2+p}) [28,45], damage and softening (δ_{o+r}) [45, 56] and viscoplastic (δ_{o+vp}) [49] models, based on corrections of the basic δ_o-model that defines behavior of isotropic material, hardening isotropically with associative plasticity [45].

Single Surface Models with Lade's Function

Schreyer and Babcock Model

Schreyer [97] proposed a modification of the function developed by Lade [72, 73] so as to lead to a single yield surface similar to that in Fig. 2-36.

Lade and Kim Model

Recently Lade and Kim [67, 74, 75] have proposed what is referred to as the single hardening constitutive models. Figure 2-38 shows the yield and plastic potential surfaces in

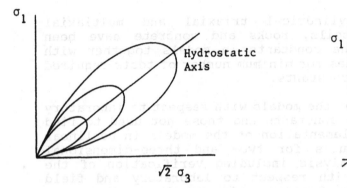
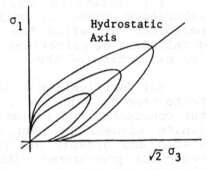

(a) Yield Surfaces in Triaxial
Plane

(b) Plastic Potential Surfaces
in Triaxial Plane

Figure 2-38 Yield and Plastic Potential Surfaces: Lade and Kim [67,74,75]

those models. In view of the above review, this approach is considered similar to the hierarchical single surface (HISS) approach developed previously by Desai and co-workers, Fig. 2-36.

The yield function F by Lade and Kim is expressed as the product of two functions as

$$F = f_1 \cdot f_2 \tag{2-60a}$$

where

$$f_1 = (\psi_1 \frac{J_1^3}{J_3} - \frac{J_1^2}{J_2})$$

$$f_2 = (\frac{J_1}{p_a})^h / e^q$$

and the plastic potential function as

$$Q = (\psi_1 \frac{J_1^3}{J_3} - \frac{J_1^2}{J_2} + \psi_2) (\frac{J_1}{p_a})^\mu \tag{2-60b}$$

where ψ_1, ψ_2, h, q and μ = material response functions or parameters. Note that F in Eq. (2-60a) has the same form as F in Eq. (2-59a), and Q in Eq. (2-60b) can be expressed as Q = F + h as in Eq. (2-59b), and Fig. 2-38 is similar to Fig. 2-36.

Lade and Kim model is based on a number of above factors that were considered by Desai and co-workers in their HISS approach that integrates both theoretical and experimental aspects. There are some (minor) differences, however. For example, (1) Lade and Kim approach defines a separate failure surface, which in the HISS approach is a subset of the compact single surface function, and (2) the number of constants required is greater than that in the HISS approach. For instance, the total number of elastic and plastic constants in the δ_1-nonassociative model in the HISS approach is 8, whereas for the comparable nonassociative model in Lade and Kim model is 11 [Ref. 75], Appendix 1, and (3) the damage/softening response in the HISS approach is based on a consistent definition of damage function that leads to incorporation of damage due to microcracking and discontinuous nature of the material damage and subsequent softening [56],

whereas in the Kim and Lade approach, softening is
incorporated in the continuum sense involving contraction of
the yield surface.

Finally, the author believes that Lade and co-workers
have performed valuable and comprehensive laboratory testing
of geomaterials, and have proposed improved constitutive
models. However, it is also felt that in their recent work
on single hardening models [67, 74, 75], it would be
appropriate to provide references to previously published and
essentially identical works on the subject.

Table 2-1. Hierarchical Models for Soils, Rocks, Concrete and Joints/Interfaces

Model	Hardening	Flow Rule	Special Forms	Constants (in addition to 2 elastic constants)	References
δ_0	Iso	A	Straight ultimate envelope	ultimate m, γ, β phase change n hardening a_1, n_1	Desai [13] Baker and Desai [4] Desai and Faruque [24] Desai, Somasundaram and Frantziskonis [45]
			Curved ultimate envelope ($f (J_1)$)	ultimate $m, \gamma, \beta, \beta_1$ phase change n hardening a_1, n_1	Desai and Salami [39, 40] Desai and Varadarajan [46]
δ_1	Iso	NA	Straight ultimate envelope and constant κ	ultimate m, γ, β phase change n hardening a_1, n_1 nonassociative κ	Desai and Faruque [24] Frantziskonis, Desai and Somasundaram [57]
			Straight ultimate envelope, constant κ and split hardening (i.e., $\alpha = f (\xi, \xi_d)$)	ultimate m, γ, β phase change n hardening b_1, b_2, b_3, b_4 nonassociative κ	Desai and Hashmi [30]
			curved ultimate envelope, variable κ	ultimate $m, \gamma, \beta, \beta_1$ phase change n hardening a_1, n_1 nonassociative κ_1, κ_2	Varadarajan and Desai [106]
δ_2	Aniso	NA	Straight ultimate envelope Plastic unloading and reloading	ultimate m, γ, β phase change n hardening (virgin) a_1, n_1 interpolation h_1, h_2 and h_3 translation h_4	Desai, Somasundaram and Frantziskonis [45] Somasundaram and Desai [100] Somasundaram [99]

Table 2-1. (Cont'd.)

Model	Hardening	Flow Rule	Special Forms	Constants (in addition to 2 elastic constants)	References
δ_{2+p}	Aniso	NA	Same as δ_2 above with modified stress for pore pressure	constants for δ_2 + porosity $\bar{n}$ + bulk modulus of water, K_f	Galagoda [58], Desai, Galagoda and Wathugala [29]
δ_{1+r}	Same as δ_1	Same as δ_1	$\delta_0 + r$	constants for δ_0 + damage r_u, R, $\bar{\kappa}$	Desai, Somasundaram and Frantziskonis [45], Frantziskonis and Desai [56]
δ_{1+v}	Same as δ_1	Same as δ_1	$\delta_0 + v$ viscoplastic	constants for δ_1 + visco $\bar{Y}$, N	Desai and Zhang [49]
δ_{1+r}^j	Same as δ_1	Same as δ_1	δ_{1+r}^j for Discontinuities: Joints and Interfaces	ultimate Y, phase change n, hardening a_1, η_1, nonassociative κ	Fishman and Desai [54], Desai and Fishman [25], Ma [78], Navayogarajah [81]

Note: A = Associative, NA = Nonassociative, Iso = Isotropic hardening, Aniso = Anisotropic Hardening

Basic Models

δ_0 = isotropic hardening, associative
δ_1 = isotropic hardening, nonassociative
δ_2 = anisotropic hardening, nonassociative
δ_{1+r} = δ_1 model with damage (r) (strain-softening)
δ_{1+v} = δ_1 model with viscoplastic modifications
δ_{1+r}^j = models for discontinuities (joints/interfaces)

Table 2-2. Definitions and Physical Meanings for Material Constants

Constants for δ_o Model				Additional Constants for δ_1 Model
Ultimate Parameters		Phase Change Parameter (constants)	Hardening Parameters (constants)	Nonassociative Parameter
γ	β	n	a_1 and η_1	κ
C → Compression S → Simple Shear E → Extension *tan θ = $\sqrt{\gamma F_s}$ *slope of the ultimate envelope depends on γ *γ is related to the classical friction angle (φ), Eq. (20) *For φ = 0, γ = 0 *When φ increases γ increases	*β determines the yield surface on the octahedral plane *β is related to friction angle φ, Eq. (19a) *β increases with φ *For φ = 0, β = 0 *F_s could be expressed with β and β_1 to include the effect of J_1 on the shape of ultimate yield surface	P → point at which $d\varepsilon_v^P = 0$ * n determines the stress at which the material starts dilating *$n = \dfrac{2\gamma F_s}{\gamma F_s - (\dfrac{J_{2D}}{J_1^2})}$ * n > 2 * n has an effect on the shape of yield surfaces in $J_1 - \sqrt{J_{2D}}$ space.	*$\alpha = \dfrac{a_1}{\xi^{\eta_1}}$ *Lower a_1 and higher η_1 denotes a stiffer soil	* κ restrains dilation * when κ increases S_u decreases * κ can be a function of stress path (Varadarajan and Desai)

Table 2-2. (Cont'd.)

Additional for δ_2 Model		Additional for δ_{1+r} Damage/Softening	Additional for δ_{1+r} Viscoplastic	ζ^*_{1+r} (Discontinuities/ Joints)
Interpolation Constants h_1, h_2 and h_3	Translation Constant h_4	r_u, $\bar{k}$, R	$\bar{\Gamma}$, N	γ, n, a_1, η_1, κ

$*h_1$, h_2 and h_3 define the interpolation formula to calculate λ_{nv}

$*d\varepsilon^p_{ij} = \lambda_{nv} \dfrac{\partial Q}{\partial \sigma_{ij}}$ for nonvirgin loadings

*Ref. Desai, Somasundaram and Frantziskonis for details

$*\Pi = h_4 \, a_n = $ a measure of induced anisotropy

*Ref. Desai, Somasundaram and Frantziskonis for details

$*r = r_u \, [1 - \exp(-\bar{k} \, \bar{\xi}_D^R)]$

$*r = \dfrac{\text{Damaged Volume}}{\text{Total Volume}}$

$*r_u$ determines the residual strength

$*\bar{k}$, R determines the shape of the peak in the stress-strain curve.

*Ref. Frantziskonis and Desai

$*\dot{\varepsilon}_1 = k \, t^{0.4} \, \sigma_\alpha \, 3.0$

$*\dot{\varepsilon}_{vp} = \bar{\Gamma} \left(\dfrac{F}{F_o}\right)^N \dfrac{\partial F}{\partial \bar{\sigma}}$

$*\bar{\Gamma}$ Fluidity Parameter

*N Exponent

*Ref. Desai and Zhang for details

*All the constants have the similar physical meanings as that of corresponding constants in the model for solids.

*Ref. Fishman and Desai for details

*For δ_{2+r} model additional parameters $(\bar{n}, k_f)$ are required

Table 2-3. Relationship Between Elastic Constants and Unloading/Reloading Slopes for Different Stress Path Tests

Test	E	ν
CTC RTE	$\dfrac{3}{\sqrt{2}} S_1$	$\dfrac{2\lvert S_1\rvert}{\lvert S_2\rvert + \lvert S_3\rvert}$
CTE RTC	$\dfrac{3\sqrt{2}\lvert S_1\rvert(\lvert S_2\rvert + \lvert S_3\rvert)}{(4\lvert S_1\rvert + \lvert S_2\rvert + \lvert S_3\rvert)}$	$\dfrac{\lvert S_2\rvert + \lvert S_3\rvert}{4\lvert S_1\rvert}$
TC TE	$\dfrac{\sqrt{2}}{3}(1+\nu)(\lvert S_1\rvert + \lvert S_2\rvert + \lvert S_3\rvert)$	-
SS	$\dfrac{2}{\sqrt{3}}(1+\nu)(\lvert S_1\rvert + \lvert S_3\rvert)$	-

where S_i = slope of the unloading/reloading curve, τ_{oct} vs. ε_i.

Table 2-4.   Constants for Different Sands

(All constants, except where indicated, are dimensionless)

Constants	Munich Sand	Leighton Buzzard Sand	Fuji River Sand	
Elastic:				
E,(unloading), psi (kPa)	8.49×10^4 (5.85×10^5)	15.0×10^4 (10.34×10^5)	6.52×10^4 (4.49×10^5)	
ν	0.25	0.29	0.30	
Ultimate:				
γ	0.105	0.0965	0.046	
β	0.747	0.612	0.558	For Model, δ_o
m	-0.5	-0.5	-0.5	
Shape:				
n	3.2	2.5	3.0	
Growth:				
a_1 $\Big\}(\alpha_o = 1 \text{ psi})$	8.5×10^{-6}	5.21×10^{-4}	2.62×10^{-7}	
η_1	0.761	0.495	1.764	
Interpolation:				
h_1	0.230	0.63	0.50	
h_2	3.40	0.95	1.27	Additional for Model, δ_2
h_3	0.70	2.00	2.52	
Translation:				
h_4 psi (kPa)	0.40	0.55	-0.60	
	(2.76)	(3.80)	(-4.13)	

Table 2-5. Values of a_n and R for Series 1 to 5 for HC
Loading, σ_o = 15 to 20 psi

| Series | a_n | | $R = \sigma_1/\sigma_o$ | | Remarks |
	Starting	End	Compression	Extension	
1	0.05	0.04			Pure HC Loading
2	2.09	2.03	4.0		Fig. 8(a)
3	2.21	2.19		0.133	Fig. 8(b)
4	1.86	1.77	4.0		Fig. 8(c)
*{ 5a	1.51	1.45	4.0		Fig. 8(d)
5b	2.40	2.39		0.133	Fig. 8(d)

*In this series, the last HC loading (3-4), Fig. 8(d), after the com-
pression cycle 1-2-3 with R = 4.0 was unloaded to 15 psi (4-3), then the
extension cycle 5-6-7 with R = 0.133 was applied, and the HC load
(3-4) was applied again. Then 5a denotes path 1-2-3-4, and 5b denotes
5-6-7-4.

Table 2-6. Analogous Quantities

Solid	Joint
J_1	σ_n
$\sqrt{J_{2D}}$	τ
$\varepsilon_{ii} = \varepsilon_v$	v_r
$e_{ij} = \varepsilon_{ij} - \dfrac{\varepsilon_{kk}}{3}\,\delta_{ij}$	u_r
$\xi = \int (d\varepsilon^p_{ij}\, d\varepsilon^p_{ij})^{1/2}$	$\int [(dv^p_r)^2 + (du^p_r)^2]^{1/2}$
$\xi_v = \dfrac{1}{\sqrt{3}} \int d\varepsilon^p_{kk}$	$\int [(dv^p_r)^2]^{1/2}$
$\xi_d = \int (de^p_{ij}\, de^p_{ij})^{1/2}$	$\int [du^p_r]$

where

ε_{ij} = strain tensor

ε_{ii} = volumetric strain

e_{ij} = deviatoric component of strain tensor

ε^p_{ij} = plastic strain tensor

ε^p_{kk} = volumetric plastic strain

e^p_{ij} = deviatoric component of plastic strain tensor

ξ = plastic strain trajectory

ξ_v = volumetric part of ξ

ξ_d = deviatoric part of ξ

v_r = relative normal displacement

u_r = relative shear or tangential displacement

v^p_r = plastic relative normal displacement

u^p_r = plastic relative shear displacement

Note: Here u^p_r and v^p_r include both the inelastic or plastic deformations and the motion due to slippage. They are computed on the basis of the elastic shear and normal stiffnesses, k_s and k_n, respectively.

Table 2-7(a) **Constants for δ_o and δ_1 Models**
Quasi-static Forward Loading

Constant	Surface i				Average Constants
	Flat	5°	7°	9°	
γ	0.36	0.42	0.78	0.81	-
n	2.00	2.20	3.00	2.90	2.5
a	0.023	0.011	0.031	0.047	0.028
b	-0.116	-0.293	-0.223	-0.162	-0.199
κ	0.70	0.70	0.70	0.55	0.66

Table 2-7(b) **Constants for Reverse Loading**
i = 9°

Constant

$\gamma = 0.45$, n = 2.50, a = 0.024, b = -0.15, $\kappa = 1.50$

Table 2-8.   REVIEW OF VARIOUS MODELS

Model	Yield Function, F Potential Function, Q	Number of Constants and Comments
1. Classical Plasticity	$F = F (J_1, J_{2D}, c, \phi)^* = 0$ $Q \equiv F$	Constants: 3 or 4 One failure surface defines plastic behavior.
2. Critical State	$F_y = F_y [J_1 \text{ (or } p), J_{2D} \text{ (or } q), \lambda, \kappa, M, e_o] = 0$ $F_c = F_c [J_1 \text{ (or } p), J_{2D} \text{ (or } q), M] = 0$ $Q \equiv F$	Constants: 6 F_y (below F_c) defines continuous yielding and F_c defines critical state. Later modifications consider nonassociative behavior.
3. Cap Models	$F_y = F_y (J_1, J_{2D}, R, D, W, Z) = 0$ $F_f = F_f (J_1, J_{2D}, \alpha, \gamma, \beta) = 0$ $Q \equiv F$	Constants: 9 F_y (below F_f) defines continuous yielding and F_f defines failure surface.
4. Matsuoka and Nakai	$F = \dfrac{J_1 J_2}{J_3} - \text{constant} = 0$	F is open failure surface.
5. Lade	$F = \dfrac{J_1^3}{J_3} - \text{constant} = 0$ $F_c = \text{spherical cap}$ $Q = J_1^3 - \kappa J_3$	Constants: 13 F is open (failure) surface.
6. Desai (HISS)	$F = (\dfrac{J_{2D}}{P_a})^2 - F_b F_s = 0$	Constants: 7 for δ_o 8 for δ_1 Single surface F for continuous yielding and ultimate, including failure, peak and critical state.
7. Schreyer	$F = \gamma \dfrac{J_1}{3} - L - \sigma_s = 0$	Constants: 10 Single surface F defines continuous yielding approaching the failure or stationary state.

8. Lade and Kim $F = f_1 \cdot f_2 = 0$

Constants: 11
Single surface F defines continuous yielding. Failure is defined by a separate function.

$$f_1 = (\psi_1 \frac{J_1^3}{J_3} - \frac{J_1^2}{J_2})$$

$$f_2 = (\frac{J_1}{P_a})^h \, e^q$$

$$Q = q_1 \cdot q_2$$

$$q_1 = (\psi_1 \frac{J_1^3}{J_3} - \frac{J_1^2}{J_2} + \psi_2)$$

$$q_2 = (\frac{J_1}{P_a})^\mu$$

---

*Definition of symbols not defined in text:

λ, κ	= slopes of consolidation and swelling curves
M	= slope of critical state line in the p-q space
p	= mean pressure
q	= shear stress
R, D, W, Z	= parameters associated with F and hardening
α, γ, β	= parameters associated with F_f
κ	= parameter in potential function
γ	= material parameter
L	= $[(\sigma_1 - \sigma_s)(\sigma_2 - \sigma_s)(\sigma_3 - \sigma_s) + \sigma_s^3]^{1/3}$
σ_s	= shift stress

PART 3.   DYNAMICS OF AND TRANSIENT FLOW IN POROUS MEDIA

3.1   DYNAMIC COUPLED RESPONSE OF POROUS MEDIA

General formulations of the problem of dynamics of porous media, based on the Biot's theory [6-8], have been proposed by various investigators [59, 89, 93, 113]. The authors and co-workers [12, 28, 42, 58, 98] have developed similar formulations in the context of the finite element method. Some of the main new contributions are inclusion of nonlinear behavior of solids (soils) and interfaces in the formulation. Here the HISS models allowing for anisotropic hardening (δ_2, δ_{2+p} models) and softening have been implemented, and the procedure is applied for solution of a number of practical problems involving closed-form solutions and laboratory and field observations. A brief description of the theoretical developments is given below.

Theoretical Considerations

A brief review of Biot's theory is presented here. Let u_i denote the components of displacement of solid, U_i denote components of displacement of fluid, n the porosity of the mixture. Then, average relative displacement of fluid with respect to solid is given by

$$w_i = \bar{n} \, (U_i - u_i) \qquad\qquad (3-1)$$

The tensor of strain components is given by

$$\varepsilon_{ij} = \frac{1}{2} \, (u_{i,j} + u_{j,i}) \qquad\qquad (3-2)$$

The change of fluid volume per unit volume of solid skeleton is given by

$$\zeta = w_{i,i} \qquad\qquad (3-3)$$

Coupled consitutive relations of the porous medium with linear elastic solid skeleton, as defined by Biot [6-8], are given by

$$\sigma_{ij} = (C_{ijk\ell} + a^2 M \, \delta_{ij} \, \delta_{k\ell}) \, d\varepsilon_{k\ell} + aM\zeta \, \delta_{ij} \qquad (3-4a)$$

$$p = M \, [a\varepsilon_{ii} + \zeta] \qquad\qquad (3-4b)$$

where σ_{ij} are the components of total stress tensor, p is the fluid pressure, and a and M are defined as

$$a = 1 - \frac{\delta_{ij} C_{ijk\ell} \delta_{k\ell}}{9K_s} \qquad (3\text{-}5a)$$

$$\frac{1}{M} = \frac{\overline{n}}{K_f} + \frac{a-\overline{n}}{K_s} \qquad (3\text{-}5b)$$

in which $C_{ijk\ell}$ is the constitutive tensor for solid skeleton under drained conditions, and K_s and K_f are bulk moduli of solid grains and fluid, respectively. For incompressible soil grains, Eqs. (3-5) reduce to

$$a = 1.0 \qquad (3\text{-}6a)$$

$$M = \frac{K_f}{\overline{n}} \qquad (3\text{-}6b)$$

The term $(1-a)$, Eq. (3-5a), can be regarded as the stiffness of solid skeleton relative to stiffness of solid grains, whereas M can be regarded as effective bulk modulus of fluid. It can be shown that under undrained conditions, the bulk modulus of the mixture is equivalent to $(K + a^2M)$ where $K = (\delta_{ij} C_{ijk\ell} \delta_{k\ell})/9$.

For nonlinear solid skeleton, the incremental form of Eq. (3-4) can be written as

$$d\sigma_{ij} = (C^t_{ijk\ell} + a^2 M \, \delta_{ij} \delta_{k\ell}) \, d\varepsilon_{k\ell} + aMd\zeta \, \delta_{ij} \qquad (3\text{-}7a)$$

$$dp = aM \, [ad\varepsilon_{ii} + d\zeta] \qquad (3\text{-}7b)$$

where superscript 't' refers to tangent quantity.

Specializing of Eqs. (3-7) to the undrained condition results in

$$d\sigma_{ij} = [C^t_{ijk\ell} + a^2 M \, \delta_{ij} \delta_{k\ell}] \, d\varepsilon_{k\ell} \qquad (3\text{-}8a)$$

$$dp = aM \, d\varepsilon_{ii} \qquad (3\text{-}8b)$$

Note that behavior of the solid skeleton is governed by the effective stress $\sigma'_{ij} = \sigma_{ij} - p \, \delta_{ij}$.

Governing Equations

Two different equilibrium equations can be derived for the two phases involved. Then they can be expressed as

$$\sigma_{ij,j} + \rho \, b_i - \rho_f \, \ddot{w}_i = 0 \qquad (3\text{-}9a)$$

$$p_i + \rho_f \, b_i - \rho_f \, \ddot{u}_i + \frac{\rho_f}{n} \, \ddot{w}_i + k_{ij}^{-1} \, w_j = 0 \qquad (3\text{-}9b)$$

where overdot denotes time derivative.

Finite Element Equations

The finite element equations corresponding to Eqs. (3-9a) and (3-9b) are derived as

$$[M_{uu}] \, \ddot{\bar{u}} + [M_{uw}] \, \{\ddot{w}\} + \int_V [B_u]^T \, \{\sigma\} \, dV = \{f_u\} \qquad (3\text{-}10a)$$

$$[M_{uw}]^T \, \{\ddot{u}\} + [M_{ww}] \, \{\ddot{w}\} + [C_{ww}] \, \{\dot{w}\} + \\ \int_V [H_w]^T \, p \, dV = \{f_w\} \qquad (3\text{-}10b)$$

where $[M_{uu}]$ is the mass matrix for solids, $[M_{ww}]$ is the mass matrix for fluid, $[M_{uw}]$ is the coupled mass matrix for solid and fluid, $[B_u]$ is the strain-displacement relation for solid, $[C_{ww}]$ is the flow properties matrix, $\{\sigma\}$ is the vector of total stress components, $[H_w]$ is the flow-gradient relation for fluid, $\{u\}$ is the vector of nodal displacements of solids, p is pore water pressure, $\{f_u\}$, $\{f_w\}$ = load vectors for solid and fluid, respectively, and the overdot denotes time derivative.

Time Integration and Iterative Scheme

The time integration of Eq. (3-10) can be performed using Newmark's family of time integration schemes given by

$$\{x\}_{n+1} = \{x\}_n + \Delta t \, \{\dot{x}\}_n + \frac{\Delta t^2}{2} \, \{(1\text{-}2\theta_1) \, \{\ddot{x}\}_n + 2\theta_1 \, \{\ddot{x}\}_{n+1} \qquad (3\text{-}11a)$$

$$\{\dot{x}\}_{n+1} = \{\dot{x}\}_n + \Delta t \, (1\text{-}\theta_2) \, \{\ddot{x}\}_n + \theta_2 \, \{\ddot{x}\}_{n+1} \qquad (3\text{-}11b)$$

where $\{x\}$ = vector of unknown variables, Δt is the time step size, n is the step number, θ_1 and θ_2 are the parameters which govern accuracy and stability of scheme. With Eqs. (3-11), a Newton-Raphson type iterative procedure is developed for the solution of Eq. (3-10) as

$$[K^*] \, \{\Delta x\}_{n+1}^{(i)} = \{f^*\}_{n+1}^{(i)} \qquad (3\text{-}12)$$

where $[K^*]$ and $\{f^*\}$ are equivalent stiffness and load vectors, respectively, and $\{\Delta x\}$ incremental displacement vector.

The matrices $[K_{uu}]$, $[K_{uw}]$ and $[K_{ww}]$ are tangent stiffness matrices given by

$$[K_{uu}] = \int_V [B_u]^T [[C^{ep}] + a^2 M \{m\} \{m\}^T] [B_u] \, dV \qquad (3\text{-}13a)$$

$$[K_{uw}] = \int_V [B_u]^T aM \{m\} [H_w] \, dV \qquad (3\text{-}13b)$$

$$[K_{ww}] = \int_V [H_w]^T \{M\} [H_w] \, dV \qquad (3\text{-}13c)$$

where $\{m\}$ is the vector form of δ_{ij}, and $[C^{ep}]$ is the elastic-plastic constitutive matrix, described later. The displacements $\{u\}$ and $\{w\}$ are updated at the end of every iteration as

$$\{x\}_{n+1}^{(i)} = \{x\}_{n+1}^{(i-1)} + \{\Delta x\}_{n+1}^{(i)} \qquad (3\text{-}14)$$

Iterations can be carried out until convergence criterion, for instance

$$\frac{||\{f^*\}^{(i)}||}{||\{f^*\}^0||} \leq \varepsilon \qquad (3\text{-}15)$$

where ε is a tolerance factor of the order of 0.01.

MODEL FOR CYCLIC PLASTICITY [28, 45, 58, 100, 107]

The complex nonlinear, inelastic characteristics of the soil skeleton are difficult to simulate, especially if the loading involves multidimensional stress paths and their reversals through cyclic loading and rotation of principal stresses. A model for these conditions should account for factors such as state of stress and strain, stress path, deformation history, volume change under shear, and initial and induced anisotropy. Desai, Somasundaram and Frantziskonis [45] presented a hierarchical approach whereby a basic isotropic hardening model is modified to yield models of progressively higher grades that account for initial and induced anisotropy, as well as nonassociativeness and softening, Part 2. One member of the family is anisotropically hardening model (δ_2) which allows for elastic-plastic behavior during unloading and reloading. This model

possesses general and fundamental basis, yet at the same time
it is conceptually simple compared to other available models
with similar capabilities.

Implementation of the Model in Finite Element Procedure

The initial stress iterative procedure for static
problems in finite element method is expressed as

$$[K] \{\Delta u\}^{(k)} = \{r\}^{(k)} \tag{3-16a}$$

$$\{u\}^{(k)} = \{u\}^{(k-1)} + \{\Delta u\}^{(k)} \tag{3-16b}$$

where [K] is the tangent stiffness matrix, {r} is the residual
load vector, Δ denotes incremental quantities, k is the
iteration counter.

The residual load vector at any iteration is written as

$$\{r\}^{(k)} = \{f\} - \int_V [B]^T \{\sigma\}^{(k-1)} \, dV \tag{3-17}$$

where {f} is the vector of applied loads, {σ} is the vector
of total stress components, [B] is the strain-displacement
transformation matrix.

In order to proceed with the iterative procedure, it is
necessary to update the vector {σ} at each Gauss point. For
this, it is necessary to convert the incremental strains
obtained through finite element calculations to corresponding
incremental stresses. In general, this is achieved as
follows:

$$\Delta p = M \{m\}^T \{\Delta \varepsilon\} \tag{3-18a}$$

$$\{d_\varepsilon\} = \{\Delta \varepsilon\}/N \tag{3-18b}$$

$$\{\Delta \sigma'\} = [C^{ep}] \{d_\varepsilon\} \tag{3-18c}$$

$$\{\Delta \sigma\} = \{\Delta \sigma'\} + \{m\}^T \Delta p \tag{3-18d}$$

where {$\Delta \varepsilon$} is the vector of strain increments, and N is the
number of (small) subincrements.

It has been found [Potts and Gens, 87] that accurate
estimate of {$\Delta \sigma$} could be achieved by using $\|\{d_\varepsilon\}\|$ of the
order of 0.0005. If larger values of {$d \varepsilon$} are used, the
consistency (dF = 0) may not be satisfied, and the stress
point may deviate from the yield surface. However, use of

smaller increments of the order of 0.0005 can be very costly in large finite element analysis. Alternative (drift) correction procedures have been proposed, and an evaluation of such algorithms has been reported by Potts and Gens [87]. One of these algorithms with a number of modifications is adopted in this study, and is described below.

Drift Correction

Virgin Loading. In the case of virgin loading, the consistency conditions on both F and Q have to be satisfied. At first, the consistency on F is satisfied, and then Q is moved in the stress space, as described previously. However, due to the incremental nature of this procedure, the consistency on Q would not be satisfied. So a correction would be applied to the location of Q until $Q \cong 0$. These aspects are described below.

Figure 3-1 illustrates schematically the drift from the yield surface. The initial state of stress $\{\sigma'\}$ lies on the yield surface F $(\{\sigma'\}, \xi_A) = 0$. Application of stress increment $\{\Delta\sigma'\}$, calculated on $[C^{ep}]_A$ $\{\Delta\varepsilon\}$ would move the stress point to B, whereas the yield surface would move to the position as shown by the dashed line corresponding to updated value of $\xi = \xi_B$. Then, point B would not necessarily lie on F. So the correction to $\{\sigma'\}_B$ is applied as follows:

$$F (\{\sigma'\}_B + \{d\sigma'\}^c, \xi_B) = 0 \qquad (3\text{-}19)$$

where $\{d\sigma'\}^c$ is the correction. Using the Taylor series expansion on Eq. (3-19), the following relation is obtained:

$$F (\{\sigma'\}_B, \xi_B) + \{\frac{\partial F}{\partial \sigma'}\}^T \{d\sigma'\}^c \ldots = 0 \qquad (3\text{-}20)$$

In order to obtain the magnitude of $\{d\sigma'\}^c$, it is necessary that direction of $\{d\sigma'\}^c$ is known. Following the work of Potts and Gens [87], this direction is chosen to be perpendicular to the plastic potential Q. Then the following correction is obtained:

$$\{d\sigma'\} = \bar{k} \ \{\frac{\partial Q}{\partial \sigma'}\} \qquad (3\text{-}21a)$$

$$\bar{k} = - \ \frac{F (\{\sigma'\}, \xi_B)}{\{\frac{\partial F}{\partial \sigma'}\}^T \ \{\frac{\partial Q}{\partial \sigma'}\}} \qquad (3\text{-}21b)$$

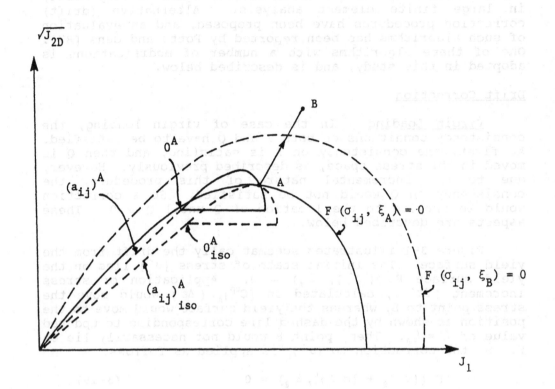

Figure 3-1 Correction of Yield Surface During Virgin Loading

Once the stress point is corrected for the drift from the yield surface, it is necessary to locate Q in the stress space appropriately. As described earlier, this could be achieved by using Eq. (3-20) and the consistency

Additional details of theory, constitutive models and applications are given elsewhere [12, 28, 42, 98, 107, 109].

3.2 TRANSIENT FREE SURFACE SEEPAGE

Since 1969, Desai and co-workers [5, 11, 21, 34, 43, 77, 101] have developed a distinct procedure called the Residual Flow Procedure (RFP) for steady and transient free surface seepage problems. Westbrook [110] has shown that this procedure is similar to the recently developed variational inequality methods. The procedure has been applied extensively for solution of two- and three-dimensional problems including its verifications with respect to analytical, laboratory experiments and field observations. In view of the length limitations, only a brief description is provided.

Some of the attributes and advantages of the RFP are: (1) it involves invariant mesh, Fig. 3-2(b), which avoids the need of mesh modifications, Fig. 3-2(a), (2) it permits use of arbitrary shapes of (isoparametric) elements, whereas the inequality methods may often be found difficult to implement for such shapes, (3) it can accommodate variable material properties whereas in the variable mesh schemes they can involve considerable difficulties, (4) it can incorporate partially saturated zones and (5) it is found to be relatively economical; e.g., in comparison with the inequality methods.

MATHEMATICAL BACKGROUND

For incompressible fluid flow through a porous medium, the governing equations of flow can be written as

$$\text{div } (k \text{ grad } \phi) - \overline{Q} = S_s \frac{\partial \phi}{\partial t} \qquad (3-22)$$

where ϕ = total fluid potential or head = $p/\gamma + z$, p = pressure, z = elevation head, γ = density of fluid, k = coefficient of permeability of the medium, $\overline{Q}$ = applied source or sink, and S_s = specific storage.

For the flow domain (dam) shown in Fig. 3-3, Eq. (3-22) is assumed to hold in both Ω and $D - \Omega$ by introducing a definition of k as follows:

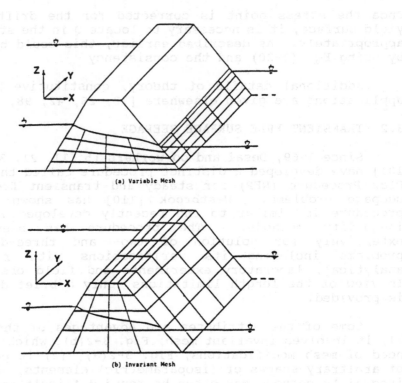

Figure 3-2 Variable and invariant mesh procedures.

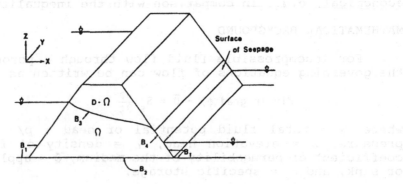

Figure 3-3 Schematic of flow domain and boundaries.

$$k(p) = \begin{cases} k_s & \text{on } \Omega \\ k_{us} = k_s - f(p) & \text{on } D - \Omega \end{cases} \qquad (3-23a)$$

and

$$S_s = \begin{cases} S_s & \text{on } \Omega \\ S_{us} = S_s - g(p) & \text{on } D - \Omega \end{cases} \qquad (3-23b)$$

where 's' and 'us' denote saturated and unsaturated zones, and
$f(p)$ and $g(p)$ are functions of pressure p. Typical pressure-
permeability relations are depicted in Fig. 3-4(a). If such
relations are not available from laboratory and/or field
tests, for the implementation of the RFP it can be appropriate
to assume simplified relations shown in Fig. 3-4(b). For S_s,
it may be appropriate to ignore the terms related to S_{us} by
assuming, as an approximation, that the specific storage of
the unsaturated zone above the free surface is negligible.

The associated boundary conditions are expressed as

$$\phi = \bar{\phi} \text{ on } B_1 \qquad (3-24)$$

and

$$k_x \frac{\partial \phi}{\partial x} \ell_x + k_y \frac{\partial \phi}{\partial y} \ell_y + k_z \frac{\partial \phi}{\partial z} \ell_z = \bar{q}_n \qquad (3-25)$$

on part B_2 of the boundary where $\bar{q}_n$ is the specified intensity
of flow. On the free surface, B_3, the boundary conditions are

$$\phi = z \qquad (3-26a)$$

and

$$k_x \frac{\partial \phi}{\partial x} \ell_x + k_y \frac{\partial \phi}{\partial y} \ell_y + k_z \frac{\partial \phi}{\partial z} \ell_z = 0 \qquad (3-26b)$$

The following boundary conditions exist on the surface of
seepage, B_4:

$$\phi = z \qquad (3-27a)$$

$$k_x \frac{\partial \phi}{\partial x} \ell_x + k_y \frac{\partial \phi}{\partial y} \ell_y + k_z \frac{\partial \phi}{\partial z} \ell_z < 0 \qquad (3-27b)$$

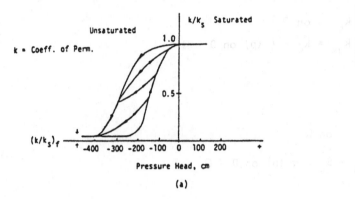

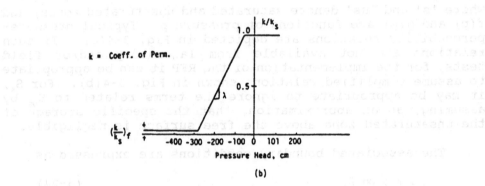

Figure 3-4 Pressure-permeability curves. (a) Saturated-unsaturated flow, (b) Idealized.

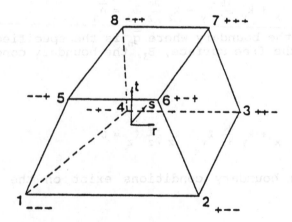

Figure 3-5 Three-dimensional element.

Here, k_x, k_y and k_z = coefficients of permeabilities in x, y and z directions, respectively, and ℓ_x, ℓ_y, ℓ_z = direction cosines of outward normal to the boundaries.

Transient Free Surface Flow

Development of the RFP is first described for the transient case; then the steady free surface case is obtained as a special case. The pseudo-functional, U, for the transient case can be expressed as [5, 21, 34, 77, 82, 101]

$$U\ (\phi,\ f,\ t)\ =\ \frac{1}{2}\ \int_\Omega\ \{(k_s\ -\ f)\ [(\frac{\partial\phi}{\partial x})^2\ +\ (\frac{\partial\phi}{\partial y})^2\ +\ (\frac{\partial\phi}{\partial z})^2\ ]$$

$$(3\text{-}28)$$

$$-\ 2\ [\overline{Q}\ -\ (S_s\ -\ g)\ \frac{\partial\phi}{\partial}]\ \phi\}\ d\Omega\ -\ \int_{B_2}\ \overline{q}_n\ \phi\ dB$$

by taking the variation of the functional in Eq. (3-28) and finding the stationary value, the following element equations can be obtained:

$$[k_s]\ \{q\}\ -\ [k_{us}]\ \{q\}\ +\ [p_s]\ \{\dot{q}\}\ -\ [p_{us}]\ \{\dot{q}\}\ =\ \{Q\} \qquad (3\text{-}29a)$$

or

$$[k_s]\ \{q\}\ +\ [p_s]\ \{\dot{q}\}\ =\ \{Q\}\ +\ [k_{us}]\ \{q\}\ +\ [p_{us}]\ \{\dot{q}\}$$

$$(3\text{-}29b)$$

$$=\ \{Q\} \qquad \{Q_r\}\ =\ \{Q^*\}$$

where

$$[k_s]\ \{q\}\ =\ \int_\Omega\ [B]^T\ [R]\ [B]\ d\Omega,$$

$$[k_{us}]\ =\ \int_\Omega\ [B]^T\ [f]\ [B]\ d\Omega,$$

$$\{Q\}\ =\ \int_\Omega\ [N]^T\ \{\overline{Q}\}\ d\Omega\ +\ \int_{B_2}\ [N]^T\ \{\overline{q}_n\}\ dB,$$

$$[p_s]\ =\ \int_\Omega\ S_s\ [N]^T\ [N]\ d\Omega,$$

$$[p_{us}]\ =\ \int\ g\ [N]^T\ [N]\ d\Omega,$$

{q} = vector of nodal heads,

{Q} = vector of forcing function,

[B] = gradient-head transformation matrix,

$$[R] = \begin{bmatrix} k_x & 0 & 0 \\ 0 & k_y & 0 \\ 0 & 0 & k_z \end{bmatrix}$$

is the matrix of (principal) permeabilities,

$$[f] = \begin{bmatrix} f_x & 0 & 0 \\ 0 & f_y & 0 \\ 0 & 0 & f_z \end{bmatrix}$$

is the matrix of (principal) values of f in Eq. (3-23a), and the overdot denotes time derivative. The residual flow or load vector $\{Q_r\}$ is composed as

$$\{Q_r\} = [k_{us}] \{q\} + [p_{us}] \{\dot{q}\} \qquad (3\text{-}30)$$

Finite Element Procedure

An isoparametric hexahedral element, Fig. 3-5, with 8 nodes, is used. The fluid head describing the approximate variation of   within an element is expressed as

$$\phi = [N] \{q\} \qquad (3\text{-}31)$$

where $[N] = [N_1\ N_2\ N_3\ N_4\ N_5\ N_6\ N_7\ N_8]$ is the matrix of interpolation functions, N_i, $\{q\}^T = [\phi_1\ \phi_2\ \phi_3\ \phi_4\ \phi_5\ \phi_6\ \phi_7\ \phi_8]$ is the vector of nodal fluid heads, and the functions N_i are given by $N_i = 1/8\ (1 + rr_i)\ (1 + ss_i)\ (1 + tt_i)$, $i = 1, \ldots, 8$. Here r, s, t = local coordinates and r_i, s_i and t_i are (+) or (-), as shown in Fig. 3-5.

Time Integration

For convenience, Eq. (3-29a) can be written as

$$[c] \{\dot{q}\} + [k'] \{q\} = \{Q\} \qquad (3\text{-}32)$$

where

$$[k'] = [k_s] - [k_{us}]$$

$$[c] = [p_s] - [p_{us}]$$

At time $(t + \Delta t)$, Eq. (3-32) can be expressed as

$$[c] \{\dot{q}\}_{t+\Delta t} + [k'] \{q\}_{t+\Delta t} = \{Q\}_{t+\Delta t} \qquad (3-33)$$

Now $\{q\}_{t+\Delta t}$ is written as

$$\{q\}_{t+\Delta t} = \{q\}_t + \Delta t \{(1-\theta) \{\dot{q}_t\} + \theta \{\dot{q}\}_{t+\Delta t}\} \qquad (3-34)$$

where   is a parameter. Solving for $\{\dot{q}\}_{t+\Delta t}$,

$$[\dot{q}]_{t+\Delta t} = \frac{\{q\}_{t+\Delta t} - \{q\}_t - \Delta t (1-\theta) \{\dot{q}\}_t}{\theta \Delta t} \qquad (3-35)$$

Different values of θ can be used, each representing a different method, e.g., $\theta = 1$.... Backward Euler; $\theta = 2/3$.... Galerkin; and $\theta = 1/2$.... Crank-Nicholson. It should also be noted that for $\theta > 1/2$, all schemes are unconditionally stable; and when $0 < \theta < 1/2$, stability is conditional. Based on a parametric study for the problems herein, the backward Euler scheme ($\theta = 1$) is employed.

Now substitution of Eq. (3-35) with $\theta = 1$ in Eq. (3-33) leads to element equations at $t + \Delta t$ as

$$[k] \{q\}^i_{t+\Delta t} = \{Q\}_{t+\Delta t} + \{Q_r\}^{i-1} \qquad (3-36)$$

where

$$[\bar{k}] = [k_s] + \frac{1}{\Delta t} [p_s]$$

and

$$\{Q_r\}^{i-1} = [k_{us}] \{q\}^{i-1}_{t+\Delta t} + \frac{1}{\Delta t} [p_s] \{q\}_t$$

$$+ \frac{1}{\Delta t} [p_{us}]^{i-1} (\{q\}^{i-1}_{t+\Delta t} = \{q\}_t)$$

where $i = 1, 2, 3 \ldots$ denote iterations for a given $t + \Delta t$, and Δt is the time step.

For the first iteration, $\{Q_r\}$ is assumed to be zero, and the solution of

$$[\bar{k}] \{q\}^1_{t+\Delta t} = \{Q\}_{t+\Delta t} \qquad (3-37a)$$

provides $\{q\}^1_{t+\Delta t}$; here the superscript denotes iteration. Now $\{Q_r\}$ is found as

$$\{Q_r\}^1 = [k_{us}] \{q\}^t_{t+\Delta t} + \frac{1}{\Delta t} [p_s] \{q\}_t$$

$$+ \frac{1}{\Delta t} [p_{us}]^1 (\{q\}^1_{t=\Delta t} - \{q\}_t) \qquad (3\text{-}37b)$$

Hence, for the second iteration,

$$[\overline{k}] \{q\}^2_{t+\Delta t} = \{Q\}_{t+\Delta t} + \{Q_r\}^1 \qquad (3\text{-}37c)$$

and, in general,

$$[\overline{k}] \{q\}^i_{t+\Delta t} = \{Q\}_{t+\Delta t} + \{Q_r\}^{i-1} \qquad (3\text{-}37d)$$

The following convergence criterion is used to terminate iterations·

$$\frac{\left| (\sum_{j=1}^{m} \phi j)^i - (\sum_{j=1}^{m} \phi j)^{i-1} \right|}{(\sum_{j=1}^{m} \phi j)^{i-1}} \leq \varepsilon \qquad (3\text{-}38)$$

where m = number of nodes on the free surface and ι = small nonnegative number; a value of ε = 0.005 is used herein. The average number of iterations required for the problems considered here is 3.

After every iteration, based on the computer nodal heads, the separation of fully and unsaturated zones is obtained by locating along nodal planes, the points of zero pressure; details are given in the next section. The negative values of pressure head ($p/\gamma = \phi - z$) in the unsaturated zone allows evaluation of $[k_{us}]$ in Eq. (3-37) from Fig. (3-4), where γ is the fluid density. As stated before, if pressure vs. S_s relation is not available, the term related to $[p_{us}]$ can be ignored. This assumption is considered satisfactory for situations involving negligible capillary zones and distinct free surface, and for small time steps, Δt.

Steady State Case

The above procedure for transient free surface can be specialized for the steady free surface case. This is achieved essentially by deleting the term [c] in Eq. (3-32).

Further details for locating of free surface and surface of seepage, mathematical analyses and applications are given elsewhere [5, 11, 21, 34, 43, 77, 101].

3.3   TIME INTEGRATION

Most dynamic and field problems in geomechanics are nonlinear. The integration schemes for transient analysis of these problems are most often relevant for linear problems. Their application to nonlinear problems often needs modification or development of alternative and improved schemes.

The subject of time integration schemes, which is vital for reliable and robust solutions, is wide in scope and it is not intended to review it here. Brief statements with relevant references of the recent research by the author and co-workers for time integration schemes for field and dynamic problems are given below.

3.3.1   Nonlinear Field Problems

The author and co-workers have considered the question of time integration schemes for free surface seepage and consolidation problems. An alternative direction explicit procedure (ADEP) in conjunction with the Saulev schemes was used for solution of finite difference equations for one- and two-dimensional transient flow through porous media [43], and variable time step procedures were proposed for the flow problems [92]. A number of finite element and finite difference schemes were evaluated for stability and convergence for the solution of parabolic equation governing one-dimensional consolidation [31, 32, 36]; here closed-form stability criteria were derived for the finite element schemes.

Recently Kujawski and Desai [70, 71] have proposed quasi-explicit two time level schemes for solution of nonlinear equations governing a wide range of field problems [1, 61-63]. Here stable and accurate quadratic families of schemes are derived for the generalized trapezoidal method (GTM) and the generalized midpoint method (GMM). The study involved analysis of accuracy, stability and osciallatory behavior of the schemes. The proposed schemes provide significant improvements in terms of stability and computational time compared to those for previously available methods.

3.3.2 Nonlinear Dynamic Problems

An algorithm based on the generalized time finite element method (GTFEM) for the time integration of dynamic nonlinear equations has been proposed by the author and co-workers [33, 68, 69]. It involves two free parameters, θ_1 and θ_2, which are optimized for improved accuracy and computational effort. The proposed algorithm is compared with the commonly used Newmark and Wilson procedures.

The algorithm involves three time levels and is based on weighted nonlinear stiffness matrix which contains the two free parameters θ_1 and θ_2. For linear problems, both $\theta_1 - \theta_2 = 0$, and the proposed algorithm reduces to the Newmark method ($\gamma = \frac{1}{2}$ and $\beta \neq 0$). Based on a number of numerical experiments for nonlinear problems, it is found that the GTFEM procedure can lead to improved results in comparison to those from the Newmark, Wilson and other methods [33, 68, 69].

PART 4. APPLICATIONS

The hierarchical single surface models (δ_0, δ_1, δ_{2+p}, δ_{o+up}, δ_{o+r}) have been implemented in nonlinear finite element procedures for static and dynamic analysis of geomechanical structures. Few examples of these applications are presented below. Subincrementation and modified drift correction techniques have been used in the incremental analysis with the elastic-plastic models involving the foregoing versions of the models. Also presented is an example of recent application of the RFP for three-dimensional free surface seepage.

4.1. FOOTING ON SAND - δ_0, δ_1 MODELS

The associative and nonassociative models (δ_0 and δ_1) are implemented in a two-dimensional finite element procedure. The numerical procedure is used to evaluate a number of schemes for handling nonassociative equations [30, 60]. It is also used to verify the models with respect to laboratory tests on a model footing.

Model Footing. Figure 4-1 shows details of model footing with width and depth of the soil box 54.0 x 18.0 inches (137 x 10 x 46 cms), respectively. Walls of the box were constructed using two 0.25 in (0.6 cm) thick pieces of glass. The Leighton Buzzard Sand at a relative density of about 95% was placed in the box. A smooth steel footing 4.0 x 4.0 x 0.75 inches (15.3 x 10.8 x 1.2 cms) was used.

To measure contact pressures, three circular 1.5 inch (3.8 cm) diam. and 0.25 inch (0.64 cm) thick cells were embedded in cavities at the bottom of the steel footing, with thin outer surface of the cell flush with the bottom surface of the footing. These stress cells were placed at the center, 2.5 inches (6.5 cms) from the center along the center line, and at the corner along the diagonal. A stress cell was also embedded at the bottom of the soil box, 18 inches (48.0 cms) from the surface of the sand. To measure stress against the side of the glass wall, a stress cell was also mounted on the side, 2.0 inches (5.0 cms) below the surface of sand. Spring Return Linear Position Sensor Modules (SRLPSM) were used to record the displacements of the footing, and on the surface of the sand at 5.0 inches (13.0 cms), 8.0 inches (20.0 cms) and 12.0 inches (30.5 cms) from center of the footing.

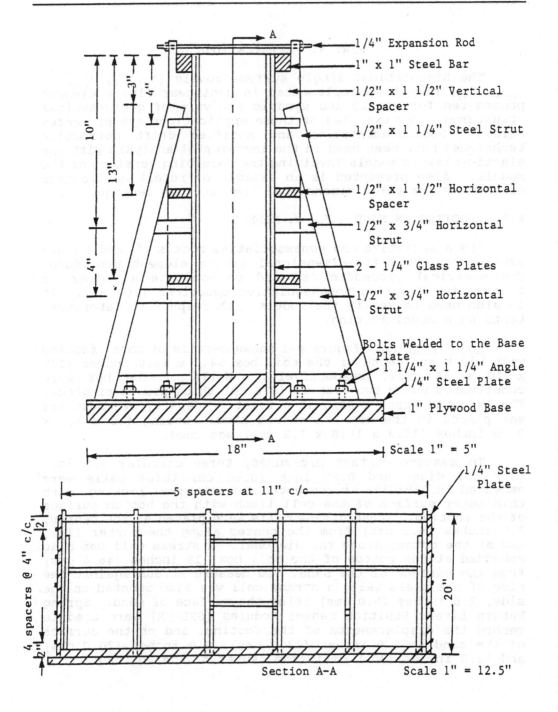

Figure 4-1 Details of Test Box for Laboratory Footing

The load on the footing was applied in increments by using an MTS test frame. A total load of 25.0 psi (172 kPa) was applied in steps of 2.50 psi (17 kPa), and the resulting displacements and stresses were recorded.

Analysis. Due to the symmetry, only half of the domain is discretized, Fig. 4-2. The finite element mesh consists of 207 nodes and 58, eight-noded isoparametric elements. Both associative and nonassociative models were used to analyze the response of the system.

Comparison of Displacements

Figure 4-3 shows comparisons of load-displacement response of the center of the footing (Node 15) as observed in the laboratory and predicted by the finite element method. Based on these and other similar comparisons, it can be concluded that the nonassociative model provides better predictions of the observed response.

Comparison of Stresses

Figure 4-4 shows comparison of the normal stress measured against the side of the soil box. The vertical stress predictions are similar from both models, while the nonassociative model shows improved predictions for horizontal stress normal to the wall.

The finite element procedure with the nonassociative model shows improved predictions for the stress-deformation response of a model footing in Leighton Buzzard Sand. In lieu of analytical solutions for the nonlinear system, it is considered appropriate to use experimental results from well instrumented tests as a reference for the comparisons.

Based on the results of this study, it can be concluded that (1) for cohesionless soils, it is often necessary to use nonassociated plasticity, (2) solution of the nonsymmetric system involved in the nonassociative model may be a better strategy than use of special schemes to produce equivalent symmetric system; this observation may need modifications for larger sized problems than those considered herein; (3) the nonassociative model provides more realistic predictions of stresses and deformations of a model footing, and (4) with the drift correction scheme used, no instability in computations was observed for the problems considered in this study.

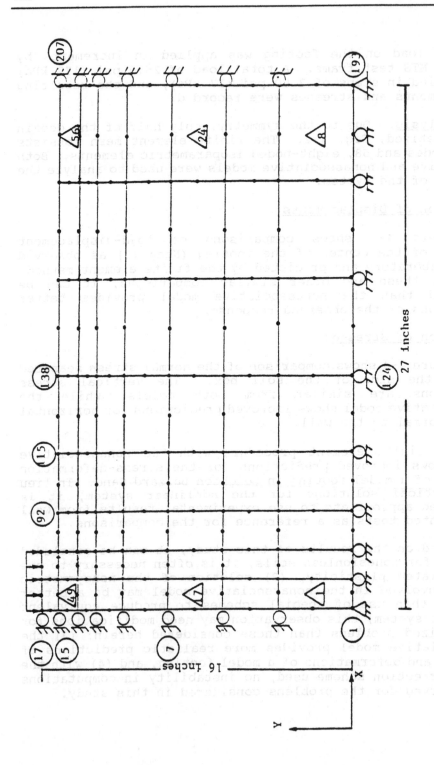

Figure 4-2 Finite Element Mesh of Soil-Footing System
(1 in = 2.54 cm)

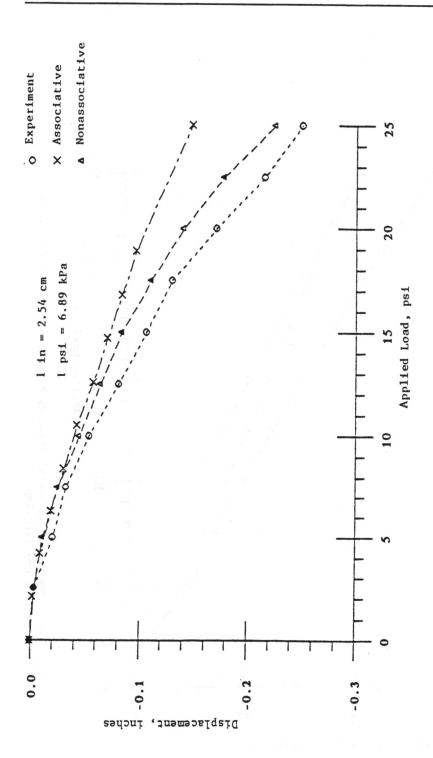

Figure 4-3 Comparison for Load-Displacement Response at Center
of Footing (Node 15)

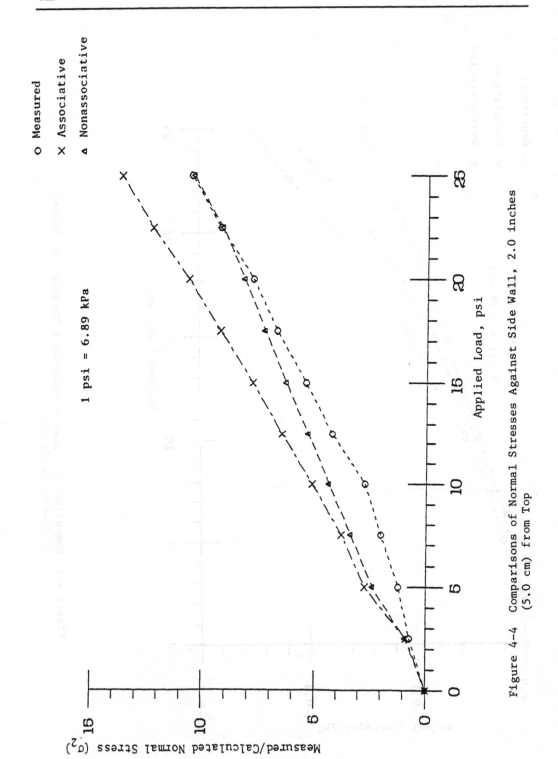

Figure 4-4 Comparisons of Normal Stresses Against Side Wall, 2.0 inches
(5.0 cm) from Top

4.2 PILE IN MARINE CLAY UNDER CYCLIC LOADING

The finite element procedure for dynamics of porous media [28, 109] was used to analyze response of instrumented pile segments tested in the field near Sabine, Texas by Earth Technology Corporation who were the industrial collaborators for this research [53, 107, 109]. Cylindrical and square specimens of the marine clay, at a depth of about 50' in which the test piles were installed, were obtained and tested in the laboratory using cylindrical triaxial and multiaxial devices. A modified anisotropic hardening model was used to simulate the behavior of clay. The computer simulation involved various phases: insitu stress, driving, consolidation, tension tests and cyclic loading. Results of preliminary analysis are given below.

Testing Procedure

Cyclic and quasi-static load tests on instrumented model pile segments of 3 inch diameter and 1.72 inch diameter (X-probe), were performed at Sabine Pass, Texas, in collaboration with Earth Technology Corporation [53]. Preliminary backpredictions for the results of the cyclic load tests carried out at (near) the end of consolidation on X-probe are presented here.

A bore hole of 6.0 inch diameter was drilled up to the uniform clay layer of about 10-12 feet thickness occurring at a depth of about 50 ft. It was cased before the probe was installed by pushing. The test setup is illustrated in Fig. 4-5. At about 97% consolidation (54.5 hours after installation), a two-way cyclic load test was performed.

Finite Element Analyses

Displacement-time history of the pile segment measured in the field is used to simulate the two-way cyclic pile load test using the finite element procedure. Undrained conditions are assumed because the test was fast and permeability of the clay is considered to be low.

Predicted shear stresses vs. time for the first two cycles of the load test for X-probe are presented in Fig. 4-6 along with the field observations. Shear stress at the soil-pile interface (τ_i) is approximately estimated from the computed stresses at the Gauss point closest to the interface, τ_g, using the following formula:

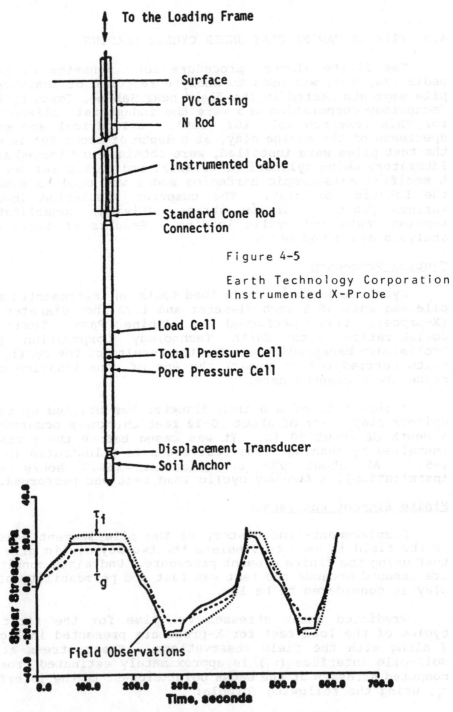

To the Loading Frame

Surface

PVC Casing

N Rod

Instrumented Cable

Standard Cone Rod
Connection

Figure 4-5

Earth Technology Corporation
Instrumented X-Probe

Load Cell

Total Pressure Cell

Pore Pressure Cell

Displacement Transducer

Soil Anchor

τ_i

τ_g

Field Observations

Figure 4-6 Variation of Shear Stress at Soil-Pile
            Interface with Time

$$\tau_i = \tau_g \frac{r_g}{r_i} \qquad\qquad (4-1)$$

where r_i is the radius of the pile, and r_g is the radial distance to the Gauss point from the pile center line.

It is observed from Fig. 4-6 that the test results lie in between τ_i and τ_g. Results for τ_i vs. (relative) displacements are plotted in Fig. 4-7, and show good agreement with test data.

Conclusions

Proposed procedure shows good promise of improved and detailed predictions for cyclic behavior of axially loaded piles in saturated clays. With the use of only one consolidated undrained triaxial test, the method provided satisfactory predictions of the shear stresses, from which pile capacity can be evaluated. Predictions of the pore pressure generation still need further research, and are underway at the University of Arizona. When all the laboratory test data, cylindrical and multiaxial, are available, it is anticipated that improved predictions of pore pressures will be available.

4.3   ANCHORS IN SAND [37]

The δ_0 and δ_1 models were used to simulate behavior of the 'Munich' Sand that was tested extensively using the multiaxial device. These models were implemented in a three-dimensional nonlinear finite element procedure, which was used to backpredict the field behavior of anchors installed in the Munich Sand near Munich, West Germany [65, 84, 95].

Field Tests

A series of full-scale field tests on grouted anchors in sand was performed near Munich, West Germany, and details are reported by Jelinek and Scheele [65], Ostermeyer and Scheele [84], and Scheele [95]. The dimensions of the test pit were about 5m x 10m x 10m (depth x length x width); details of the anchor components are shown in Fig. 4-8. In one of the test series, the existing soil in the pit was replaced by a cohesionless soil, called herein "Munich" sand. This material is a coarse- to medium-well-graded sand with approximately round grains and about 3% silty particles. The soil's specific gravity $G_s = 2.578$; coefficient of uniformity $C_u =$

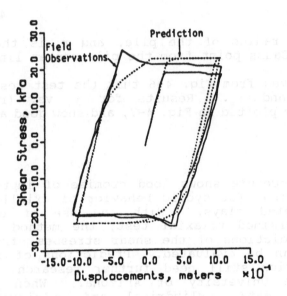

Figure 4-7   Variation of Shear Stress at Soil-Pile Interface
with Pile Displacement

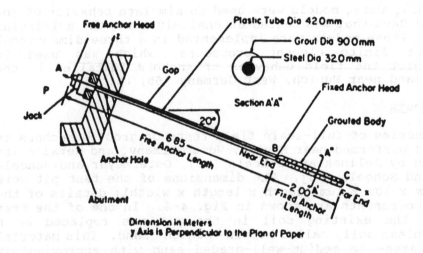

Figure 4-8 Details of Components of Anchor

$D_{60}/D_{10} = 9.0$; coefficient of gradation $C_c = (D_{30})^2/D_{10}D_{60} = 0.8$; maximum and minimum dry densities = 2.07 and 1.69 t/m³, respectively; and natural water content ≅ 4.0%. The sand was compacted in layers of 30.0 cm to the desired density using a vibrator. The compacted sand carried a surcharge of gravel with a height of about 2.0m.

Procedure and Elements

Details of the formulation of the three-dimensional finite element procedure are available in a number of publications [20, 37, 44]. Eight-noded isoparametric solid element is used for soil, and grout (concrete) and steel in the anchor (Fig. 4-9), whereas the thin-layer interface element is used to simulate the interaction behavior. The thin-layer interface can account for various modes of deformation such as stick or no slip, slip or slide, debonding or separation, and rebonding. Details of this element are given elsewhere [48]. The thin-layer element is treated essentially like a solid isoparametric element (Fig. 4-9), but with different constitutive properties, described subsequently. Figure 4-10 shows the finite element mesh.

Tensile Stress Redistribution.
Tensile strength of the sand is assumed to be zero. The computed tensile stresses in the interface and the soil around the anchor are identified and evaluated at every stage of incremental loading. They are converted into an equivalent or residual load. This load is then redistributed to the surrounding medium during the iterative procedure.

Loading

In the field test, a total pullout (axial) load, P = 250 kN, was applied at the free anchor head point A (Fig. 4-8) in five increments by using an annular jack. Three finite element analyses were performed by assuming the sand behavior to be linear, nonlinear (elastic-plastic) without interface, and nonlinear with interface. For the linear elastic analysis, the total load was applied in one increment. For the nonlinear analysis, a maximum load equal to 120% of 250 kN was applied in six equal increments.

Comparisons

Typical comparisons between computed and field results are shown in Figs. 4-11 and 4-12. Figure 4-11 shows comparisons of load displacement curves at the fixed anchor

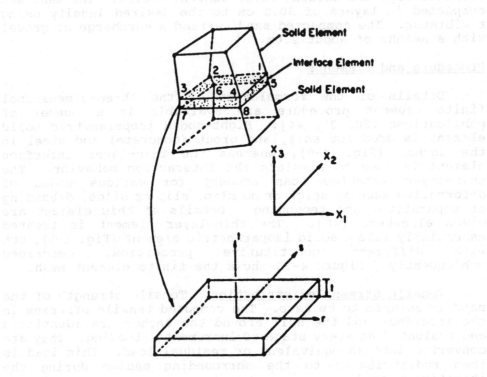

Figure 4-9   Schematic of Solid and Interface Elements

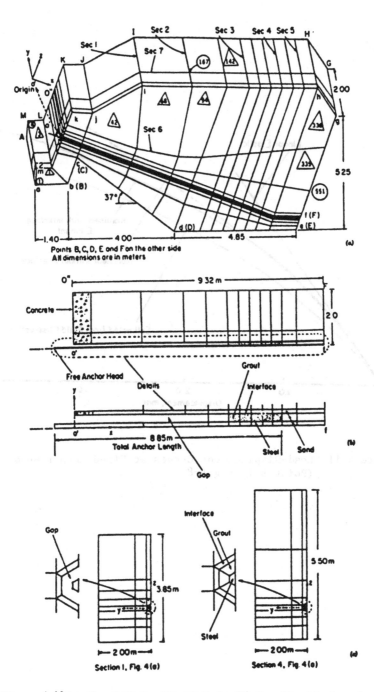

Figure 4-10 Details of Finite Element Mesh: (1) Overall; (b) along Grouted
Anchor; and (c) across Grouted Anchor.

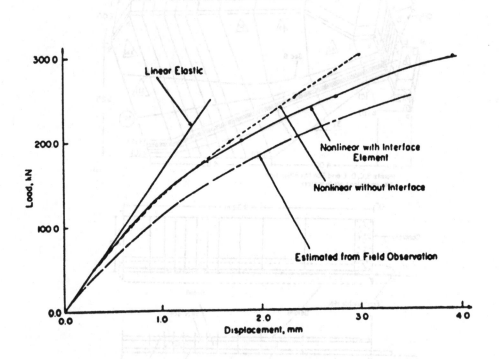

Figure 4-11   Load-Displacement Curves at Fixed Anchor Head
              (Point B in Fig. 4-8

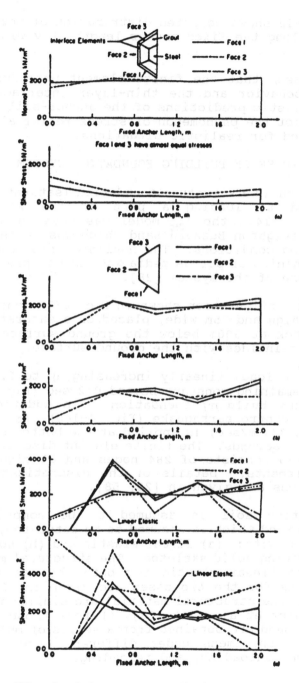

Figure 4-12 Normal and Shear Stress Distributions in Interface Elements on Grouted
Anchor: (a) At P = 50 kN; (b) at P = 150 kN; and (c) at P = 250 kN.

head; and Fig. 4-12 shows computed distribution of normal and
shear stresses along the fixed anchor length at various load
levels.

It can be seen that the finite element procedure with
nonlinear soil behavior and the thin-layer interface model
provide more realistic predictions of the anchor-soil system.
It can also predict the phenomenon of stress relief along the
anchor as expected for realistic situations.

4.4   DYNAMIC RESPONSE OF BUILDING FOUNDATION SYSTEM

Figure 4-13 shows a structure foundation (saturated sand)
system   subjected   to   loading   a   part   of   the   El   Centro
earthquake, Fig. 4-14.   the δ_{2+p} model was used to simulate
behavior of the Leighton Buzzard Sand which was tested under
multiaxial loading conditions.   The nonlinear finite element
procedure for dynamics of porous media was used to predict the
transient response of the system [28].

The problem configuration shown in Fig. 4-13 consists of
a building, 75m high and 75m wide, placed on a saturated soil
mass with bed rock at 300m below the ground surface [113].
The water table coincides with the ground surface.

A distributed load, linearly increasing up to 0.13 MN/m
in 0.2 sec and remaining constant beyond 0.2 sec (Fig. 4-14a)
is applied at the building foundation.   In addition, the
motion of the El Centro earthquake (Fig. 4-14b), (El Centro,
1940 NS Max. Acc. 3/4 m/sec) is applied at the bottom boundary
during the first 5 seconds.   The finite element discretization
shown in Fig. 4-15 consists of 284 nodes and 81 eight-noded
isoparametric elements.   Details of the discretization and
boundary conditions are given in [28, 58].

The   foundation   soil   is   assumed   to   be   comprised   of
Leighton Buzzard Sand which is assumed to be saturated, porous
two-phase material with: (a) linear elastic and (b) nonlinear
anisotropic hardening solid skeleton.   The structural material
is assumed to be linear elastic for both the cases.   The
material constants for the two cases are given in [28, 58].
The constants for the structural material are similar to those
used by Zienkiewicz et al. [112,113], whereas those for the
Leighton Buzzard Sand are obtained from a series of multiaxial
tests on cubical specimens under different stress paths,
including loading, unloading, and reloading.

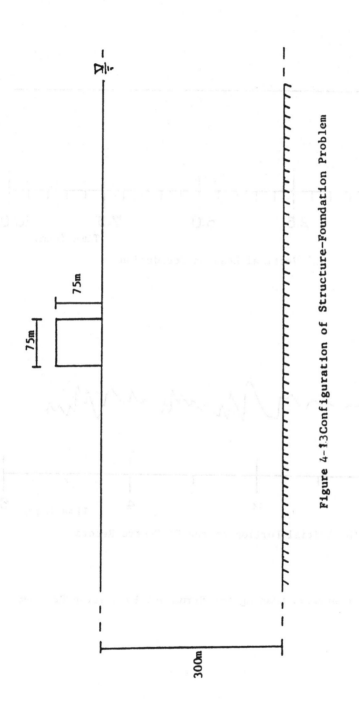

Figure 4-13Configuration of Structure-Foundation Problem

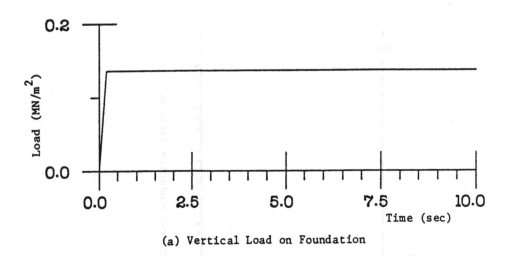

(a) Vertical Load on Foundation

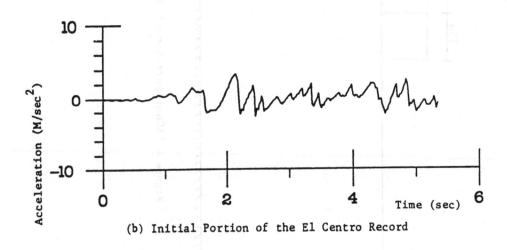

(b) Initial Portion of the El Centro Record

Figure 4-14 Applied Loading for Structure-Foundation Problem

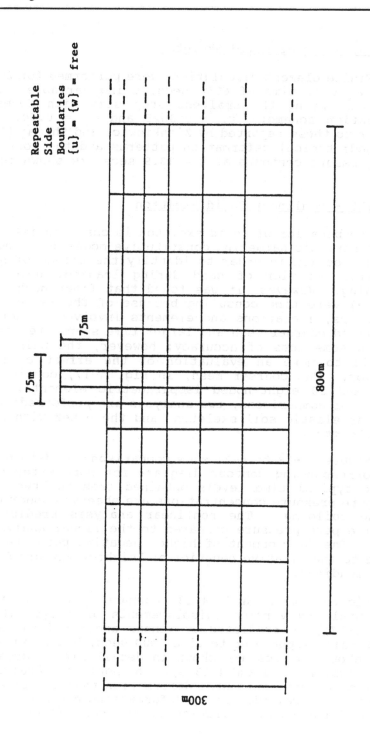

Figure 4-15 Finite Element Discretization for Structure-Foundation Problem

Case (a) - Linear Solid Skeleton

Finite element calculations were performed for 250 steps, up to a total time of 6354 secs, using variable time steps with 0.02 being the smallest step size with Newmark time integration scheme: $\theta_1 = 0.3025$ and $\theta_2 = 0.60$. Results similar to those reported by Zienkiewicz and Shiomi [113] were obtained; typical deformation patterns and the corresponding pore pressure contours at t = 13.9 secs are shown in Fig. 4-16.

Case (b) - Nonlinear Solid Skeleton

The behavior of solid skeleton is characterized by using the anisotropic hardening constitutive model described in the earlier section in order to identify the effect of nonlinear inelastic behavior of sand during loading, unloading and reloading. However, it was found that finer meshes used in case (a) were time consuming because of the large number of time steps, iterations and elements involved. Therefore, a simplified coarser mesh and larger time steps are used. This entails some loss of accuracy; however, the main intention here is to make an evaluation of the effect of nonlinear behavior. The coarser mesh, in Fig. 4-17, consists of 171 nodes and 30 eight-noded isoparametric elements. For the purpose of comparisons, two analyses are performed: one with a linear elastic soil skeleton, and the other with nonlinear soil skeleton.

Figures 4-18 to 4-21 show comparisons of deformed shapes and corresponding contour diagrams for pore water pressures at two typical time levels obtained from the two analyses. High pore pressure concentrations are observed under the edges of the building. The nonlinear analysis predicts higher negative pore pressure compared to the linear analysis (Fig. 4-20). The development of higher negative pore pressure may be due to the dilative behavior of Leighton Buzzard Sand under shear loadings.

Figures 4-22 and 4-23 compare the variations of horizontal and vertical displacements of a typical node 36 (Fig. 4-17), as predicted by linear and nonlinear analyses; for overall response up to time 7500 secs; the initial results up to about 40 secs are shown in detail with a greater time scale, Figs. 4-22(a) and 4-23(a). The nonlinear analysis shows higher permanent horizontal deformation than linear analysis; however, vertical (downward) deformations are much smaller in the nonlinear case. Dilation of the sand layer due to the

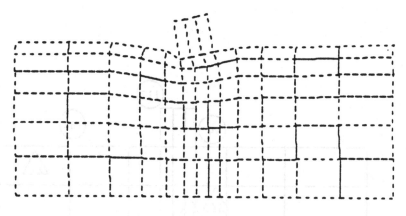

magnification factor 225
scale 1 cm = 66.6m

(a) Deformation Pattern

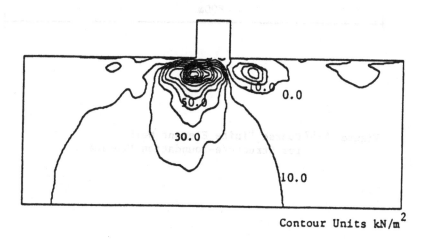

Contour Units kN/m^2

(b) Pore Pressure Contours

Figure 4-16 Deformation Pattern and Pore Water Pressure
Contours at 13.9 sec.

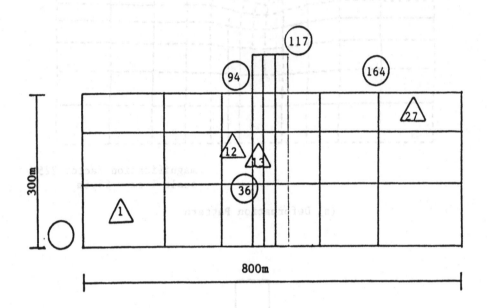

Figure 4-17 Coarse Finite Element Mesh
              for Structure-Foundation Problem

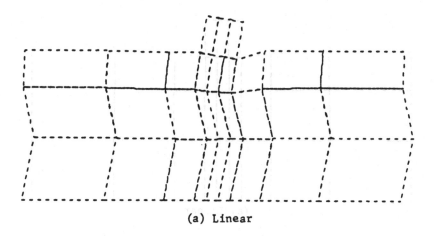

(a) Linear

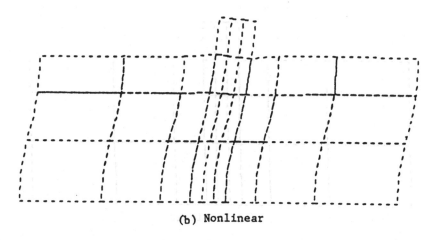

(b) Nonlinear

Figure 4-18 Comparison of Deformed Shapes at Time 5.8 secs
(Scale: 1 cm = 66.6m)

Magnification Factor 225

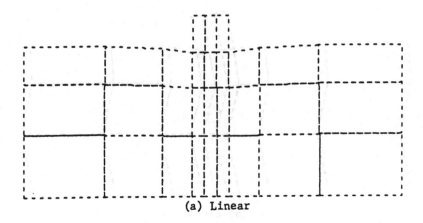

(a) Linear

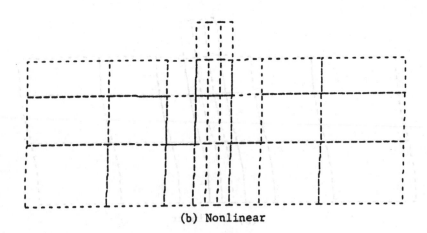

(b) Nonlinear

Figure 4-19 Comparison of Deformed Shapes at Time 6351.4 secs
(Scale: 1 cm = 66.6m)

Magnification Factor 225

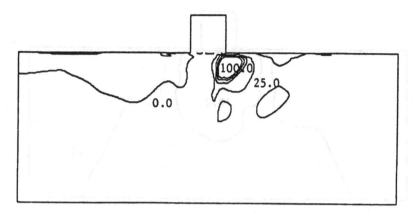

(a) Linear

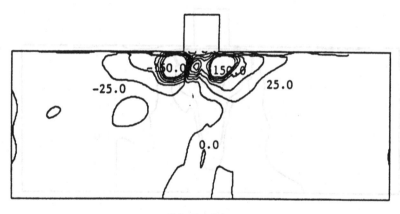

(b) Nonlinear

Contour Units kN/m^2

Figure 4-20 Comparison of Countour Diagrams for
Pore Pressure at Time 5.8 secs

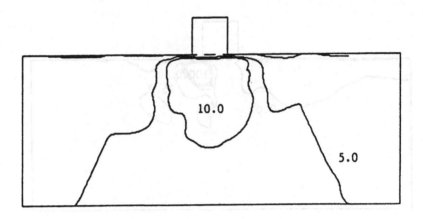

(a) Linear

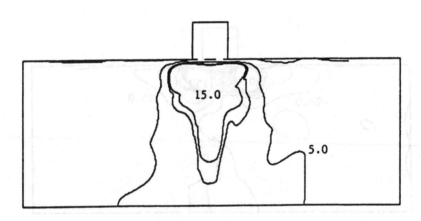

(b) Nonlinear

Contour Units kN/m^2

Figure 4-21 Comparison of Contour Diagrams for
Pore Pressure at Time 6351.4 secs

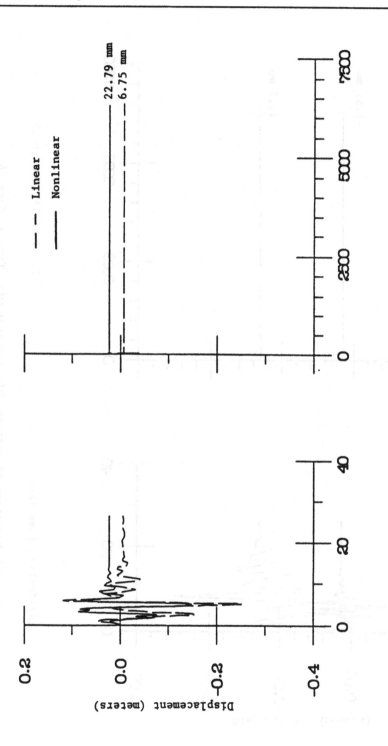

(a) Initial Time (sec)

(b) Overall Time (sec)

Figure 4-22 Variation of Horizontal Displacement with Time at Node 36

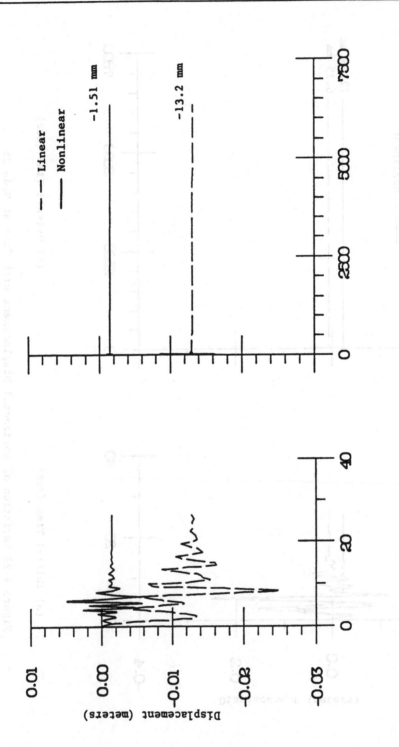

(a) Initial Time (sec)

(b) Overall Time (sec)

Figure 4-23 Variation of Vertical Displacement with Time at Node 36

earthquake induced shear loading may have reduced the downward displacement since dilation may be essentially reflected in upward displacements.

The time variations of pore pressures and shear stresses in a typical element 12 are presented in [28, 58]. The magnitude of shear stresses and the pore pressure are different in the two cases. At higher time levels, nonlinear analysis shows about ten times higher shear stress than the linear analysis does. Even during the initial period of shaking, higher shear stresses are predicted by the nonlinear analysis. The trend of development and dissipation of pore pressure is considerably different for the two analyses.

Comments

The above comparisons between the predictions from the linear and nonlinear material responses indicate that the consideration of the latter can involve significant influence on the dynamic behavior of the structure-foundation system. The anisotropic hardening model allows for various factors such as plastic deformations during virgin and nonvirgin loading and volume changes including dilatation. These appear to be among the factors that lead to significant differences between linear and nonlinear analysis. For instance, the nonlinear response shows (1) increased magnitudes and zones of concentration of pore water pressures, (2) increased magnitudes of horizontal displacements, (3) decreased magnitudes of vertical displacements, and (4) increased magnitudes of shear stresses.

4.5 THREE-DIMENSIONAL TRANSIENT FREE SURFACE SEEPAGE

The Residual Flow Procedure (RFP) has been applied for solution of a number of two- and three-dimensional laboratory and field problems. Here, a recent application involving comprehensive laboratory tests with a three-dimensional glass bead model is described [5, 21].

Although various three-dimensional formulations have been proposed, there are hardly any laboratory experimental results involving three-dimensional flow available in the literature. Hence, it was proposed to design and construct a laboratory glass-bead model that can simulate three-dimensional free surface seepage in dams. The model is capable of simulating both homogeneous and nonhomogeneous domains with steady and transient (rise and drawdown) fluctuations on the upstream. The comprehensive series of tests conducted using the model provides a unique capability to verify the proposed RFP for the three-dimensional free surface flow.

General Description of the Model

The schematic of the model is shown in Fig. 4-24. The model consists of two components: an outside box of Plexiglas panels, and wire meshes that are placed inside these panels to simulate the sloping sides of a model dam section and also provide barriers between different sizes of glass beads to simulate nonhomogeneous domains. Glass beads of various mean diameters (3mm and 1mm) are used to simulate granular soil media. These beads are coated with silicon by using a commerical spray in order to reduce capillary effects, and then they are placed in the model dam with given densities of packing. The permeabilities of the glass beads are obtained using a constant head test. The fluid used is glycerin mixed with about 20% water in order to reduce capillarity.

The fluid level on the upstream side of the model can be raised or lowered by means of a motorized unit, thus creating rise, drawdown and steady-state conditions. An overflow tank is also provided in order to keep the upstream head steady. The time dependent locations of the free surface are recorded photographically for various sections of the three-dimensional model.

Determination of Permeabilities and Storage Coefficients

The values of coefficients of permability among specific storage coefficients were obtained by using laboratory tests and empirical methods [5, 21].

Experiments and Verifications

The 3-D laboratory model (Fig. 4-24) is used with different size glass beads to perform three different experiments: homogeneous, and nonhomogeneous with two types of zoned conditions. The homogeneous case is labeled as Experiment 1, while two nonhomogeneous experiments with Zone I and Zone II are called Experiments 2 and 3, respectively. For each case, the behavior of the free surface is recorded photographically for the rise, steady-state and drawdown conditions. These results are then compared with the numerical predictions from the 3-D RFP. In order to clarify the presentation of the results, different sections of the

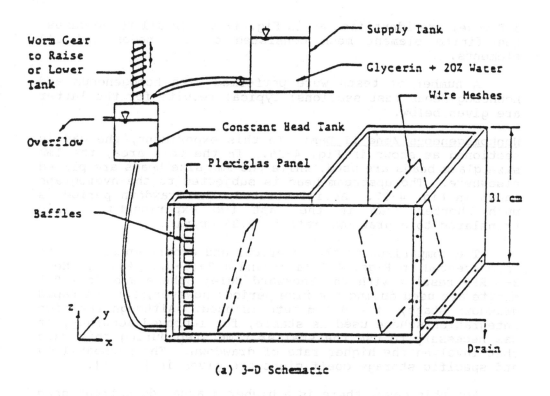

(a) 3-D Schematic

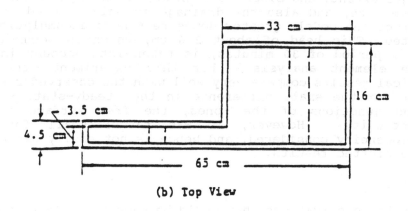

(b) Top View

FIGURE 4-24 CONFIGURATION OF THE 3-D LABORATORY MODEL

3-D model are identified as in Fig. 4-25. In all three cases, the finite element mesh consisted of 624 nodes and 348 elements.

A number of tests were performed with homogeneous and nonhomogeneous test sections; typical results for the latter are given below.

Nonhomogeneous Zone II Dam: In this experiment, the dam is sectioned as shown in Fig. 4-26. In the core area, the 1mm size glass beads are used, while the 3mm size beads are placed elsewhere. The upstream head is subjected to the hydrograph shown in Fig. 4-27. As illustrated, the drawdown period is much shorter than in the last two experiments. This translates to a drawdown rate of 9.33 cm/min.

The comparison of the numerical and experimental results is presented in Figs. 4-28(a-d) and 4-29(a-d). Again, the 3-D RFP results with the Backward Euler Scheme and t = 0.5 minute are used during the rise period; however, for the rapid drawdown stage, t = 0.1 minute is used. Although the time integration scheme used is stable, for improved accuracy, it was necessary to use the smaller time step during this case that involved the higher rate of drawdown. The permeability and specific storage coefficient are given in [5, 21].

For this case, there is a higher gradual downstream head accumulation as compared to the previous two experiments. This is due to the fact that the 3mm size beads are used to a larger extent; and since these beads are more previous than the 1mm size, and also the drainage is only allowed at one corner of the model, greater downstream head accumulation is expected This head of about 2.6 cm, which is accumulated after a period of 30 minutes, is taken into account in the finite element analysis. For this experiment, the 3-D numerical results compare very well with the observed results except for the small difference in the steady-state stage. At the junctions of the zones, the free surface showed discontinuity. However, it was rather small for the nonhomogeneity considered, and hence is not easily discerned in the plotted results.

Comments

Other observations that could be made are related to the free surface behavior. As presented in the results, in case of both nonhomogeneous dams, and at the junctions of the zones, the free surface shows discontinuity. In the case of

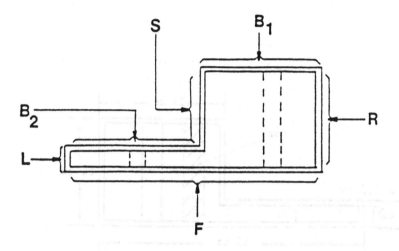

(top view)

FIGURE 4-25 IDENTIFICATION OF DIFFERENT SECTIONS
OF THE 3-D LABORATORY MODEL

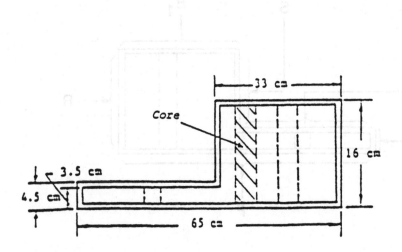

Figure 4-26   SCHEMATIC OF THE NONHOMOGENEOUS ZONE II DAM

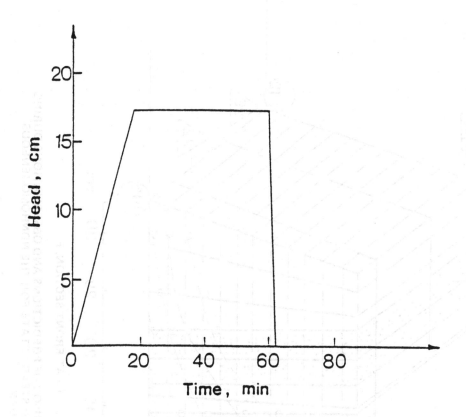

Figure 4-27   **EXPERIMENT 3: VARIATION OF HEAD WITH TIME**

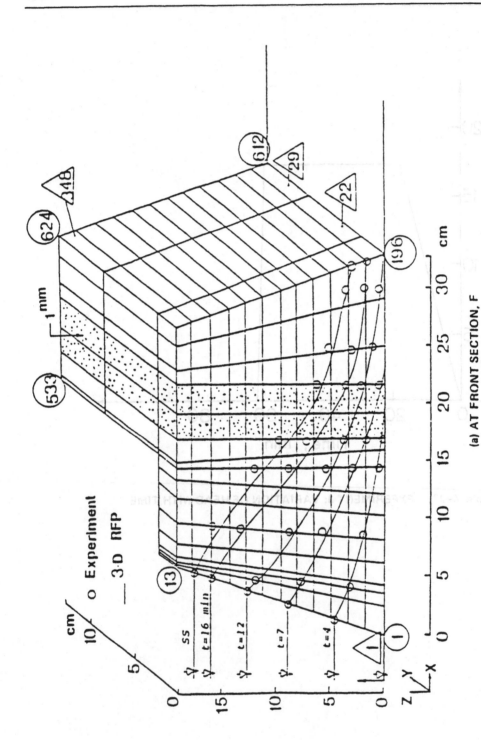

Figure 4-28 **COMPARISON OF PREDICTIONS AND OBSERVATIONS DURING RISE AND STEADY-STATE FOR THE NONHOMOGENEOUS ZONE II DAM**

(a) AT FRONT SECTION, F

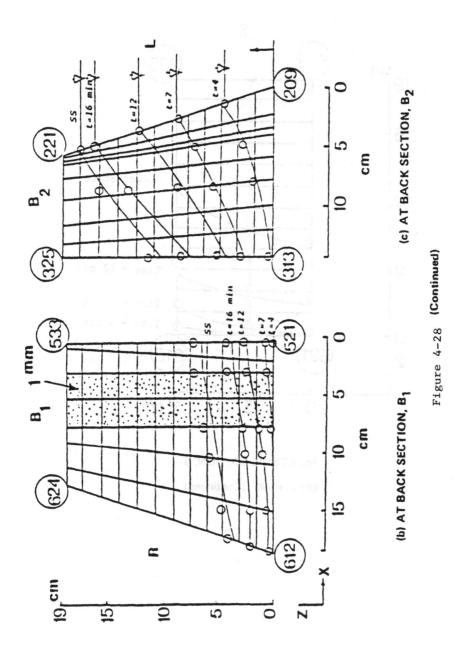

(b) AT BACK SECTION, B₁

(c) AT BACK SECTION, B₂

Figure 4-28   (Continued)

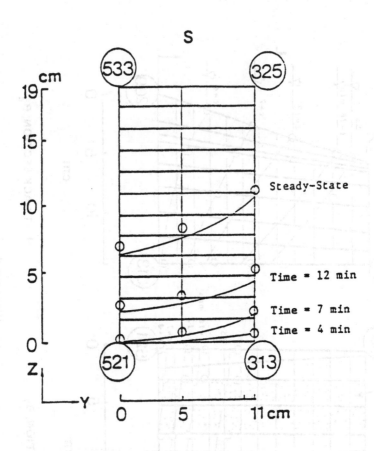

(d) AT SIDE SECTION, S

Figure 4-28   (Continued)

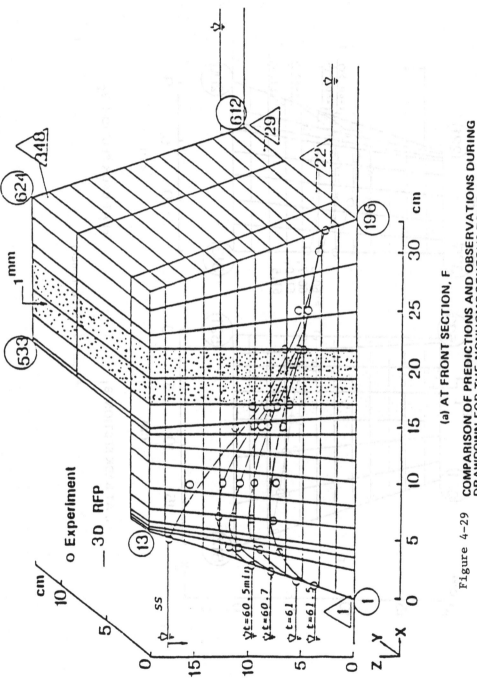

Figure 4-29 COMPARISON OF PREDICTIONS AND OBSERVATIONS DURING
DRAWDOWN FOR THE NONHOMOGENEOUS ZONE II DAM

(a) AT FRONT SECTION, F

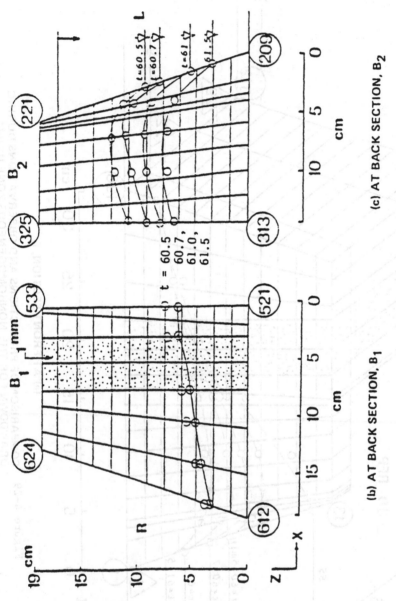

(b) AT BACK SECTION, B_1

(c) AT BACK SECTION, B_2

Figure 4-29.(Continued)

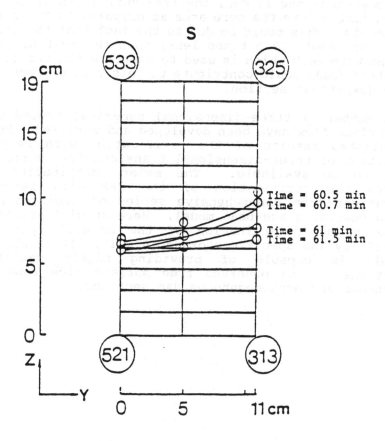

(d) AT SIDE SECTION, S

Figure 4-29   (Continued)

nonhomogeneous-Zone II dam, the free surface is higher in the section just before the core area as compared to the first two experiments. This could be due to the fact that the core area is less permeable (2.5 times less) than the rest of the dam. Also the wire mesh which is used to separate the two different size glass beads could contribute to this rise of free surface in the downstream section.

A number of three-dimensional numerical procedures for free surface flow have been developed and verified. However, no published results of the verification with respect to observation of three-dimensional transient free surface flow appear to be available. The major contribution herein involves the verification of the RFP with respect to observations from comprehensive series of laboratory tests using a special glass-bead model. Here special attention is also given to determination of the material properties, permeability and specific storage [5, 21]. It is shown that the RFP is capable of providing highly satisfactory predictions of the observed free surface flow results for homogeneous and nonhomogeneous dam sections.

PART 5:   COMPUTER CODES HISS-DOD1:
EVALUATION OF MATERIAL CONSTANTS,
VERIFICATION AND IMPLEMENTATION

Computer codes are prepared for:

(a) computation of material constants in various hierarchical models (δ_0, δ_1, δ_2, δ_{o+r}, δ_{1+vp}); at this time the code of δ_0, δ_1 versions is available. This code allows computation of all elastic and plastic constants based on typical stress-strain curves input as data points in the code.

(b) back prediction of stress-strain tests under different stress paths by using the constants found in (a) and by integration of the incremental constitutive equations, $\{d\sigma\} = [C^{ep}] \{d\varepsilon\}$.

(c) a routine that can be integrated in (available) finite element codes (or in codes based on other methods) through equation $[K] \{\Delta q\} = \{\Delta Q\}$. It is based on a modified drift correction procedure.

Complete details of the above items are available in Ref. [18]. For details regarding the availability, contact the author.

PART 6.   ACKNOWLEDGMENTS

The results reported herein are based on a number of research grants and projects supported by the National Science Foundation, Washington, D. C.; Air Force Office of Scientific Research, Bolling Air Force Base; and Office of University Research, Department of Transportation.   Acknowledgments are due to various co-workers, S. Armaleh, B. Baseghi, M. O. Faruque, K. L. Fishman, G. N. Frantziskonis, H. M. Galagoda, Q.S.E. Hashmi, J. Kujawski, A. Muqtadir, B. K. Nagaraj, N. Navayogarajah, M. R. Salami, F. Scheele, S. Somasundaram, A. Varadarajan, and G. W. Wathugala, whose contributions have been used directly in these notes.   The reference list also provides acknowledgment of the contributions of a number of other co-workers.

REFERENCES:

1. Argyris, J. H., L. E. Vaz and K. J. Willam: Higher order methods for transient diffusion analysis, Comp. Meth. Appl. Mech. Eng., 12 (1977), 243-278.

2. Armaleh, S.: Disturbed state modelling approach for cohesionless soils," Ph.D. Dissertation, University of Arizona, Tucson (1989), in progress.

3. Arthur, J.R.F., K. S. Chua, T. Dunstant and J. I. Rodriguez del C.: Principal stress rotation: a missing parameter, J. Geotech. Eng. Div., ASCE, 106, GT4 (1980), 419-433.

4. Baker, R. and C. S. Desai: Induced anisotropy during plastic straining, Int. J. Num. Analyt. Meth. in Geomech., 8, 2 (1984), 167-185.

5. Baseghi, B. and C. S. Desai: Laboratory verification and applications of residual flow procedure for three-dimensional seepage, J. Water Resources Research, in press.

6. Biot, M.A.: General theory of three-dimensional consolidation, J. App. Physics, 12 (1941), 155-164.

7. Biot, M.A.: Theory of propagation of elastic waves in a fluid saturated porous solid I & II, J. Acoustical Soc. of America, 28 (1956), 168-191.

8. Biot, M.A.: Mechanics of deformation and acoustic propagation in porous media, J. App. Physics, 33 (1962), 1482-1498.

9. Dafalias, Y.F.: Initial and induced anisotropy of cohesive soils by means of a varying nonassociated flow rule, Colloque Internationale du C.N.R.S., 319, Villard-du-Lans, Grenoble, France, June (1981).

10. Dafalias, Y.F.: On Rate Dependence and Anisotropy in Soil Constitutive Modelling, in: Constitutive Relations for Soils, (Eds. G. Duehus, F. Darve and I. Vardoulakis), A.A. Balkema Publishers, 1984, 457-462.

11. Desai, C.S.: Finite element residual schemes for unconfined flow, Int. J. Num. Meth. Engng, 10 (1976), 1415.

12. Desai, C.S.: Effect of driving and subsequent consolidation on driven piles, Int. J. Num. Analyt. Meth. in Geomech., 2 (1978), 283-301.

13. Desai, C.S.: A general basis for yield, failure and potential functions in plasticity, Int. J. Num. Analyt. Meth. in Geomech., 4 (1980), 361-375.

14. Desai, C.S.: A dynamic multi-degree-of-freedom shear device, Report No. 8-36, Department of Civil Engineering, Virginia Tech, Blacksburg, Virginia (1980).

15. Desai, C.S.: Behavior of interfaces between structural and geologic media. A State-of-the-Art Paper for Int. Conf. on Recent Advances in Geotech. Earthquake Engrg. and Soil Dyn., St. Louis, Missouri (1981).

16. Desai, C.S.: Further on unified hierarchical models based on alternative correction of "damage" approach, Report, Dept. of Civil Engineering and Engineering Mechanics, University of Arizona, Tucson, Arizona (1987).

17. Desai, C.S.: Single surface yield and potential function plasticity models: a review, Letter to Editor, J. Comp. and Geotechnics, 7 (1989) 319-335.

18. Desai, C.S.: Computer Code HISSDOD1 for Computation of constants and back predictions of hierarchical single surface models, Report, Tucson, AZ (1990).

19. Desai, C.S. and J. F. Abel: Introduction to the Finite Element Method, Van Nostrand Reinhold Co., New York (1972).

20. Desai, C.S. and G. C. Appel: 3-D analysis of laterally wooded structures, Proc., 2nd Int. Conf. on Num. Meth. Geomech., ASCE, Blacksburg, Virginia (1976).

21. Desai, C.S. and B. Baseghi: Theory and verification of residual flow procedure for 3-D free surface seepage, Ad. Water Resources, 11 (1988), 195-203.

22. Desai, C.S., E.C. Drumm and M.M. Zaman: Cyclic testing and modeling of interfaces, J. of Geotech. Engrg., ASCE, 111, 6 (1985), 793-815.

23. Desai, C.S., I.M. Eitani and C. Haycocks: An application of finite element procedure for underground structures with nonlinear analysis and joints, Proc., 5th Int. Conf. Soc. of Rock Mech., Melbourne, Australia (1983).

24. Desai, C.S. and M.O. Faruque: Constitutive model for geologic materials, J. Eng. Mech. Div., ASCE, 110, 9, September (1984), 1391-1408.

25. Desai, C.S. and K.L. Fishman: Constitutive models for rocks and discontinuities (joints), Proc., 28th Symp. on Rock Mech., Tucson, Arizona (1987).

26. Desai, C.S. and K. L. Fishman: Plasticity based constitutive model with associated testing for joints, Int. J. Rock Mech. and Min. Sc. (1988), accepted for publication.

27. Desai, C.S., G.N. Frantziskonis and S. Somasundaram: Constitutive modelling for geologic materials, Proc., 5th Intl. Conf. on Num. Meth. in Geomech., Nagoya, Japan, April (1985), 19-34.

28. Desai, C.S. and H.M. Galagoda: Earthquake analysis with generalized plasticity model for saturated soils, J. Earthquake Eng. and Struct. Dyn., 18, 6 (1989), 903-919.

29. Desai, C.S., H.M. Galagoda and G.W. Wathugala: Hierarchical modelling for geologic materials and discontinuities - joints and interfaces, Proc., 2nd Int. Conf. on Constitutive Laws for Eng. Mat., Tucson, AZ (1987), Elsevier, N.Y.

30. Desai, C.S. and Q.S.E. Hashmi: Analysis, evaluation and implementation of a nonassociative model for geologic materials, Int. J. Plasticity, 5 (1989),

31. Desai, C.S. and L.D. Johnson: Evaluation of two finite element formulations for one-dimensional consolidation, J. Computers and Structures, 2, 4 (1972).

32. Desai, C.S. and L.D. Johnson Evaluation of some numerical schemes for consolidation, Int. J. Num. Meth. Eng., 7 (1973).

33. Desai, C.S., J. Kujawski, C. Miedzialowski and W. Ryzynski: Improved time integration of nonlinear dynamic problems, Comp. Meth. in Appl. Mech. and Eng., 62 (1987), 155-168.

34. Desai, C.S. and G. C. Li: A residual flow procedure and application for free surface flow in porous media, Adv. in Water resources, 6 (1983), 27.

35. Desai, C.S. and J.G. Lightner: Mixed finite element procedure for soil-structure interaction and construction sequences, J. Num. Meth. in Eng., 21 (1985), 801-814.

36. Desai, C.S. and R.L. Lytton: Stability criteria for two finite element schemes for parabolic equations, Int. J. Num. Meth. Eng., 9 (1975).

37. Desai,C.S., A. Muqtadir and F. Scheele: Interaction analysis of anchor-soil systems, J. Geotech. Eng., ASCE, 112, 5 (1985), 537-553.

38. Desai, C.S. and B.K. Nagaraj: Modeling for cyclic normal and shear behavior of interfaces, J. of Eng. Mech., ASCE, 114 (1988), 1198-1217.

39. Desai, C.S. and M.R. Salami: A constitutive model and associated testing for soft rock, Int. J. Rock Mech. Min. Sc., 24 (1987), 299-307.

40. Desai, C.S. and M.R. Salami: A constitutive model for rocks, J. Geotech. Eng., ASCE, 113, 5 (1987).

41. Desai, C.S. and S.M. Sargand: Hybrid finite element procedure for soil-structure interaction, J. Geotech. Eng., ASCE, 110 (1984), 473-486.

42. Desai, C.S. and S.K. Saxena: Consolidation analysis of layered anisotropic foundations, Int. J. Num. Analyt. Meth. in Geomech., 1 (1977).

43. Desai, C.S. and W.C. Sherman: Unconfined transient seepage in sloping banks, J. Soil Mech. and Found. Engg. Div., ASCE, 97, SM2 (1971), 357.

44. Desai, C.S. and H.J. Siriwardane: Numerical models for track support structures, J. of Geotech. Eng., ASCE, 108, 3 (1982).

45. Desai, C. S., S. Somasundaram and G.N. Frantziskonis: A hierarchical approach for constitutive modelling of geologic materials, Int. J. Num. Meth. in Geomech., 10, 3 (1986).

46. Desai, C.S. and A. Varadarajan: A constitutive model for quasistatic behavior of rock salt, J. of Geophys. Research, Manuscript No. 6136091 (October 1987).

47. Desai,C.S. and   G. W. Wathugala: Hierarchical and unified models for solids and discontinuities (joints/interfaces), Notes for Short Course, Tucson, AZ (1987).

48. Desai, C.S., M.M. Zaman, J.G. Lightner and H.J. Siriwardane: Thin-layer element for interfaces and joints, Int. J. Num. Analyt. Meth. Geomech., 8, 1 (1984).

49. Desai, C.S. and D. Zhang: Viscoplastic model (for rocks) with generalized yield function, Int. J. Num. Analyt. Meth. Geomech., 11 (1987), 603-620..

50. DiMaggio, F.L. and I. Sandler: Material model for granular soils, J. of Eng. Mech. Div., ASCE, 97 (1971), 935-950.

51. Drucker, D.C., R.E. Gibson and D.J. Henkel: Soil mechanics and work-hardening theories of plasticity, Trans. ASCE Eng. Mech. Div., Paper No. 2864 (1955), 81.

52. Drumm, E.E. and C.S. Desai: Determination of parameters for a model for the cyclic behavior of interfaces, J. of Earthq. Engrg. and Struct. Dyn., 16, 1 (1986), 1-18.

53. The Earth Technology Corp: Pile segment tests-Sabine pass, some aspects of the fundamental behavior of axially loaded piles in clay soils, ETC Report No. 85-007, Long Beach, CA (1986).

54. Fishman, K.L. and C. S. Desai: A constitutive model for hardening behavior of rock joints, Proc., 2nd Int. Conf. on Const. Laws for Eng. Mat., Elsevier, New York (1987).

55. Fishman, K.L. and C.S. Desai: Measurements of normal deformations in joints during shear using inductance devices, Geotech. Testing J., 12, 4 (1989), 297-301.

56. Frantziskonis, G. and C.S. Desai: Constitutive model with strain softening, Int. J. of Solids Structures, 23, 6 (1987).

57. Frantziskonis, G., C.S. Desai and S. Somasundaram: Constitutive model for nonassociative behavior, J. of Eng. Mech., ASCE, 112 (1986), 932-946.

58. Galagoda, H.M.: Nonlinear analysis of porous soil media and application, Ph.D. Dissertation, University of Arizona, Tucson, AZ (1986).

59. Ghaboussi, J. and K.J. Kim: Quasi-static and dynamic analysis of saturated soils, in: Mechanics of Engineering Materials (Eds. C.S. Desai and R.H. Gallagher), John Wiley & Sons Ltd. (1984).

60. Hashmi, Q.S.E.: Modelling for nonassociative behavior and implementation, Ph.D. Dissertation, University of Arizona, Tucson, Arizona (1987).

61. Hogge, M.A.: Accuracy and cost of integration techniques for nonlinear heat transfer, in: Finite Element Methods in the Commerical Environment (Ed. J. Robinson), Robinson Assoc., Wimborne, U.K. (1978).

62. Hogge, M.A.:   A comparison of two- and three-level integration schemes for non-linear heat conduction, in: Numerical Methods in Heat Transfer (Eds. R.W. Lewis, K. Morgan and O.C. Zienkiewicz), Wiley (1981).

63. Hughes, T.J.R.: Analysis of transient algorithms with particular reference to stability behavior, in: Computational Methods in Transient Analysis (Eds. T. Belytschko and T.J.R. Hughes), North-Holland Publishing Co., Amsterdam, The Netherlands (1983).

64. Hvorslev, M.J.:   Uber die Festigkeitsei genschaffen gestorter bindiger Boden (on physical properties of disturbed cohesive soils), Ingeniorvidenskabelige Skrifter, No. 45, Austria (1937), 159 p.

65. Jelinek, R. and F. Scheele: Tragfahigkert und tragverhalten von verpressanken, Bundes Ministerium fur Raumordnung, Bauwesan und Statebau (1977).

66. Katti, D.R.:   Disturbed state approach for cyclic behavior of clays, Ph.D. Dissertation, University of Arizona, Tucson, Arizona (1989), in progress.

67. Kim, M.K. and P.V. Lade: Single hardening constitutive model for frictional materials: I. Plastic potential function, Computers and Geotechnics, 5 (1988), 307-324.

68. Kujawski, J. and C.S. Desai: Unconditionally stable time finite element family of schemes for time integration in linear and non-linear dynamic problems, Rept. 2/KD/83, Department of Civil Engineering and Engineering Mechanics, University of Arizona, Tucson, AZ (1983).

69. Kujawski, J. and C.S. Desai: Generalized time finite element algorithm for non-linear dynamic problems, Engrg. Comput., 1, 3 (1984), 247-251.

70. Kujawski, J. and C.S. Desai: A quasi-explicit modification of two time level family of schemes for nonlinear transient analysis, Int. J. Num. Analyt. Meth. in Geomech., 9, 5 (1985).

71. Kujawski, J. and C.S. Desai: Construction of highly stable explicit algorithms for transient field problems, J. of Tech. Physics, 28, 1 (1987).

72. Lade, P.V.:   Elasto-plastic stress-strain theory for cohesionless soil with curved yield surfaces, Int. J. Solids Struct., 13 (1977), 1019-1035.

73. Lade, P.V. and J.M. Duncan: Elasto-plastic stress-strain theory of cohesionless soils, J. Geot. Eng., ASCE, 101 (1975), 1037-1053.

74. Lade, P.V. and M.K. Kim: Single hardening constitutive model for frictional materials: II. Yield criterion and plastic work contours, Computers and Geotechnics, 6 (1988), 13-29.

75. Lade, P.V. and M.K. Kim: Single hardening constitutive model for frictional materials: III. Comparison with experimental data, Computers and Geotechnics, 6 (1988), 31-47.

76. Lewin, P.I.:   The deformation of soft clay under generalized stress conditions, Ph.D. Thesis, Kings College, University of London (1978).

77. Li, G. C. and C.S. Desai: Stress and seepage analysis of earth dams, J. Geotech. Eng., ASCE, 109 (1983), 946-960.

78. Ma, Y.:   Hierarchical single surface model for rock joints, Ph.D. Dissertation, University of Arizona, Tucson, Arizona (1989), under preparation.

79. Matsuoka, H. and T. Nakai: Stress-deformation and strength characteristics of soil under three different principal stresses, Soils & Found., J. Jap. Soc. of Soil Mech. & Found. Eng., 232 (1974), 59-70.

80. Mroz, Z., V.A. Norris and O.C. Zienkiewicz: An anisotropic hardening model for soils and its application to cyclic loading, Int. J. Num. Analyt. Meth. Geomech., 2 (1978), 203-221.

81. Navayogarajah, N.: Hierarchical single surface model for interfaces, Ph.D. Dissertation, University of Arizona, Tucson, AZ (1989), under preparation.

82. Norrie, D.H. and G. DeVries: Application of the pseudo-functional finite element method to nonlinear problems, in: Finite Elements in Fluids (Eds. Gallagher et al.), John Wiley & Sons, U.K. (1974), 2, Ch.3.

83. Nova, R. and D.M. Wood: An experimental program to define the yield function for sand, Soils and Foundations, 18, 4 (1978), 77-86.

84. Ostemayer, H. and F. Scheele: Research on ground anchors in noncohesive soils, Sp. Session 4, at Int. Conf. on Soil Mech. and Found. Eng., Tokyo, Japan (1977).

85. Poorooshasb, H.B.:   Deformation of sand in triaxial compression, Proc., 4th Asian Conf. SMFE, Bangkok, 1 (1971), 63-66.

86. Poorooshasb, H.B., I. Holubec and A.N. Sherbourne: Yielding and flow of sand in triaxial compression, Part 2 and 3, Can. Geot. Journal, 4 (1967), 376-397.

87. Potts, D.M. and A. Gens: A critical assessment of methods of corrections for drift from the yield surface in elasto-plastic finite element analysis, Int. J. Num. Analyt. Meth. in Geomech., 9 (1985), 149-159.

88. Prevost, J.H.: Plasticity theory for soil stress-strain behavior, J. Eng. Mech. Div., ASCE, 104, EM5, October (1978), 1177-1194.

89. Prevost, J.H.: Nonlinear transient phenomena in elastic-plastic solids, J. of Eng. Mech. Div., ASCE, 108 (1982), 1297-1311.

90. Roscoe, K.H., A.N. Schofield and A. Thurairajah: Yielding of clays in states wetter than critical, Geotechnique, 13 (1963), 211-240.

91. Roscoe, K.H., A.N. Schofield and C.P. Wroth: On yielding of soils, Geotechnique, 8 (1958), 22-53.

92. Sandhu, R.S., I.S. Rai and C.S. Desai: Variable time-step analysis of unconfined seepage, Proc. Int. Symp. on Finite Element Meth. in Flow Problems, Univ. of Wales, Swansea, U.K. (1974).

93. Sandhu, R.S. and E.L. Wilson: Finite element analysis of seepage in elastic media, J. Eng. Mech. Div., ASCE, Proc., No. EM3 (1969) 641-652.

94. Scheele, F. and C.S. Desai: Laboratory Behavior of Munich Sand, Report, Dept. of Civil Eng. & Eng. Mech., Univ. of Arizona, Tucson, AZ (1983).

95. Scheele, F.: Tragfaigkeit von verpressanken in nichtbinddigen boden, Neue Erkemnt-nisien durah dehnungsmessungen in veran-Kernungsbereich, Ph.D. Disseration, University of Munich, West Germany (1981).

96. Scheele, F., C.S. Desai and A. Muqtadir: Testing and modelling of "Munich" sand, Soils & Found. J. of JSMFE, 26 (1986), 1-18.

97. Schreyer, H.L.: A third-invariant plasticity theory for frictional materials, J. of Structural Mech., June (1983).

98. Siriwardane, H.J. and C.S. Desai: Two numerical schemes for nonlinear consolidation, Int. J. Num. Meth. in Engg., 5 (1981).

99. Somasundaram, S.: Constitutive modelling for anisotropic hardening behavior with applications to cohesionless soils, Ph.D. Thesis, University of Arizona, Tucson, AZ (1986).

100. Somasundaram, S. and C.S. Desai: Modelling and testing for anisotropic behavior of soils, J. Eng. Mech., ASCE, 114 (1988), 1177-1194.

101. Sugio, S. and C.S. Desai: Residual flow procedure for sea water intrusion in unconfined aquifers, Int. J. Num. Meth. Eng., 24 (1987), 1439-1450.

102. Tatsuoka, F. and K. Ishihara: Drained deformation of sand under cyclic stresses reversing direction, Soils and Foundations, 14, 3, December (1974), 51-65.

103. Tatsuoka, F. and K. Ishihara: Yielding of sand in triaxial compression, Soils and Foundations, 14, 2 (1974), 63-76.

104. Uesugi, M. and H. Kishida: Influence factors of friction between steel and dry sands, Soils and Foundations, 26 2 (1986) 33-46.

105. Uesugi, M. and H. Kishida: Frictional resistance at yield between dry sand mild steel, Soils and Foundations, 26, 4 (1986) 139-149.

106. Varadarajan, A. and C.S. Desai: Constitutive modelling for some sands from India, Report, Dept. of Civil Eng. & Eng. Mech., Univ. of Arizona, Tucson, AZ, (1987).

107. Wathugala, G.W.:   Dynamic soil-structure interaction
     analysis with anisotropic hardening model, Ph.D.
     Dissertation, Dept. of Civil Eng. & Eng. Mech.,
     University of Arizona, Tucson, AZ (1989), under
     preparation.
108. Wathugala, G.W. and C.S. Desai: "Damage" based
     constitutive model for soils, Proc., 12th Canadian
     Congress of Appl. Mech., Ottawa, Canada (1987).
109. Wathugala, G.W. and C.S. Desai: An analysis of piles in
     Maine clay under cyclic axial loading, Offshore
     Technology Conf., OTC paper 6002, Houston, Texas (1989).
110. Westbrook, D.R.:   Analysis of inequality and residual
     flow procedures and an iterative scheme for free surface
     seepage, Int. J. for Num. Meth. in Eng., 21 (1985), 1791.
111. Zaman, M.M., C.S. Desai and E.C. Drumm: An interface
     model for dynamic soil-structure interaction.   J.
     Geotech. Div., ASCE, 110, 4 (1984), 1257-1273.
112. Zienkiewicz, O.C., K.H. Leung and E. Hinton: Earthquake
     response behavior of soils with damage, Proc. Num. Meth.
     in Geomech., (Ed. Z. Eizenstein), Edmonton (1982).
113. Zienkiewicz, O.C. and T. Shiomi: Dynamic behavior of
     saturated porous media: the generalized Biot formulation
     and its numerical solution, J. Num. Analyt. Meth. in
     Geomech., 8 (1984), 71-96.

MECHANICS OF PARTIALLY SATURATED POROUS MEDIA

B.A. Schrefler, L. Simoni
Università di Padova, Padova, Italy

Li Xikui, O.C. Zienkiewicz
University College of Swansea, Swansea, UK

Abstract

In the framework of the volume fraction theories, the governing equations of problems of mechanics of partially saturated porous media are derived. The use of general averaging principles provides the definition of averaged field variables, which allow the connection with experimental data. A finite element discretization of the governing equations is subsequently presented.

1. INTRODUCTION

A partially saturated porous medium is a portion of space occupied partly by a solid phase (solid skeleton) and partly by a void space filled with fluid phases, i.e. water and air. The description of the mechanical deformation of the solid and of the flux of the fluid phases in such a multiphase system is a problem of increasing interest. An excellent survey of the theories for modelling such problems may be found in [1].

The first attempts to describe the physical behaviour relied on models of a more or less intuitive nature (engineering or phenomenological approach) [2].

A subsequent approach is referred to as continuum theory of mixtures. In this method all components of the multiphase system are supposed to be simultaneously present everywhere and to occupy the whole domain (overlapping continua). Continuous field variables are defined which account for the behaviour of each phase and for the interaction of the phases. The balance equations, which involve these field quantities, are postulated as an intuitive extension of the classical single phase continuum equations.

The third approach to describe the problem employs the technique of local volume averaging: the system is supposed to consist, as in reality, of interpenetrating continua, each occupying only part of the space. Irregular interfaces exist which separate the components of the system. The usual field quantities are associated with each phase. These variables are continuous within each phase, but discontinuous over the entire domain. The balance laws of the continuum mechanics are integrated (averaged) over some local representative element of volume. In this way interfacial terms appear, which account for the interaction of the phases.

The averaging technique seems to provide the best framework for the description of the problem, but the connection with the traditional experimental environment is quite difficult and may only be obtained by introducing many simplifications. In the present work, we refer to the last technique, but in a short note also the phenomenological or engineering approach will be addressed. From the practical point of view this last one is always useful.

2. AVERAGING PRINCIPLES

For our purpose we introduce the following definitions:

microscopic level: we consider the real nonhomogeneous structure of the porous medium domain (fig. 1). The scale of inhomogeneity is of the order of magnitude of the dimensions of a pore or a grain, say d. Attention is focussed on what happens at a mathematical point within a single phase and the field variables describing the status of a phase are defined only at the points occupied by that phase. For the general description of the process taking place in a porous medium, this level is of no practical value.

macroscopic level: the real multiphase system that occupies the porous medium domain is replaced by a model in which each phase is assumed to fill up the entire domain. This means that at every point all phases are supposed present at the same time (overlapping continua). At this level, we usually

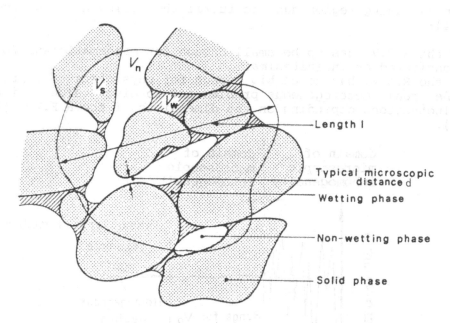

Fig. 1- Typical averaging volume (two phase flow).

deal with homogeneous media, but nonhomogeneities may still be present, e.g. strata. Their scale is of the order of magnitude comparable with the order of magnitude of the entire domain, say L.

megascopic level: at this level the conditions are similar to those of the previously defined level. The difference depends on the fact that some macroscopic inhomogeneities are eliminated by some type of averaging and/or on the fact that the mathematical model is stated in a domain which has less dimensions than the real domain, e.g. 2-D problem with field values averaged over the thickness [3].

At each level, the value of a field variable α is averaged over a certain representative volume element V_0 (R.E.V.), which contains all the present phases, according to

$$\bar{\alpha}\left(\bar{\underline{x}}, t\right) = \frac{1}{V_0} \int_{V_0} \alpha\left(\underline{x}, t\right) dV \qquad (1)$$

where $\underline{x}$ and $\bar{\underline{x}}$ represent the coordinate system at the lower and upper level.

Averaged quantities have to be independent on the size of the averaging volume and continuous in space and time. Thus

an averaging region has to fulfil the following requirements
[4]:

- the R.E.V. has to be small enough for mathematicians to be
considered as infinitesimal;
- the R.E.V. has to be big enough for physicists to represent
the real heterogeneous problem and to give values without
fluctuations depending on the dimensions of the R.E.V. (fig.
2).

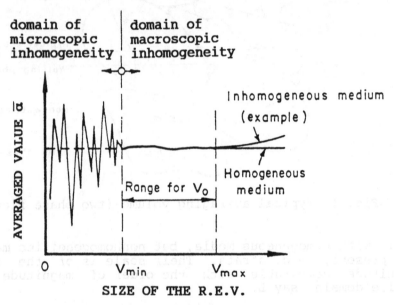

Fig. 2- Possible fluctuations of an averaged quantity as
          function of the dimensions of the R.E.V..

To obtain meaningful average values the characteristic length
of the averaging volume l must satisfy the inequality

$$d \ll l \ll L \tag{2}$$

l is dependent on the specific material which constitutes the
multiphase medium. Here are some values [5]:

metals	0.1	mm
plastics	1.	mm
wood	10.	mm
concrete	100.	mm

Usually there are no boundary conditions for the R.E.V.,
except in the case of the megascopic level. In this last

case attention has to be used in defining the averaged values
[3].
We have by definition:

$$V = \Sigma_\pi \, V_\pi \tag{3}$$

where π refers to the generic π-phase.
The following notation is used in the sequel:

$$\text{phase average} \langle \alpha_\pi \rangle = \frac{1}{V} \int_V \alpha_\pi \, dV \tag{4}$$

$$\text{intrinsic phase average} \langle \alpha_\pi \rangle^\pi = \frac{1}{V_\pi} \int_{V_\pi} \alpha_\pi \, dV \tag{5}$$

Due to the fact that $\alpha_\pi = 0$, in $\pi' \neq \pi$, we obtain

$$\langle \alpha_\pi \rangle^\pi = \frac{1}{V_\pi} \int_V \alpha_\pi \, dV = \frac{V}{V_\pi} \langle \alpha_\pi \rangle \tag{6}$$

We recall that $\langle \alpha_\pi \rangle$ is the mean value over the R.E.V. and it
is defined at a generic point even though at that point the
phase π may not be present.

3. GENERAL HYPOTHESES

The partially saturated porous medium sample is
treated as a multiphase system undergoing mechanical
deformation of the solid and with flux of the fluid phases
(water and gas). The constituents are assumed to be
immiscible and chemically non reacting. All constituents are
at the same temperature, which is assumed constant in time
and in space.
For the macroscopic behaviour of the single constituents,
the following hypotheses are made:

solid phase: - infinitesimal strain is assumed;
                - the stress-strain relationship is nonlinear
                  (i.e. nonlinear elastic or elastoplastic);
                - creep effects and strains not directly
                  associated to stress changes are accounted
                  for;
                - the stress is defined as tension positive.

fluid phases: - both fluid phases are in contact with the
                solid phase;

- the pore pressure is defined as compressive positive;
- for deriving the effective stress, shear stresses are neglected in both phases; of course they exist at microscopic level and their effect is accounted for by viscous drag forces exerted on the solid phase;
- the saturation S_π of the π phase is a known function of the capillary pressure p_c:
$$S_\pi = S_\pi (p_c)$$
where
$$p_c = p_g - p_w$$
and the subscripts g and w refer to gas (air) and water respectively;
- Darcy's law is assumed valid for the transport of both water and gas; however it will be derived from general mechanical principles;
- the permeability of the medium to each fluid is defined by the product of the intrinsic permeability times the relative permeability $k_{r\pi}$ of the π phase;
- the relative permeabilities are known functions of the degree of saturation:
$$k_{r\pi} = k_{r\pi} (S_\pi).$$

Under these hypotheses the conservative equations of the mechanical quantities may be stated using the principles of the continuum mixture theory. This step is performed using averaged values of the field variables. This fact allows for the constitutive model to be easily related to experimental models.
As it will be shown in the following, the validity of Darcy's law follows from a linearized or simplified momentum balance equations for each fluid.

4. STRESS AND PRESSURE

4.1 The stress field: averaged solution

An important question in mechanical problems concerning partially saturated porous media is the definition of the effective stress, i.e. the stress controlling both changes in volume and the strength of the medium [6]. For this purpose, we apply the previous definition of averaged quantities to the stress field in a multiphase system with a solid phase, a liquid phase (water) and a gaseous phase (air).

$$< \underline{\sigma} > = \frac{1}{V} \int_V \underline{\sigma} \, dV = \frac{1}{V} \left[\int_{V_S} \underline{\sigma} \, dV + \int_{V_f} \underline{\sigma} \, dV \right]$$

$$= \frac{V_S}{V} < \underline{\sigma}_S >^S + \frac{V_f}{V} \left[\frac{V_w}{V_f} < \underline{\sigma}_w >^w + \frac{V_g}{V_f} < \underline{\sigma}_g >^g \right]$$

$$= (1 - n) < \underline{\sigma}_S >^S + n \left[S_w < \underline{\sigma}_w >^w + S_g < \underline{\sigma}_g >^g \right] \qquad (7)$$

where n is the porosity and σ_π the stress of the π-phase. For the fluid phases, the stress vector is given by

$$\underline{\sigma}_\pi = \underline{\tau}_\pi - \underline{m} \, p_\pi \qquad (8)$$

where p_π is the pressure of the π phase, τ_π is the shear stress and $\underline{m} = [1, 1, 1, 0, 0, 0]^T$. Under the assumption that the shear stress τ_π is negligible, we obtain

$$< \underline{\sigma} > = \left[1 - n \right] < \underline{\sigma}_S >^S - n \, \underline{m} \left[S_g < p_g >^g + S_w < p_w >^w \right] \qquad (9)$$

The pore pressures p_π produce a stress state in the grains, but, for the moment, we assume they do not undergo any deformation due to this stress, i.e. we introduce the hypothesis of incompressible grains. The deformation of the solid skeleton, which depends on the effective stress, consists in the rearrangement of the grains only.
In presence of only one fluid phase, accounting for the definition of the intrinsic averaged value, equation (9) results in

$$<\underline{\sigma}> = (1-n) <\underline{\sigma}_S>^S - n \, \underline{m} <p_\pi>^\pi$$

$$= (1-n)(<\underline{\sigma}_S>^S + \underline{m} <p_\pi>^\pi) - (1-n) \, \underline{m} <p_\pi>^\pi - n \, \underline{m} <p_\pi>^\pi \qquad (10)$$

$$= \underline{\sigma}' - \underline{m} <p_\pi>^\pi$$

The stress tensor is split in two components, the pore pressure effect and the part which deforms the solid skeleton, i.e. the effective stress. This latter is given by

$$\underline{\sigma}' = (1 - n) (< \underline{\sigma}_S >^S + \underline{m} < p_\pi >^\pi) \qquad (11)$$

Equation (10) results in a splitting of the stress vector similar to that of the Terzaghi's principle. In presence of

several fluid phases, the above averaging procedure applied
to the pore pressure gives

$$< p_f >^f = S_g < p_g >^g + S_w < p_w >^w \qquad (12)$$

or

$$\bar{p} = S_g \, p_g + S_w \, p_w \qquad (12a)$$

Under more general hypotheses, a corrective term, known
as Biot's constant, has to be introduced to account for the
deformability of the grains. We assume the more general
expression of the effective stress as

$$\underline{\sigma}' = \underline{\sigma} + \underline{m} \, p \, \alpha \qquad (13)$$

where the corrective coefficient α will be determined in the
following.

4.2 Biot's constant (Experimental formulations)

We consider first the fully saturated case and derive the
expression for the above introduced coefficient α. It will be
shown that it has not the same value for shear strength and
for volume changes. We use the traditional splitting of the
'total' stress vector in fluid pore pressure and 'effective
stress' (fig. 3) as in eq. (10):

$$\underline{\sigma}' = \underline{\sigma} + \underline{m} \, p \qquad (14)$$

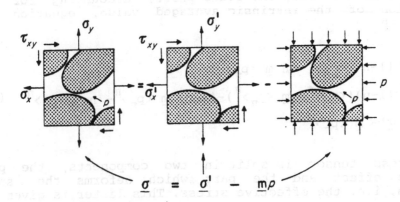

$$\underline{\sigma} = \underline{\sigma}' - \underline{m}\rho$$

Fig. 3. Total and effective stress in porous media.

Let's first consider shear strength problems. The intrinsic

shear strength in solid materials is given by [6]

$$\tau_i = k + \sigma \tan \psi \qquad (15)$$

where ψ is the angle of intrinsic friction.
       k is the intrinsic cohesion.
    Real materials present $\psi \neq 0$ with values in the following range:

$$\psi \quad < 5° \quad \text{for metals}$$
$$3° < \psi < 10° \quad \text{for minerals.}$$

    In an ideal porous material, which, by definition has $\psi = 0$, and in case of $p=0$, the shear strength, accounting for the grain deformation, is given by [6]

$$\tau_d = c' + \sigma \tan \phi' \qquad (16)$$

where ϕ' is the angle of shearing resistence;
       c' is apparent cohesion.
    In presence of pore water pressure p, which supports the sides of the particles, we have

$$\tau_s = c' + (\sigma + p) \tan \phi' \qquad (17)$$

By comparing equations (16) and (17) it follows that the effective stress is given by

$$\sigma' = \sigma + p \qquad (18)$$

which is Terzaghi's equation.
For soils, concrete and rocks the angle of intrinsic friction however differs significantly from zero. In this case, the strength at failure can be shown to be [6]

$$\tau_f = c' + \left[\sigma + \left(1 - \frac{a \tan \psi}{\tan \phi'} \right) p \right] \tan \phi' \qquad (19)$$

where a is the contact area per unit gross area (fig. 4).
Consequently, the effective stress is given by

$$\sigma' = \sigma + \left(1 - \frac{a \tan \psi}{\tan \phi'} \right) p \qquad (20)$$

This is the most general expression of the effective stress and includes, as special case ideal materials.
As it is well known, for soils within the range of pore pressure encountered in practice, Terzaghi's equation involves no significant errors in problems concerning the strength of fully saturated materials. On the other hand, experimental tests show that

$$\frac{\tan \ \psi}{\tan \ \phi} \ ' \approx \ \frac{0.3}{0.25} \quad \begin{array}{l} (\ \text{FOR SANDS} \ ) \\ (\ \text{FOR CLAYS} \ ) \end{array} \tag{21}$$

Hence the validity of Terzaghi's equation in shear strength problems for sands must depend upon the small magnitude of the contact area ratio at normal engineering pressures. Triaxial tests on limestone and marble have shown that equation (19) is correct.

As next step we consider compressibility , i.e. volume change problems, of saturated porous materials. We distinguish two limiting cases:

a)   zero pore pressure (drained or jacketed tests). We have:

$$- \left(\frac{\Delta \ V}{V} \right)_d = - \ C \ \Delta\sigma' \tag{22}$$

where: $\Delta\sigma'$ is the increase of the effective stress;
         C is the coefficient of compressibility of the material (skeleton).
The coefficient C decreases with increasing effective stress. At sufficiently high effective stress, C tends to the compressibility of the solid particles C_s.

b) pore pressure different from zero. In such a case a change in the total stress has to be split in a change of effective stress and of pore pressure.
As usual, we assume the 'perfect solid' to have zero values of the angle of intrinsic friction and of the coefficient C_s (incompressible grains). Thus a change in the applied stress with an equal change in pore pressure will cause neither volume change nor deformation of the grains nor changes in the contact area and shear strength. Changes in volume depend exclusively on increasing of the term ($\Delta \sigma + \Delta p$), i.e.

$$- \ \frac{\Delta V}{V} = - \ C \left[\Delta\sigma \ + \ \Delta p \right] \tag{23}$$

Hence the effective stress is

$$\Delta\sigma' = \Delta\sigma \ + \ \Delta p \tag{24}$$

In real problems, we have to account for the grain compressibility. When the porous material is subjected to equal increases in applied total stress and pore pressure, each grain undergoes cubical compression under hydrostatic pressure Δp. Hence

$$- \left(\frac{\Delta V}{V} \right)_s = C_s \, \Delta p \tag{25}$$

If we assume for the moment $\psi = 0$, i.e. the rearrangement of the particles is not prevented by internal friction, then a net stress increment will cause the total volume change of the porous material

$$- \left(\frac{\Delta V}{V} \right) \doteq - C \left[\Delta \sigma + \Delta p \right] + C_s \, \Delta p$$

$$\doteq - C \left[\Delta \sigma + \left(1 - \frac{C_s}{C} \right) \Delta p \right] \tag{26}$$

where the contact area is not involved. Hence the effective stress is represented by

$$\Delta \sigma' = \Delta \sigma + \left(1 - \frac{C_s}{C} \right) \Delta p \tag{27}$$

When, as in reality, $\psi \neq 0$, there will be a small increase in resistance to particle rearrangement and deformation associated with the increase of pore pressure. A more general expression for the volume changes may be used, but equation (26) gives still quite acceptable results [6]. Experimental tests confirm that the effective stress controlling volume changes is represented by equation (27).
For soils $C_s/C = 0$ and equation (27) reduces to Terzaghi's equation. For rocks and concrete we have

$$\alpha = \left(1 - \frac{C_s}{C} \right) \approx 0.4 \,:\, 0.6 \tag{28}$$

resulting in remarkable reduction for the contribution of the pore pressure in the effective stress relationship.

4.3 Bishop's equation

Terzaghi's equation or equation (27) are satisfactory in fully saturated cases, but lead to appreciable errors when applied to problems of partially saturated media. We consider here the corrections to the effective stress principle introduced by Bishop [7].
The saturation S_π was already defined in equation (7). When several phases are present, we have by definition

$$\Sigma_i \, S_i = 1 \tag{29}$$

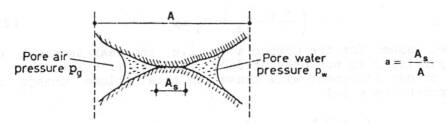

Fig. 4- Two grains contact in partially saturated case.

If the pores (fig. 4) are filled with water and air, due to
the surface tension we have

$$p_w < p_g \tag{30}$$

If the degree of saturation is rather low, the water is
present as menisci. It is assumed that the water pressure
acts over an area χ per unit gross area [8].
Consequently, the equivalent pore pressure is expressed by

$$\chi\, p_w + (1 - \chi)\, p_g \tag{31}$$

We may write the equivalent pore pressure also as

$$p_g - \chi\, (p_g - p_w) \qquad \text{or} \qquad p_w + (1 - \chi)\, (p_g - p_w) \tag{32}$$

By comparison with the fully saturated case, Bishop suggested
the following expression for the effective stress:

$$\sigma' = \sigma + [p_g - \chi\, (p_g - p_w)] \tag{33}$$

The coefficient χ is not the same in problems of shear
strength and in consolidation. For a given degree of
saturation the coefficient χ must be determined
experimentally in both types of problems.
When p_a is equal to the atmospheric pressure, since all
pressures are normally expressed with reference to the
atmospheric one, p_a is zero. Hence

$$\sigma' = \sigma + \chi\, p_w \tag{34}$$

The comparison between equations (9) and (33) is noteworthy.
The former may be written in function of the degrees of
saturation as

$$\sigma = \sigma' - [S_w\, p_w + S_g\, p_g] = \sigma' - [p_g - S_w\, (p_g - p_w)] \tag{35}$$

and coincides with equation (33) if $\chi = S_w$.

The coefficient χ is related to the area of contact between solid and fluids, whereas the degrees of saturation S_π depend on the volume occupied by the π phase. According to Morland's theory [9], we define the volume fraction as

$$\varphi_\pi = \frac{V_\pi}{V} \tag{36}$$

and cross sectional area fraction as

$$\alpha_\pi = \frac{A_\pi}{A} \tag{37}$$

In the case of microstructurally isotropic constituents, such that α_π are independent of the orientation of the surface, under the assumption that

$$\alpha_\pi = \alpha_\pi (\varphi_\pi) \tag{38}$$

it can be shown that

$$\alpha_\pi = \varphi_\pi \tag{39}$$

If this assumption on area and volume fractions is assumed as valid, only little difference exists between the definition of Bishop's coefficient χ and the degree of saturation. As a consequence, Bishop's definition of the effective stress may be directly obtained by an averaging process applied to the stress vector.

4.4 Experimental validation of Bishop's equation

An experimental validation of Bishop's equation has been obtained through a triaxial test on a soil sample [6]. The cell pressure is increased with σ_3, p_g and p_w held constant before failure. When failure conditions are reached all pressures are varied, but the following conditions are satisfied:

$$(\sigma_3 + p_g) = \text{constant}$$
$$(\sigma_3 - p_w) = \text{constant}$$

If Bishop's equation is correct, no change in effective stress and no change in the stress-strain curve should occur. As can be seen from fig. 5 regarding the results of a triaxial test on partially saturated silt, the prediction is validated within the limits of experimental accuracy.

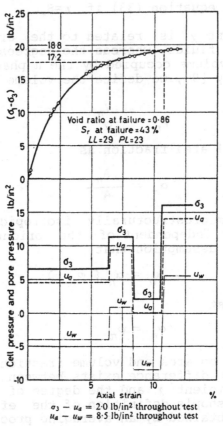

Fig. 5- Results of a triaxial test on partially saturated
       silt (Bishop and Donald) [6].

It has to be mentioned that the validity of the effective
stress concept is questioned in certain situations such e.g.
in case of collapse of the solid skeleton structure on first
wetting and rewetting if a compressive mean stress state
exists [10]. However this effect decreases on subsequent
wetting/drying cycles. In principle it is possible to include
it by a slight modification of the constitutive relation
without changing the effective stress principle. Hence in the
following we assume that the relationship (35) holds together
with an appropriate, general constitutive law, valid in both
saturated and non-saturated zones.

4.5 Correction of Bishop's equation

Equations (33) or (34) do not account for the grain

compressibility.   Operating in the same manner as for fully saturated porous media in the case when C_s and ψ are different from zero, we obtain:

$$\sigma' = \sigma + \left[1 - \frac{a \tan \psi}{\tan \phi'}\right] \left[P_g - \chi \left[P_g - P_w\right]\right]$$

$$= \sigma + \left[1 - \frac{C_s}{C}\right] \left[P_g - \chi \left[P_g - P_w\right]\right] \qquad (40)$$

Again, when the air is at atmospheric pressure, we have

$$\sigma' = \sigma + \left[1 - \frac{C_s}{C}\right] \chi \, P_w$$

$$= \sigma + \alpha \, \chi \, P_w \qquad (41)$$

As already indicated, Bishop's coefficient χ has to be determined experimentally and appears to be a nonlinear function of the degree of saturation S_w or of the pore water pressure.   However,   in accordance to what discussed previously, we assume in the sequel

$$\chi \approx S_w \qquad (42)$$

This assumption is acceptable for many materials, as can be seen e.g. in fig. 6 for two types of clay.

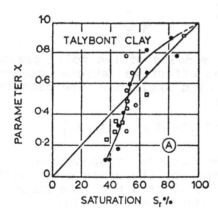
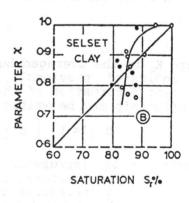

o   w CONSTANT, VARIOUS $(\sigma_3 - U_a)$
● $\sigma_3 - U_a = 0$, VARIOUS w VALUES
□ $\sigma_3 - U_a = 30$ PSI, VARIOUS w VALUES

Fig. 6- Relationship between χ and S_w [7].

As a consequence, since

$$S_w + S_g = 1$$

we express the effective stress in a partially saturated porous medium, accounting for the grain compressibility as

$$\underline{\sigma}' = \underline{\sigma} + \underline{m} \, [\, p_g - S_w(p_g - p_w)] \, \alpha \qquad (43)$$

or

$$\underline{\sigma}' = \underline{\sigma} + \underline{m} \, [\, S_w \, p_w + S_g \, p_g] \, \alpha \qquad (44)$$

as obtained through the averaging techniques. For Biot's constant α we assume the values given by eq. (27).

5. CONSTITUTIVE EQUATIONS

5.1 Solid phase

The averaged pore pressure $\bar{p}$ of the fluids occupying the void space introduces in the solid phase a hydrostatic stress distribution. The ensuing deformation is the purely volumetric strain

$$\varepsilon_v^P = - \frac{\bar{p}}{K_s} \qquad (45)$$

or, in incremental form

$$d \, \underline{\varepsilon}^P = - \underline{m} \, \frac{d \, \bar{p}}{3 \, K_s} \qquad (46)$$

where K_s is the averaged bulk modulus of the solid grains. As indicated previously, the effective stress causes all relevant deformation of the solid skeleton. The constitutive relationship may be written as

$$d \, \underline{\sigma}' = \underline{D}_T \left(d \, \underline{\varepsilon} - d \, \underline{\varepsilon}^c - d \, \underline{\varepsilon}^P - d \, \underline{\varepsilon}^\circ \right) \qquad (47)$$

where: - $d\underline{\sigma}'$ represents the stress responsible of all deformations, except for the grain compressibility;
- $d\underline{\varepsilon}$ represents the total strain of the solid;
- $\underline{D}_T$ is the tangent constitutive matrix:

$$\underline{D}_T = \underline{D}_T (\underline{\sigma}', \underline{\varepsilon}, \underline{\dot{\varepsilon}}) \qquad (48)$$

- $d\underline{\varepsilon}^c = g(\underline{\sigma}') \, dt$
  accounts for the creep strain;

- $d\underline{\varepsilon}^0$ represents all other strains in the solid skeleton not direcly dependent on effective stress.

We introduce now a modified effective stress $\underline{\sigma}''$ to also account for grain compression. This allows us to obtain again Biot's constant α. By substituting equation (46) in (47) and omitting for brevity $d\underline{\varepsilon}^0$ and $d\underline{\varepsilon}^c$ we obtain

$$d\underline{\sigma}' = \underline{D}_T \ (\ d\underline{\varepsilon} - d\underline{\varepsilon}^P \ )$$

$$= \underline{D}_T \ d\underline{\varepsilon} + \underline{D}_T \ \underline{m} \ \frac{d\bar{p}}{3K_s} \qquad (49)$$

Consequently the effective stress equation (47) becomes

$$\qquad\qquad\qquad\qquad\qquad\qquad\qquad\qquad (50)$$

where $\underline{\sigma}''$ represents the stress responsible for all deformations in the solid. In the case of an isotropic material, we have [11]

$$d\underline{\sigma} = d\underline{\sigma}'' - \frac{1}{3} \left(\underline{m}^T \underline{m} - \underline{m}^T \underline{D}_T \underline{m} \cdot \frac{1}{3 K_s} \right) \underline{m} \ dp$$

$$= d\underline{\sigma}'' - \left(1 - \frac{\underline{m}^T \underline{D}_T \underline{m}}{9 K_s} \right) \underline{m} \ dp \qquad (51)$$

where

$$\alpha = 1 - \frac{\underline{m}^T \underline{D}_T \underline{m}}{9 K_s} = 1 - \frac{K_T}{K_s} \qquad (52)$$

and K_T is the bulk modulus of the overall skeleton. The resultant expression of the constitutive relationship is therefore

$$d\underline{\sigma}'' = \underline{D}_T \ d\underline{\varepsilon} \qquad (53)$$

5.2 Liquid phase

The intrinsic pore pressure p_π of the π fluid phase causes in its phase a purely volumetric strain, which is represented by

$$\varepsilon_v^\pi = - \frac{p_\pi}{K_\pi} \qquad (54)$$

where K_π is the averaged bulk modulus of the phase.

In addition to the constraint of eq. (29) on the saturation of the two fluid phases, the pressures of air and water are subject to the constraint that capillary pressure $p_c = p_a - p_w$ is a function of S_w. The relationship of $p_c(S_w)$ is generally known from laboratory experiments.

6. GOVERNING EQUATIONS

To obtain the governing equations for the mechanics of partially saturated porous media, we use the continuum theory of mixtures, i.e. we suppose that all constituents of the porous medium occupy at the same time every point of the domain.
Due to the assumed immiscibility of the constituents [1]:
- in some local sense each constituent will obey the constitutive relation for that constituent alone;
- the constituents intrinsically have a microstructure, i.e. interfaces are present which separate the constituents.
As a consequence, the mechanical behaviour of the mixture is described by the usual thermodynamical variables, as if there would be the components occupying at the same time all the domain, but in addition the volume fraction of the constituents is introduced to account for the discontinuous distribution.
The volume fraction is defined as the volume of each constituent per unit volume of the material.
In the context of the continuum approach used, two types of theories have evolved:
- postulated theories, in which the thermodynamical balance equations are written in terms of postulated mean values of the variables;
- averaged theories, in which the thermodynamical balance equations are written in local form and are then integrated (averaged) in space, time or space and time [12,13,14].
Postulated and spatially averaged theories will be used in the sequel.

We recall now two definitions:

- Material time derivative: given a function $F(x_1, x_2, x_3, t)$ that is attributable to a moving particle, the material time derivative of F is defined by

$$\frac{DF}{Dt} = \frac{\partial F}{\partial t} + \frac{\partial F}{\partial x_1} \frac{\partial x_1}{\partial t} + \frac{\partial F}{\partial x_2} \frac{\partial x_2}{\partial t} + \frac{\partial F}{\partial x_3} \frac{\partial x_3}{\partial t} \qquad (55)$$

or, by using the summation over repeated indices, by

$$\frac{DF}{Dt} = \frac{\partial F}{\partial t} + \frac{\partial F}{\partial x_i} \dot{u}_i \tag{56}$$

-Material time derivative of volume integral: consider a volume integral

$$I = \int_V A \, dV \tag{57}$$

where $A(\underline{x},t)$ denotes a property of the continuum, the boundary S of the volume V is a moving surface and the integral is evaluated at an instant of time t. The material time derivative of I is defined as

$$\frac{D}{Dt} \int_V A \, dV = \int_V \left(\frac{DA}{Dt} + A \frac{\partial \dot{u}_i}{\partial x_i} \right) dV \tag{58}$$

with the summation over repeated indices being used. The same notation will be used in the following.

6.1 Mass balance equations

6.1.1 Solid phase

The mass of the soil skeleton in a representative elementary volume may be written as

$$M_S = \int_V \rho^S \, dV \tag{59}$$

where ρ^S is the local value of the density of the solid. We express the same quantity using intrinsic averaged density ρ_s and the volume fraction of the solid as

$$M_S = \int_V \left(1 - n \right) \rho_s \, dV \tag{60}$$

where n is the porosity of the medium, which may vary during the phenomenon.

The mass conservation requires that

$$\frac{D}{Dt} \int_V (1-n)\rho_s dV = \int_V \left[\frac{D}{Dt} [(1-n) \, \rho_s] + (1-n) \, \rho_s \frac{\partial \dot{u}_{s,i}}{\partial x_i} \right] dV$$

$$= \int_V \left[\frac{\partial (1-n)\rho_s}{\partial t} + \frac{\partial (1-n)\rho_s \, \dot{u}_{s,i}}{\partial x_i} \right] dV = 0 \tag{61}$$

The term $\dot{u}_{\pi,i}$ represents the i-th component of the absolute velocity of the π-phase.

Then, by localization, we obtain the equation of continuity of the solid which is expressed in terms of intrinsic averaged quantities:

$$\frac{\partial(1-n)\rho_s}{\partial t} + \frac{\partial(1-n)\rho_s \dot{u}_{s,i}}{\partial x_i} = 0 \tag{62}$$

In the volume preserving case, eq. (62) reduces to

$$\frac{\partial(1-n)\rho_s}{\partial t} = 0 \tag{63}$$

6.1.2 Fluid phases (water and air)

The mass of water in a R.E.V., in partially saturated conditions may be expressed in terms of intrinsic averaged quantities using the pertinent volume fraction as

$$M_w = \int_V n \, S_w \, \rho_w \, dV \tag{64}$$

where ρ_w is the intrinsic averaged density of water. External supply is not accounted for in the following.

The water mass conservation requires the material time derivative with respect to the flowing water to be zero:

$$\frac{DM_w}{Dt} = \frac{D}{Dt} \int_V n \, S_w \, \rho_w \, dV$$

$$= \int_V \left[\frac{\partial(n \, S_w \, \rho_w)}{\partial t} + \frac{\partial(n \, S_w \, \rho_w \, \dot{u}_{w,i})}{\partial x_i} \right] dV = 0 \tag{65}$$

Then by localization we obtain the continuity equation for water:

$$\frac{\partial(n \, S_w \, \rho_w)}{\partial t} + \frac{\partial(n \, S_w \, \rho_w \, \dot{u}_{w,i})}{\partial x_i} = 0 \tag{66}$$

For the gaseous phase we have in a similar way:

$$\frac{\partial(n \, S_g \, \rho_g)}{\partial t} + \frac{\partial(n \, S_g \, \rho_g \, \dot{u}_{g,i})}{\partial x_i} = 0 \tag{67}$$

It is worth to remark that all equations are expressed in terms of intrinsic averaged quantities and contain the volume fraction as variable. Moreover $\dot{u}_{\pi,i}$ represents, by definition of the material time derivative, the i-th component of the absolute velocity of the π phase. It is costumary to describe the flow of the fluid phases through the porous medium using Darcy's type relationships. These are expressed in terms of volume averaged velocities relative to the moving solid, i.e. the velocities measurable in experimental tests. To define the necessary quantities $\dot{\alpha}_i$, we apply in the R.E.V. the averaging process to the previously defined intrinsic phase averaged quantities:

$$\dot{\alpha}_i = \frac{1}{V} \int_{V_\pi} \left(\dot{u}_{\pi,i} - \dot{u}_{s,i} \right) dV = \frac{1}{V} \int_{V_\pi} \left(\dot{u}_{\pi,i} - \dot{u}_{s,i} \right) dV$$

$$= \frac{V_\pi}{V} \left(\dot{u}_{\pi,i} - \dot{u}_{s,i} \right) \tag{68}$$

In this way we obtain the water (w_i) and gas (v_i) velocities relative to the moving solid:

$$\dot{w}_i = n \, S_w \, (\dot{u}_{w,i} - \dot{u}_{s,i})$$
$$\dot{v}_i = n \, S_g \, (\dot{u}_{g,i} - \dot{u}_{s,i}) \tag{69}$$

From a practical point of view, it is useful to express the balance equations in terms of the solid skeleton velocity ($\dot{u}_i$), where the subscript s has been omitted, and in terms of fluid phases velocity relative to the moving solid. This yields

$$\frac{(1-n)}{\rho_s} \frac{D\rho_s}{Dt} - \frac{Dn}{Dt} + (1 - n) \, \dot{u}_{i,i} = 0 \tag{70}$$

$$\frac{n}{S_w \rho_w} \frac{D(S_w \rho_w)}{Dt} + \frac{Dn}{Dt} + n\dot{u}_{i,i} + \frac{1}{S_w} \dot{w}_{i,i} + \frac{\dot{w}_i}{S_w \rho_w} \frac{\partial \rho_w}{\partial x_i} = 0 \tag{71}$$

$$\frac{n}{S_g \rho_g} \frac{D(S_g \rho_g)}{Dt} + \frac{Dn}{Dt} + n\dot{u}_{i,i} + \frac{1}{S_g} \dot{v}_{i,i} + \frac{\dot{v}_i}{S_g \rho_g} \frac{\partial \rho_g}{\partial x_i} = 0 \tag{72}$$

respectively for solid, water and air, where the notation

$$(*)_{i,i} = \frac{\partial (*)_i}{\partial x_i} = \text{div} \, (*) \tag{73}$$

is used, with summation over repeated indices. All material time derivatives are performed with respect to the moving solid, i.e.

$$\frac{D \rho_s}{Dt} = \frac{\partial \rho_s}{\partial t} + \frac{\partial \rho_s}{\partial x_i} \dot{u}_i \tag{74}$$

$$\frac{D n}{D t} = \frac{\partial n}{\partial t} + \frac{\partial n}{\partial x_i} \dot{u}_i \tag{75}$$

$$\frac{D(S_w \rho_w)}{Dt} = \frac{\partial(S_w \rho_w)}{\partial t} + \frac{\partial(S_w \rho_w)}{\partial x_i} \dot{u}_i \tag{76}$$

$$\frac{D(S_g \rho_g)}{Dt} = \frac{\partial(S_g \rho_g)}{\partial t} + \frac{\partial(S_g \rho_g)}{\partial x_i} \dot{u}_i \tag{77}$$

A further simplification is obtained by summing the mass balance equation of each fluid phase with that of the solid, obtaining respectively for water and air

$$\frac{(1-n)}{\rho_s} \frac{D\rho_s}{Dt} + \frac{n}{\rho_w} \frac{D\rho_w}{Dt} + \frac{n}{S_w} \frac{DS_w}{Dt} + \dot{u}_{i,i} + \frac{1}{S_w} \dot{w}_{i,i}$$

$$+ \frac{\dot{w}_i}{S_w \rho_w} \frac{\partial \rho_w}{\partial x_i} = 0 \tag{78}$$

$$\frac{(1-n)}{\rho_s} \frac{D\rho_s}{Dt} + \frac{n}{\rho_g} \frac{D\rho_g}{Dt} + \frac{n}{S_g} \frac{DS_g}{Dt} + \dot{u}_{i,i} + \frac{1}{S_g} \dot{v}_{i,i}$$

$$+ \frac{\dot{v}_i}{S_g \rho_g} \frac{\partial \rho_g}{\partial x_i} = 0 \tag{79}$$

The material derivatives of the intrinsic phase averaged densities, which appear in equations (78) and (79) are obtained next. For the fluid phases we recall that these are assumed barotropic. Consequently the equations of state may be directly obtained from the constitutive equations. From the definitions of the bulk modulus of water K_w and of the air K_a

$$\frac{1}{\rho_\pi} \frac{\partial \rho_\pi}{\partial p_\pi} = \frac{1}{K_\pi} \tag{80}$$

we obtain

$$\frac{1}{\rho_w} \frac{D \rho_w}{Dt} = \frac{1}{K_w} \frac{D p_w}{Dt} \tag{81}$$

$$\frac{1}{\rho_g} \frac{D \rho_g}{Dt} = \frac{1}{K_g} \frac{D p_g}{Dt} \tag{82}$$

As far as the solid skeleton is concerned, this is usually assumed as microscopically incompressible, i.e. grains do not undergo any deformation. As a consequence

$$\frac{D \rho_s}{Dt} = 0 \tag{83}$$

To account for the effect of the grain deformation on the solid density we use the solid mass conservation equation in differential form:

$$\frac{D(\rho_s v_s)}{Dt} = 0 \tag{84}$$

which yields

$$\frac{1}{\rho_s} \frac{D \rho_s}{Dt} = - \frac{1}{V_s} \frac{D V_s}{Dt} \tag{85}$$

The change of the volume of solid V_s depends on the intrinsically averaged fluid pressure and on the mean effective stress σ'_0, which is a volume averaged quantity:

$$- \frac{1}{V_s} \frac{D V_s}{Dt} = - \frac{1}{V_s} \left[- \frac{V_s}{K_s} \frac{D \bar{p}}{Dt} + \frac{1}{K_s} \frac{D \sigma'_0}{Dt} \right]$$

$$= - \frac{1}{V_s} \left[- \frac{V_s}{K_s} \frac{D \bar{p}}{Dt} + \frac{1}{K_s} \left(K_T \frac{D \varepsilon_v}{Dt} + \frac{K_T}{K_s} \frac{D \bar{p}}{Dt} \right) \right]$$

$$= \frac{1}{1-n} \left[\frac{\alpha - n}{K_s} \frac{D \bar{p}}{Dt} - (1 - \alpha) \frac{D \varepsilon_v}{Dt} \right] \tag{86}$$

where ε_v is the volumetric strain and α is defined in eq. (52).

Substituting equations (81), (82) and (86) in (78) and (79), we express the mass balance equation of the mixture solid+air and solid+water as follows:

$$\left[\frac{\alpha - n}{K_s} \frac{D \bar{p}}{Dt} - (1 - \alpha) \dot{u}_{i,i}\right] + \frac{n}{K_w} \frac{D p_w}{Dt}$$

(87)

$$+ \frac{n}{S_w} \frac{D S_w}{Dt} + \dot{u}_{i,i} + \frac{1}{S_w} \dot{w}_{i,i} + \frac{\dot{w}_i}{S_w \rho_w} \frac{\partial \rho_w}{\partial x_i} = 0$$

$$\left[\frac{\alpha - n}{K_s} \frac{D \bar{p}}{Dt} - (1 - \alpha) \dot{u}_{i,i}\right] + \frac{n}{K_g} \frac{D p_g}{Dt}$$

(88)

$$+ \frac{n}{S_g} \frac{D S_g}{Dt} + \dot{u}_{i,i} + \frac{1}{S_g} \dot{v}_{i,i} + \frac{\dot{v}_i}{S_g \rho_g} \frac{\partial \rho_g}{\partial x_i} = 0$$

We recall the definition of the averaged fluid pressure

$$\bar{p} = S_w p_w + S_g p_g$$

(89)

and the constraint relative to the degrees of saturation:

$$S_w + S_g = 1$$

From the definition of the time material derivative, it follows

$$\frac{D S_g}{Dt} = - \frac{D S_w}{Dt}$$

(90)

and the mass balance equations result in

$$\left[\frac{\alpha - n}{K_s} S_w^2 + \frac{n S_w}{K_w}\right] \frac{D p_w}{Dt} + \frac{\alpha - n}{K_s} S_g S_w \frac{D p_g}{Dt} + \alpha S_w \dot{u}_{i,i}$$

(91)

$$+ \left[\frac{\alpha - n}{K_s} S_w (p_w - p_g) + n\right] \frac{D S_w}{Dt} + \dot{w}_{i,i} + \frac{\dot{w}_i}{\rho_w} \frac{\partial \rho_w}{\partial x_i} = 0$$

$$\left[\frac{\alpha - n}{K_s} S_g^2 + \frac{n S_g}{K_g}\right] \frac{D p_g}{Dt} + \frac{\alpha - n}{K_s} S_w S_g \frac{D p_w}{Dt} + \alpha S_g \dot{u}_{i,i}$$

(92)

$$+ \left[\frac{\alpha - n}{K_s} S_g (p_g - p_w) + n\right] \frac{D S_g}{Dt} + \dot{v}_{i,i} + \frac{\dot{v}_i}{\rho_g} \frac{\partial \rho_g}{\partial x_i} = 0$$

6.2 Linear momentum balance equations

The microscopic (or local) momentum balance equation for the π phase

$$\frac{D}{Dt}\left(\rho_\pi \, \underline{v}_\pi \right) - \underline{\nabla} \cdot \underline{\sigma}_\pi - \rho_\pi \, \underline{g}_\pi = 0 . \qquad (93)$$

has to be integrated over the region occupied by the phase and then expressed in terms of quantities averaged over the R.E.V.. In this equation $\underline{v}_\pi$ is the intrinsic velocity of the phase. In doing so, attention has to be paid to the fact that the limiting surface of the volume V_π is different from that of the R.E.V. and mechanical interactions arise at the interfaces. We assume that the external momentum supply is $\underline{g}$, no phase change takes place and the temperature is stationary in space and in time.
To account for these conditions, we use the averaging theorem by Hassanizadeh and Gray [12], by which we obtain the following form of the momentum balance equation:

$$\varepsilon_\pi \, \rho_\pi \, \frac{D^\pi \, v_1^\pi}{Dt} - \sigma_{kl,k}^\pi - \varepsilon_\pi \, \rho_\pi \, g_1^\pi - \varepsilon_\pi \, \rho_\pi \, T_1^\pi = 0 \qquad (94)$$

where ε_π represents the volume fraction for the phase. The material derivative is performed with respect to the flowing phase. $\sigma_{,k}^\pi$ represents the gradient of the intrinsic stress vector for the π phase and $\underline{T}^\pi$ accounts for the exchange of momentum due to mechanical interactions with the other phases.

The constitutive relationships have to be introduced in eq. (94). In particular for the momentum exchange term, a simplified constitutive equation, which is however thermodynamically consistent, may be assumed as

$$\varepsilon_f \, \rho_t \, \underline{T}^f = - \, \underline{R}^f \, \underline{v}_f^d \qquad (95)$$

This term represents the dissipative part of the fluid-solid exchange of momentum and is related to the fluid relative velocity $\underline{v}^d$. In this way we account for local effects, depending also on the viscosity of the fluid, assuming the fluid to be macroscopically inviscid.

6.2.1 Fluid phases

By recalling the initial assumption for the fluid stress vector and substituting in eq. (94) the volume fraction as function of the porosity and the saturation, we obtain

$$n \ S_\pi \ \rho_\pi \ \frac{D^\pi \ v_1^\pi}{Dt} = -n \ S_\pi \ P_{\pi,k} + n \ S_\pi \ g_1^\pi \ \rho_\pi - R_{kl}^f \ v_1^d = 0 \qquad (96)$$

Then, if inertial effects are neglected, the fluid phase momentum equation, in matrix form, becomes

$$\underline{v}^d = -\left(\ \underline{R}^f \right)^{-1} \left[\underline{\nabla} \ P_\pi - \underline{g}_\pi \ \rho_\pi \right] = - \ \underline{K} \ \left[\underline{\nabla} \ P_\pi - \rho_\pi \ \underline{g}_\pi \right] \qquad (97)$$

where $\underline{K}$ is the permeability matrix of the medium.
This equation represent Darcy's law, which therefore is valid as first approximation for the slow flow of a macroscopically inviscid fluid through the porous medium. To obtain eq. (97) the resistivity matrix $\underline{R}^\pi$ is assumed invertible, i.e. whenever $\underline{v}^\pi=0$ the pore pressure will be hydrostatic and vice versa.
When several fluid phases are present, dissipative terms arise at the fluid-solid and fluid-fluid interfaces. To express the constitutive equations of these terms, we assume the permeability of the fully saturated case (intrinsic permeability), which depends on the geometry of the solid skeleton, modified by the presence of other fluid phases.
This modification depends on the volume fraction of the other phases. As a consequence we assume the resulting permeability as the product of the intrinsic permeability times a coefficient $k_{r\pi}$ (relative permeability), which is a function of the degree of saturation S_π. In isotropic conditions we have

$$v_{i\pi}^d = k_\pi \left(- P_{\pi,i} + \rho_\pi \ g_i^\pi \right) \qquad (98)$$

where k_π is evaluated by the following expression:

$$k_\pi = \frac{k_{a\pi}}{\mu_\pi} \ k_{r\pi} \left(S_\pi \right) \qquad (99)$$

In this equation $k_{a\pi}$ represents the absolute permeability, μ_π is the viscosity of the π-phase.

6.2.2 Solid phase

By operating in the same way as in the previous case, we obtain the momentum balance equation for the solid as

$$\left[1-n\right]\rho_s \frac{D^s \ v_1^s}{Dt} = \left[1-n\right]\sigma_{kl,k}^s + \left[1-n\right]\rho_s \ g_1^s + R_{kl}^f \ v_1^d \qquad (100)$$

6.2.3 Whole mixture

By summing the momentum balance equation for each component, we obtain the equilibrium equation for the whole mixture. We remark however that:
- the exchanges of momentum due to mechanical interactions are subjected to the following condition:

$$\sum_\pi \varepsilon_\pi \rho_\pi T_1^\pi = 0 \qquad (101)$$

- the external supply for each phase comes from gravity effects, consequently

$$\underline{g}^s = \underline{g}^f = \underline{g}^g = \underline{b} \qquad (102)$$

- the averaged density of the mixture is obtained by an averaging process as

$$\rho = (1 - n)\,\rho_s + n\,S_g\,\rho_g + n\,S_w\,\rho_w \qquad (103)$$

- the averaged stress for the mixture is given by eq. (8). This results in the following expression of the balance equation:

$$-\left(1-n\right)\rho_s\frac{D^s v_1^s}{Dt} - nS_w\rho_w\frac{D^w v_1^w}{Dt} - nS_g\rho_g\frac{D^g v_1^g}{Dt} + \rho g_1 + \sigma_{kl,k} = 0 \qquad (104)$$

where the material time derivatives are performed with respect to the velocity of each phase.

If the inertial effects are neglected, we obtain:

$$\sigma_{kl,k} + \rho\,b_1 = 0 \qquad (105)$$

A more convenient form of the Eulerian equation of motion for the mixture may be obtained in the following way:
- multiply the momentum balance equation (in vectorial form) by a vector function $\delta\underline{u}$ which has continuous first partial derivatives with respect to the coordinates x_i and satisfies displacement boundary conditions where prescribed (virtual displacement). Integrate over the body volume:

$$-\int_\Omega \left(1-n\right)\rho_s \delta\underline{u}^T \frac{D^s \underline{v}^s}{Dt}\,d\Omega -\int_\Omega n\,S_w\rho_w\delta\underline{u}^T\frac{D^w \underline{v}^w}{Dt}\,d\Omega -\int_\Omega n\,S_g\rho_g\delta\underline{u}^T\frac{D^g \underline{v}^g}{Dt}\,d\Omega$$

$$+\int_\Omega \rho\,\delta\underline{u}^T\,\underline{g}\,d\Omega + \int_\Omega \delta\underline{u}^T\,\underline{\sigma}\,d\Omega = 0 \qquad (106)$$

- by integrating the last term by parts and owing to the Gauss' theorem, we obtain

$$-\int_{\Omega}\left(1-n\right)\rho_s\delta\underline{u}^T\,\frac{D^s\underline{v}^s}{Dt}\,d\Omega-\int_{\Omega}nS_w\rho_w\delta\underline{u}^T\,\frac{D^w\underline{v}^w}{Dt}\,d\Omega\,-\int_{\Omega}nS_g\rho_g\delta\underline{u}^T\,\frac{D^g\underline{v}^g}{Dt}\,d\Omega$$

$$+\int_{\Omega}\rho\,\delta\underline{u}^T\underline{b}\,\,d\Omega\,-\int_{\Omega}\delta\underline{\varepsilon}^T\,\underline{\sigma}\,\,d\Omega\,+\int_{\Gamma}\delta\underline{u}^T\underline{t}\,\,d\Gamma\,=\,0 \qquad (107)$$

where $\delta\underline{\varepsilon}^T$ is the deformation relative to the virtual displacement and $\underline{t}$ are the boundary tractions. Equation (107) is the weak form of the differential equation of motion, i.e. the principle of virtual work. This is very useful numerically. Moreover there are many relevant mathematical theorems regarding the weak form of a differential equation and this form is still valid under conditions for which differential equation (104) do not necessarily make sense [15].

7 SUMMARY OF THE GOVERNING EQUATIONS

For a more compact form of the governing equations, we define the following coefficients [16]:

$$C_{ww} = \frac{S_w^2(\alpha-n)}{K_s} + \frac{S_w\,n}{K_w}$$

$$C_{gg} = \frac{S_g^2(\alpha-n)}{K_s} + \frac{S_g\,n}{K_g} \qquad (108)$$

$$C_{wg} = C_{gw} = \frac{(\alpha-n)S_w\,S_g}{K_s}$$

and use a short notation for the material derivative

$$\dot{F} = \frac{DF}{Dt}$$

Furthermore the relative velocities of equation (98) are substituted in the mass balance equations (91) and (92) and the total stress of the mixture is expressed by eq. (51). Recalling the definitions of capillary pressure of the fluid and the constitutive relationship for S_w, we have

$$\dot{p}_g = \dot{p}_c + \dot{p}_w \qquad (109)$$

and

$$\dot{S}_w = \frac{d\,S_w}{d\,p_c}\,\dot{p}_c \tag{110}$$

Hence the final form of the governing equation is the following:

- equilibrium equation for the mixture:

The equilibrium condition for the mixture is given by eq. (107), where the effective stress relationship (44) and the constitutive equation (53) have to be introduced. Furthermore the mean pore pressure $\bar{p}$ has to be expressed by eq. (12). However, when dealing with quasi static problems, a more appropriate form for numerical applications may be obtained by differentiating this equation with respect to time. In such a formulation, variations in the density of the mixture is accounted for by the variation of the body forces. As a consequence we obtain the following form for the equilibrium equation, which is still defined in the time domain:

$$-\int_\Omega \delta\,\underline{\varepsilon}^T \underline{D}_T \frac{\partial\,\underline{\varepsilon}}{\partial t}\,d\Omega + \int_\Omega \delta\,\underline{\varepsilon}^T \alpha\,\underline{m}\,\frac{\partial\,p_w}{\partial t}\,d\Omega - \int_\Omega \delta\,\underline{\varepsilon}^T \alpha\,\underline{m}\,p_c\,\frac{dS_w}{dp_c}\,\frac{\partial p_c}{\partial t}\,d\Omega$$

$$-\int_\Omega \delta\underline{u}^T\,\frac{\partial(\rho\underline{b})}{\partial t}\,d\Omega - \int_\Gamma \delta\underline{u}^T\,\frac{\partial\,\underline{t}}{\partial t}\,d\Gamma = 0 \tag{111}$$

- continuity equation for water:

$$\left[C_{ww} + C_{wg}\right]\dot{p}_w + C_{wg}\dot{p}_c - \frac{C_{wg}}{S_g}\,p_c\,\frac{dS_w}{dp_c}\,\dot{p}_c + n\,\frac{dS_w}{dp_c}\,\dot{p}_c + \alpha\,S_w\,\dot{u}_{i,i}$$

$$\tag{112}$$

$$+ \left[k_w\left(-p_{w,i} + \rho_w\,b_i\right)\right]_{,i} + \frac{1}{\rho_w}\left[k_w\left(-p_{w,i} + \rho_w\,b_i\right)\right]\frac{\partial\,\rho_w}{\partial x_i} = 0$$

- continuity equation for air:

$$\left(C_{gw}+ C_{gg}\right)\dot{p}_w+ C_{gg}p_c - \frac{C_{gw}}{S_w}\ p_c\frac{d\ S_w}{d\ p_c}\ \dot{p}_c -n\ \frac{d\ S_w}{d\ p_c}\ \dot{p}_c+ \alpha\ S_g\ \dot{u}_{i,i}$$

$$+\left(k_g\left(-p_{g,i}+ \rho_g\ b_i\right)\right)_{,i}+ \frac{1}{\rho_g}\left(k_g\left(-p_{g,i}+ \rho_g\ b_i\right)\right)\frac{\partial \rho_g}{\partial x_i}\ =\ 0 \tag{113}$$

The boundary value problem for the mechanical behaviour of partially saturated sample requires that the governing equations (111)-(113) be satisfied within all points of the domain Ω and that the boundary conditions are satisfied on the boundary Γ of the domain. Such type of conditions are either imposed displacements or boundary tractions for the displacement field, imposed pressures or fluxes for the pressure fields.

7.1 The case $p_g=0$.

Sometimes phenomena regarding desaturating porous media occur with the air or gas at atmospheric pressure, which represents the reference pressure. Consequently $p_a=0$, $p_c=-p_w$ and the averaged pressure of the fluid phase is simply

$$\bar{p}\ =\ S_w\ p_w \tag{114}$$

We define the specific moisture C_s content as

$$C_s\ =\ n\ \frac{d\ S_w}{d\ p_w}\ =\ -\ \frac{d\ S_w}{d\ p_c} \tag{115}$$

The mass balance equation for water (91) results therefore in

$$\left[\frac{\alpha-n}{K_s}\ S_w^2\ +\ \frac{n\ S_w}{K_w}\ +\ \frac{C_s}{n}\ \frac{\alpha-n}{K_s}\ S_w\ p_w\ +\ C_s\right]\frac{D\ p_w}{Dt}$$

$$+\ \alpha\ S_w\ \dot{u}_{i,i}\ +\ \dot{w}_{i,i}\ +\ \frac{\dot{w}_i}{\rho_w}\ \frac{\partial \rho_w}{\partial x_i}\ =\ 0 \tag{116}$$

In equation (116) the terms

$$\frac{\dot{w}}{\rho_w}\ \frac{\partial\ \rho_w}{\partial\ x_i}\qquad\qquad \frac{1}{K_w}\ \frac{D\ p_w}{Dt} \tag{117}$$

result in a material derivative with respect to the flowing water. Moreover we introduce in the balance equation (116) the transport law (98) and, as usually done in quasi-static soil mechanics problems, we overlook the convective terms. This yields

$$- \nabla^T \left\{ \underline{k} \, \frac{k_{rw}}{\mu_w} \, \nabla \left(p_w + \rho_w \, gh \right) \right\} + S_w \, \alpha \, \underline{m}^T \, \frac{\partial \, \underline{\varepsilon}}{\partial \, t}$$

$$(118)$$

$$+ \left[c_s + n \, \frac{S_w}{K_w} + S_w^2 \left(\alpha - n \right) + \frac{\alpha - n}{K_s} \, S_w \, \frac{c_s}{n} \, p_w \right] \frac{\partial \, p_w}{\partial \, t} = 0$$

Let's consider the gas mass balance equation, which has the form

$$\frac{\alpha - n}{K_s} \, S_w \, S_g \, \frac{D \, p_w}{Dt} + \left[\frac{c_s}{n} \, \frac{\alpha - n}{K_s} \, S_g \, p_w - c_s \right] \frac{D \, p_w}{Dt} + \alpha \, S_g \, \dot{u}_{i,i}$$

$$+ \dot{v}_{i,i} + \frac{\dot{v}_i}{\rho_g} \, \frac{\partial \rho_g}{\partial x_i} = 0 \qquad (119)$$

Due to the fact that the degree of saturation S_a may be expressed as a function of the degree of saturation of the water S_w, by introducing the transport law and overlooking, as usual, convective terms, the air mass balance equation results in

$$\left[\left(1 - S_w \right) \frac{\alpha - n}{K_s} \left(S_w + \frac{c_s}{n} \right) - c_s \right] \frac{\partial \, p_w}{\partial t}$$

$$+ \left(1 - S_w \right) \alpha \, \underline{m}^T \, \frac{\partial \, \underline{\varepsilon}}{\partial \, t} + \dot{v}_{i,i} = 0 \qquad (120)$$

This equation may be used as an auxiliary equation when we want to controll the flow of air in the sample, assuming the boundary fluxes as unknowns of the resulting boundary value problem. In all other cases this equation may be neglected and the only mass balance equation is that of the solid and the water at the same time. The equilibrium statement is the same as in eq. (111), provided that definition (115) is accounted for. Then we have

$$\int_{\Omega} \delta\underline{\varepsilon}^T \underline{D}_T \frac{\partial \underline{\varepsilon}}{\partial t} d\Omega - \int_{\Omega} \delta\underline{\varepsilon}^T \underline{D}_T \frac{\partial \underline{\varepsilon}_0}{\partial t} d\Omega - \int_{\Omega} \delta\underline{\varepsilon}^T \underline{D}_T \frac{\partial \underline{\varepsilon}_c}{\partial t} d\Omega$$

$$\tag{121}$$

$$-\int_{\Omega} \delta\underline{\varepsilon}^T \underline{m} \ \alpha \left(p_w \frac{C_s}{n} + S_w \right) \frac{\partial p_w}{\partial t} d\Omega - \int_{\Omega} \delta\underline{u}^T \frac{\partial \underline{b}}{\partial t} d\Omega - \int_{\Gamma} \delta\underline{u}^T \frac{\partial \hat{\underline{t}}}{\partial t} d\Gamma = 0$$

8. FINITE ELEMENT DISCRETIZATION IN SPACE

The solution of the boundary value problem represented by the equations (111)-(113) results easier if a weak form for all equations is obtained. In this way in fact second derivatives in space do not appear and natural boundary conditions are automatically satisfied. The weak form of eqs. (112) and (113) is easily obtained using the same technique as for the equilibrium equation:

- each term of the balance equation is multiplied by a vector function δp_π which has continuous first partial derivatives with respect to the coordinates x_i and satisfy the boundary conditions where prescribed (virtual pressure). Integration is then performed over the domain Ω.

- terms involving second spatial derivatives are transformed according to the Gauss' theorem. q_π represents the flux of the π phase at the boundary Γ of the domain.

Consequently the continuity equations assume the following form for water

$$\int_{\Omega} \delta\underline{p}_w^T \left\{ \left(C_{ww} + C_{wg} \right) \dot{p}_w + \left(C_{wg} - \frac{C_{wg}}{S_g} \ p_c \frac{d \ S_w}{d \ p_c} + n \frac{d \ S_w}{d \ p_c} \right) \dot{p}_c \right.$$

$$+ \alpha \ S_w \ \dot{u}_{i,i} + \frac{k_w}{\rho_w} \left(-p_{w,i} + \rho_w \ b_i \right) \frac{\partial \ \rho_w}{\partial \ x_i} \right\} d\Omega \tag{122}$$

$$- \int_{\Omega} \nabla \left(\delta\underline{p}_w^T \right)^T \left[\underline{k}_w \left(-p_{w,i} + \rho_w b_i \right) \right] d\Omega + \int_{\Gamma} \delta\underline{p}_w^T \ q_w \ d\Gamma = 0$$

and for gas

$$\int_{\Omega} \delta p_c^T \Bigg\{ \Big(C_{gw} + C_{gg} \Big) \dot{p}_w + \Big(C_{gg} - \frac{C_{gw}}{S_w} p_c \frac{d\,S_w}{d\,p_c} - n \frac{d\,S_w}{d\,p_c} \Big) \dot{p}_c$$

$$+ \alpha\, S_g\, \dot{u}_{i,i} + \frac{k_g}{\rho_g} \Big[-\big(p_c + p_w \big)_{,i} + \rho_g\, b_i \Big] \frac{\partial\,\rho_g}{\partial\,x_i} \Bigg\}\, d\Omega \qquad (123)$$

$$- \int_{\Omega} \nabla\big(\delta p_c\big)^T \cdot \Big[k_g \big[-\big(p_c + p_w \big)_{,i} + \rho_g\, b_i \big] \Big]\, d\Omega + \int_{\Gamma} \delta p_c^T\, q_g\, d\Gamma = 0$$

The finite element discretization in space is now carried out and u, p_w and p_c are chosen as basic variables. These are expressed in terms of their nodal values u, p_w and p_c by means of the shape functions N^d and N^p:

$$u = N^d\, u, \qquad p_w = N^p\, p_w, \qquad p_c = N^p\, p_c \qquad (124)$$

The numbers and locations of the nodes are not necessarily the same for all variables [11].

By substituting the piecewise interpolation functions (124) into the governing equations, and taking into account the arbitrary choice of δu, δp_w and δp_c, we may obtain the semidiscretized nonlinear system of differential equations in time as follows

$$K\, \dot{u} + C_{sw}\, \dot{p}_w + C_{sc}\, \dot{p}_c + F_s = 0$$

$$C_{ws}\, \dot{u} + P_{ww}\, \dot{p}_w + C_{wc}\, \dot{p}_c + H_{pp}\, p_w + F_w = 0 \qquad (125)$$

$$C_{cs}\, \dot{u} + C_{cw}\, \dot{p}_w + P_{cc}\, \dot{p}_c + H_{cc}\, p_c + H_{cv}\, p_w + F_c = 0$$

where the matrices are listed in appendix II. The system (125) is nonsymmetric and coupled, i.e. we can not solve any field separately from the others nor any set of variables can be explicitly eliminated. The complete solution may now be obtained by means of appropriate time stepping algorithms. Regarding the choice of the solution procedure and the relative stability and accuracy analyses, we refer to specialized literature [17,18].

9. NUMERICAL EXAMPLE

It is very difficult to choose appropriate tests to validate the presented numerical model and its numerical implementation. This is so because there exist no analytical solutions for this type of coupled problems, where deformations of the solid skeleton are studied together with the saturated-unsaturated flow of the fluid. On the other hand, the major part of laboratory experiments on this type of phenomena are performed to check the hydraulic parameters only.

The choosen problem for a numerical test for the case of $p_g=0$ is the experiment conducted by Liakopoulos (1965) on the drainage of water from a vertical column of sand. This test was also used by other authors to check their numerical models [19,20].

A column of perspex, 1 meter high, was packed by sand and instrumented to measure continuously the moisture tension at various points within the column (fig. 7). Prior to the start of the experiment (t<0) water was continuously added from the top of the column and was allowed to drain freely at the bottom through a filter. The flow was carefully regulated until tensiometers read zero pore pressure, which corresponds to a unit vertical gradient of the potential. These are the initial conditions. The inflow of water from the top was then stopped and the upper boundary made impermeable to water. From then on the tensionmeter readings were recorded. The physical properties of the sand were measured in an independent experiment: the porosity was 29.75% and the parameters k_{rw} and S_w are presented in fig. 8 as function of the pressure of the water. For numerical purposes the column was simulated by 5 to 20 finite elements of equal size, always obtaining similar results.

The coupling of the fields results of paramount importance particularly at the onset of the phenomenon. This was also the conclusion of other authors, for instance [19]. In fact, only by accounting for this term in the flow equation it is possible to obtain numerical results similar to the experimental ones. Unfortunately in Liakopoulos' experiment the parameters defining the behaviour of the soil skeleton are not given.
For this reason a trial analysis was performed using linear elastic models. The value E=1300 KN/m² gave acceptable results as shown in fig. 9 . This value is in the range of possible values for this material.

Fig. 10 presents the calculated vertical displacements at four stations in the column. We observe that the coupling effect in the hydraulic field, which depends on S_w multiplied by the time derivative of the displacements, is important at the onset of the phenomenon. With elapsing time, the

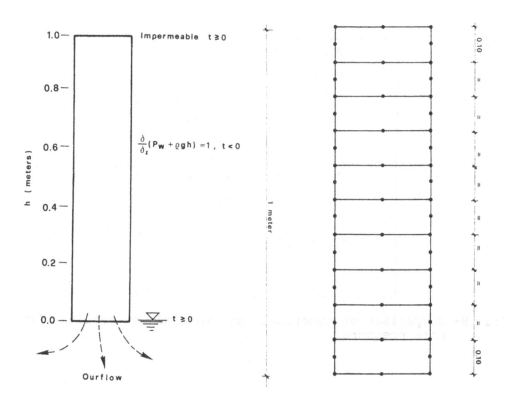

Fig. 7- Test problem and finite element discretization.

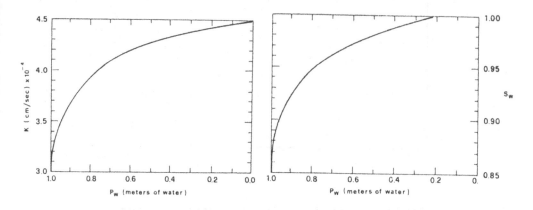

Fig. 8- Hydraulic material properties.

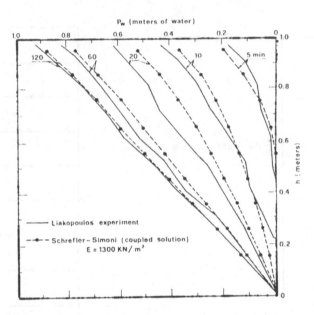

Fig. 9- Comparison of numerical and experimental results for
        pore pressure.

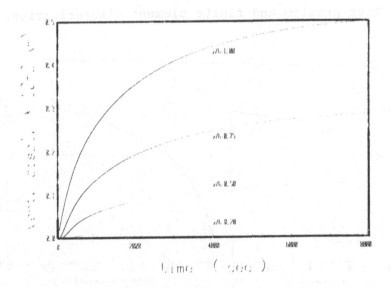

Fig. 10- Calculated vertical displacements versus time at
         four stations in the column.

equations uncouple. For this reason the obtained results using uncoupled formulations and coupled ones, independently of the deformations characteristics assumed for the solid, agree with experimental results at high time values.

10. CONCLUSIONS

An averaging technique was applied to derive the governing equations for the mechanics of multi-component continua, in the case of partially saturated soils. Mass balance equations and linear momentum balance equations of the constituents and of the whole mixture were obtained in quasi static conditions and the interaction forces between the phases were taken into account.

The special case of air pressure equal to the atmospheric pressure in the semi-saturated zone is obtained in a straightforward manner as subcase of the u (displacements) p_w (water pressure) and p_c (capillary pressure) formulation adopted.

An example was shown where numerical results based on this approach are compared with experimental results.

APPENDIX I

The simplified governing equations may be obtained using a less rigorous formulation, which is called phenomenological approach. The assumptions are the same as in the previous case and convective terms are neglected.

- equilibrium equation

In this case we use the weak formulation represented by the virtual work principle, disregarding the inertia terms. We first define the averaged stress as in eq. (8). To impose the equilibrium condition at microscopic level, we use the Hill's macro-homogeneity condition [21]. This states that: if the stress and displacement fields are admissible, i.e. satisfy the equilibrium equations and the boundary conditions, the average of the work at macroscopic level equals the work performed by the averaged values of stress and strain. Accounting for the constitutive relationships and the definition in equation (44) we obtain immediately equation (111).

- mass balance equations for fluid phases

In the same manner as in [20], we account for the following contributions to the rate of fluid accumulation:

a) rate of change of the total strain

$$\frac{\partial \varepsilon_v}{\partial t} = \underline{m}^T \frac{\partial \underline{\varepsilon}}{\partial t} \tag{A.1}$$

b) rate of change of the grain volume due to pressure changes

$$\frac{(1-n)}{K_s} \frac{\partial \bar{p}}{\partial t} \tag{A.2}$$

where $\bar{p}$ is the average effective pressure defined in (12);

c) rate of change of saturation of the π-phase

$$n \rho_\pi \frac{\partial S_\pi}{\partial t} \tag{A.3}$$

d) rate of change of fluid density

$$n S_\pi \frac{\partial \rho_\pi}{\partial t} \tag{A.4}$$

e) change of grain size due to effective stress changes

$$- \frac{1}{3K_s} \underline{m}^T \frac{\partial \underline{\sigma}'}{\partial t} \tag{A.5}$$

The continuity of the flow requires that the following expression is valid:

$$\text{rate of fluid accumulation} + \underline{\nabla} \left(- \frac{k_\pi \rho_\pi}{\mu_\pi} \underline{\nabla} \left(p_\pi + \rho_\pi gh \right) \right) \tag{A.6}$$

where Darcy's law for the transport is incorporated. By summing up all above terms contributing to fluid accumulation equations (112) and (113) are obtained directly from equation (A.6).

APPENDIX II

List of the matrices of equation (125) using the usual symbols [2]:

$$\underline{K} = - \int_{\Omega} \underline{B}^T \underline{D}_T \underline{B} \, d\Omega$$

$$\underline{C}_{SW} = \int_{\Omega} \underline{B}^T \alpha \, \underline{m} \, \underline{N}^P \, d\Omega$$

$$\underline{C}_{SC} = \int_{\Omega} \underline{B}^T \alpha \, \underline{m} \left(1 - S_w - \frac{d \, S_w}{d \, P_c} \, \underline{N}^P \, P_c \right) \underline{N}^P \, d\Omega$$

$$\dot{\underline{F}}_S = \int_{\Omega} \underline{N}^{d^T} \frac{d \, \underline{b}}{d \, t} \, d\Omega + \int_{\Gamma} \underline{N}^{d^T} \frac{d \, \underline{t}}{d \, t} \, d\Gamma$$

$$\underline{C}_{WS} = \int_{\Omega} \underline{N}^{P^T} \alpha \, S_w \, \underline{m}^T \underline{B} \, d\Omega$$

$$\underline{C}_{CS} = \int_{\Omega} \underline{N}^{P^T} \alpha \, S_g \, \underline{m}^T \underline{B} \, d\Omega$$

$$\underline{C}_{CW} = \int_{\Omega} \underline{N}^{P^T} S_g \left(\frac{\alpha - n}{K_S} + \frac{n}{K_g} \right) \underline{N}^P \, d\Omega$$

$$\underline{P}_{CC} = \int_{\Omega} \underline{N}^{P^T} \left[\frac{\alpha - n}{K_S} \left(S_g^2 - S_g \frac{d \, S_w}{d \, P_c} \, \underline{N}^P \, P_c \right) + \frac{S_g \, n}{K_g} - n \frac{d \, S_w}{d \, P_c} \right] \underline{N}^P \, d\Omega$$

$$\underline{H}_{CC} = - \int_{\Omega} \underline{N}^{P^T} \frac{\underline{k}_g}{\rho_g} \nabla \underline{N}^P \nabla^T \rho_g \, d\Omega + \int_{\Omega} \left(\nabla \underline{N}^P \right)^T \underline{k}_g \nabla \underline{N}^P \, d\Omega$$

$$\underline{H}_{CW} = - \int_{\Omega} \underline{N}^{P^T} \frac{\underline{k}_g}{\rho_g} \nabla \underline{N}^P \nabla^T \rho_g \, d\Omega - \int_{\Omega} \left(\nabla \underline{N}^P \right)^T \underline{k}_g \nabla \underline{N}^P \, d\Omega$$

$$\underline{P}_{WW} = \int_{\Omega} \underline{N}^{P^T} \left(\frac{\alpha - n}{K_S} + \frac{n}{K_W} \right) S_w \, \underline{N}^P \, d\Omega$$

$$\underline{C}_{WC} = \int_{\Omega} \underline{N}^{P^T} \left[\frac{\alpha - n}{K_S} \left(S_w \, S_g - S_w \frac{d \, S_w}{d \, P_c} \, \underline{N}^P P_c \right) + n \frac{d \, S_w}{d \, P_c} \right] \underline{N}^P \, d\Omega$$

$$\underline{H}_{PP} = - \int_{\Omega} \underline{N}^{P^T} \frac{\underline{k}_w}{\rho_w} \nabla \underline{N}^P \nabla^T . \rho_w \, d\Omega + \int_{\Omega} \left(\nabla \underline{N}^P \right)^T \underline{k}_w \nabla \underline{N}^P \, d\Omega$$

$$\dot{\underline{F}}_W = \int_{\Omega} \underline{N}^{P^T} \frac{\underline{k}_w}{\rho_w} \underline{b}_w \nabla^T . \rho_w \, d\Omega - \int_{\Omega} \left(\nabla \underline{N}^P \right)^T \underline{k}_w \underline{b}_w \, d\Omega + \int_{\Gamma} \underline{N}^{P^T} q_w \, d\Gamma$$

$$\dot{F}_w = \int_\Omega \underline{N}^{P^T} \frac{\underline{k}_w}{\rho_w} \, \underline{b}_w \nabla^T \cdot \rho_w \, d\Omega - \int_\Omega \left(\nabla \underline{N}^P \right)^T \underline{k}_w \, \underline{b}_w \, d\Omega + \int_\Gamma \underline{N}^{P^T} q_w d\Gamma$$

REFERENCES

1. Bedford, A. and Drumheller, D.S.: Theories of immiscible and structured mixtures, Int. J. Engng. Sci., 21(1983), 863-960.
2. Lewis, R.W. and Schrefler, B.A.: The finite element method in the deformation and consolidation of porous media, Wiley, Chichester 1987.
3. Simoni, L. and Schrefler, B.A.: F.E. solution of a vertically averaged model for regional land subsidence. Int. J. Num. Meth. Eng., 27(1989), 215-230.
4. Bear, J. and Bachmat, Y.: Transport phenomena in porous media - Basic equations, in: Fundamentals of transport phenomena in porous media (Bear, J. and Corapcioglu, M.Y. eds.), Nato ASI Series, Nijhoff, Dordrecht 1984, 5-61.
5. Lemaitre, J. and Chaboche J.L.: Mecanique des materiaux solides, Dunod, Paris 1988.
6. Skempton, A.W.: Effective stress in soils, concrete and rocks, in Pore pressure and suction in soils, Butterworths London 1961, 4-16.
7. Bishop, A.W. and Blight, G.E.: Some aspects of effective stress in saturated and partly saturated soils, Geotechnique, 13(1963), 177-197.
8. Aitchison, G.D. and Donald, I.B.: Effective stress in unsaturated soils, Proc. 2nd Australia-New Zealand Conf. Soil Mech., 1956, 192-199.
9. Morland, L.W.: A simple constitutive theory for a fluid saturated porous solid, Journal Geophys. Res. 77(1972), 890-900.
10. Lloret, A., Alonso, E.E. and Gens,a A.: Undrained loading and consolidation analysis for unsaturated soils, Proc. Eur. Conf. Num. Meth. Geomech., 2, Stuttgart, 1986.
11. Zienkiewicz, O.C. and Shiomi, T.: Dynamic behaviour of saturated porous media: the general Biot's formulation and its numerical solution, Int. J. Num. Anal. Meth. Geom., 8(1985), 71-96.
12. Hassanizadeh, M. and Gray W.G.: General conservation equations for multiphase systems: 1. Averaging procedure, Adv. Water Resources, 2(1979), 131-144.

13. Hassanizadeh, M. and Gray W.G.: General conservation equations for multiphase systems: 2. Mass, momenta, energy and entropy equations, Adv. Water Resources, 2(1979), 191-203.

14. Hassanizadeh, M. and Gray W.G.: General conservation equations for multiphase systems: 3. Constitutive theory for porous media flow, Adv. Water Resources, 3(1980), 25-40.

15. Marsden J.E. and Hughes, T.J.R.: Mathematical foundations of elasticity, Prentice-Hall, Englewood Cliffs, 1983.

16. Li, X., Ding, D., Chan, A.H.C. and Zienkiewicz, O.C.: A coupled finite element method for the soil-pore fluid interaction problems with immiscible two-phase fluid flow, Proc. V-th Int. Symposium on Num. Meth. Eng., Lausanne, 1989.

17. Zienkiewicz, O.C. and Taylor, R.L.: Coupled problems - A simple time stepping procedure, Comm. Appl. Num. Meth., 1(1985), 233-239.

18. Zienkiewicz, O.C., Paul, D.K. and Chan, A.H.C.: Unconditionally stable staggered solution procedure for soil-pore fluid interaction problems, Int. J. Num. Meth. Eng., 25(1988), 1093-1055.

19. Narasimhan, T.N. and Witherspoon, P.A.: Numerical model for saturated-unsaturated flow in deformable porous media. 3 Applications, Water Resour. Res., 14(1978), 1017-1034.

20. Schrefler, B.A. and Simoni, L.: A unified approach to the analysis of saturated-unsaturated elastoplastic porous media, in Numerical Methods in Geomechanics. Innsbruck 1988. (ed. G. Swoboda), Balkema, Rotterdam 1988.

21. Suquet, P.M.: Elements of homogenization for inelastic solid mechanics, in:Homogenization techniques for composite media (eds. E. Sanchez-Palencia and A. Zaoui), Springer-Verlag, Berlin 1987.

DYNAMIC ANALYSIS OF SATURATED NON LINEAR MEDIA

D. Aubry
Ecole Centrale de Paris, Paris, France

H. Modaressi
Bureau de Recherches Géologiques et Minières, Orléans, France

ABSTRACT

Some considerations are given on the modelling of wave propagations in porous media. Two models similar to the one proposed by Biot many years ago are described and the numerical implementation using a variational formulation and a finite element approach for the solid and the fluid phase are discussed. The discretization with time may be performed with a mixed implicit-explicit scheme which can combine the advantages of each technique. Finally the case of seismic loading in the form of a given propagating free field is analyzed.

1. INTRODUCTION

The numerical modelling of the wave propagation in saturated porous media has some important applications in many areas like earthquake engineering, geophysics or petroleum engineering. In the past years it was restricted to the linear case. But recent developments have shown that by using modern computational formulations and techniques together with the continuing progress of the power of computers it is possible to deal with nonlinear behaviour of the materials and to choose among more or less sophisticated models depending on the required cost of the analysis.

The purpose of this article is to review some of the basic aspects including discussions on the physical models themselves together with their numerical implementations. It is believed that the coverage although rather brief is rather selfcontained.

2. RESULTS FROM HOMOGENEIZATION THEORY

In this chapter we use some notions of homogeneization theory to get some better insight into the Biot type models which are essentially macroscopic. Several techniques of homogeneization have been developped over the recent years. Their usefullness is more or less appropriate depending on the detailed microscopic topology of the material. One of the weakest assumption relies on the existence of an representative elementary volume (REV) the size of which is such that some averaging of the microscopic quantities can be performed to obtain the macroscopic quantities which will be useful to describe the deformation of the porous medium at the length scale of the boundary value problem.

Let σ be the stress tensor, ε the strains, $\mathbf{u}$ the displacement vector, $\mathbf{D}$ the elasticity tensor, ρ the density of the matrix, μ the fluid viscosity, p fluid pressure, $\mathbf{n}$ the unit normal. The indices s et f stand respectively for the solid and fluid phases. The behaviour of each phase is assumed to be modelled itself by the concept of the mechanics of continuous media. The solid matrix fills the domain Ω_s in the REVand the fluid phase fills the domain Ω_f. The averaging process (Auriault and Sanchez Palencia (1977), Bear and Pinder (1983)) depends on the homogeneization technique at hand. Here only the simple averaging technique will be used.

The $<.>_\alpha$ symbol will stand for the phase average and the intrinsic average over the α phase defined respectively by :

$$<\phi>_\alpha = (1/V) \int_{V\alpha} \phi \ dV \ , \quad <\phi>^\alpha = (1/V_\alpha) \int_{V\alpha} \phi \ dV$$

Starting with the conservation of momentum the following equation are obtained :

$$\text{Div} <\sigma_s>_s + <\rho_s>_s \mathbf{g} + 1/\Omega \int_{Ss} \sigma_s. \ \mathbf{n}_s \ dS = < \rho_s \ \partial_{tt} u_s>_s$$

$$\text{Div} <\sigma_f>_f + <\rho_f>_f \mathbf{g} + 1/\Omega \int_{Sf} \sigma_f. \ \mathbf{n}_f \ dS = < \rho_f \ \partial_{tt} u_f>_f$$

If n stands for the porosity and θ_s and θ_f stand for the volumic fractions given by :

$$\theta_s = (1- n), \quad \theta_f = n$$

the total stress will be given by :

$$< \sigma > = < \sigma_s>_s + < \sigma_f>_f = \theta_s < \sigma_s>^s - \theta_f <p>^f \ \mathbf{I} + \theta_f < \tau_f>^f$$

In classical soil mechanics the pore pressure is given by :

$$p = <p>^f$$

According to the Terzaghi principle the so-called effective stresses are given by :

$$<\sigma'> = <\sigma> + p\ I$$

Thus the relationship between the effective stresses and the average stresses in the solid matrix is the following :

$$<\sigma'> \quad = <\sigma_s>_s + (1-n)\,p\,I + n <\tau_f>^f$$

which demonstrates the influence of the pore pressure and the deviatoric stresses in the fluid due to the viscosity have on the effective stresses. These former quantities are usually neglected in soil mechanics. In the same way the average density is given by :

$$<\rho> = <\rho_s>_s + <\rho_s>_f = (1-n) <\rho_s>^s + n <\rho_f>^f$$

The average conservation of momentum for the bulk material will thus be :

$$\text{Div} <\sigma> + <\rho>g = <\rho_s\,\partial_{tt}u_s> + <\rho_f\,\partial_{tt}u_f>$$

It is possible to introduce the effective stresses in this equation :

$$\text{Div} <\sigma'> - \text{grad } p + <\rho>g = <\rho_s\,\partial_{tt}u_s> + <\rho_f\,\partial_{tt}u_f>$$

The second ingredient to a modelling of the deformation of porous media is brought by some type of Darcy's law of filtration. The filtration law is indeed connected with the interaction forces between the fluid and the solid due the fluid flow :

$$R = 1/\Omega \int_{Sf} \sigma_f.\,n_f$$

which should be related to the relative velocity of the fluid with respect to solid :

$$< \partial_t u_f - \partial_t u_s>_f.$$

Usually a linear relationship is postulated thus introducing the permeability :

$$< \partial_t u_f - \partial_t u_s>_f = k\ .\ R$$

but using the momentum conservation R may be written :

$$R = <\rho_f\,\partial_{tt}u_f>_f - \text{Div } n<\sigma_f>^f - n<\rho_f>^f\,g$$

Very often the first term on the right hand side is approximated by :

$$<\rho_f\,\partial_{tt}u_f>_f \approx n <\rho_f>^f <\partial_{tt}u_f>_f$$

so that the following expression for $\partial_t u_{rf}$ is given by :

$$< \partial_t u_{rf} > \; = \; < \partial_t u_f - \partial_t u_s >^f$$
$$= \; k \; \{ \; grad \; (p - <\rho_f>^f g \; x \;) \; + <\rho_f>^f <\partial_{tt} u_f>_f \; - \; Div <\tau_f>^f \; \}$$

The last ingredient to the modelization is the conservation of mass for each phase. First for the soil :

$$(1-n) \; \partial_t <\rho_s>^s / <\rho_s>^s \; + div(\; (1-n)< \partial_t u_s>^s) \; - \partial_t n = 0$$

then for the fluid :

$$n \; \partial_t <\rho_f>^f / <\rho_f>^f \; + \; div(\; n< \partial_t u_f>^f) \; + \partial_t n = 0$$

So that the mass conservation for the bulk material gives :

$$(1-n) \; \partial_t <\rho_s>^s / <\rho_s>^s \; + n \; \partial_t <\rho_f>^f / <\rho_f>^f$$
$$+ \; div \; \{ \; (1-n)< \partial_t u_s>^s + n< \partial_t u_f>^f \} \; = 0$$

If K_s and K_f stand for the moduli of compressibility of each phase the following equation is obtained :

$$- (1-n) \; Tr< \partial_t \sigma_s>^s/K_s \; - \; n \; Tr < \partial_t \sigma_f>^f/K_f \; + div \; (< \partial_t u_s> + < \partial_t u_{rf}>) \; = 0$$

It can be shown that indeed due to microscopic fluid/structure interaction effects the permeability is frequency dependant. Fortunately this effect becomes important only at higher frequencies than those usually occuring for instance in earthquake engineering.

3. TWO BIOT TYPE MODELS

In the preceeding chapter the qualitative results arising from the homogeneization techniques have been presented. In the subsequent developments two macroscopic models with different ranges of complexity and applications are presented.

3.1 The (u_s - u_{rf}) model

The total stress σ is first decomposed into the effective stresses σ' and the pore pressure p with the convention of positive traction. The usual decomposition in soil mechanics :

$$\sigma = \sigma' - p \; I$$

is generalized by the following one with the parameter α which can also be used in rock mechanics :

$$\sigma = \sigma" - \alpha \, p \, \mathbf{I} \qquad \alpha \leq 1$$

According to the definition of α and using the modulus of compressibility of the matrix:

$$\partial_t \sigma" = \partial_t \sigma' - (1-\alpha) \, \partial_t p \, \mathbf{I} = \mathbf{D}: (\partial_t \varepsilon + \partial_t p \, \mathbf{I} / K_s) - (1-\alpha) \, \partial_t p \, \mathbf{I}$$

but $\sigma"$ is related to the overall strain rate :

$$\partial_t \sigma" = \mathbf{D} : \partial_t \varepsilon$$

so that α may be written :

$$\alpha = (1 - (\mathbf{I}:\mathbf{D}:\mathbf{I}) / 3K_s)$$

Several choices of the dependant variables have been proposed in the literature (Zienkiewicz et al.(1984)). Either the absolute displacements of the solid and of the fluid and the pore pressure can used : u_s, u_f, p. This choice leads to an interesting decoupling between the inertial effects. However the formulation retained here is such that the natural boundary conditions coming from the variational formulation wil be automatically satisfied. Thus the displacement of the solid phase u_s and the relative displacement of the fluid phase u_{rf} :

$$u_{rf} = n \, (u_f - u_s)$$

where n is the porosity will be kept as the primary dependant unknowns. Now in this model the conservation of momentum of the two phases is given by :

$$\mathbf{Div} \; \sigma + \rho \, g = \rho \; \partial_{tt} u_s + \rho_f \, \partial_{tt} u_{rf}$$

where the density is :

$$\rho = (1-n) \, \rho_s + n \; \rho_f$$

ρ_s and ρ_f are respectively the denisty of the solid and the fluid phases. By introducing the total stress decomposition the following expression is obtained :

$$\mathbf{Div} \; \sigma" - \alpha \, \mathbf{grad} \; p + \rho \; g = \rho \; \partial_{tt} u_s + \rho_f \, \partial_{tt} u_{rf}$$

The above conservation of momentum must of course be augmented with an equation which describes the movement of one phase with respect to the other. This equation is given by some type of Darcy's law :

$$-\mathbf{grad} \; p + \rho_f \, g = K^{-1} \, \partial_t u_{rf} + \rho_f \, (\partial_{tt} u_s + \partial_{tt} u_{rf} / n)$$

where $\mathbf{K}$ is the permeability tensor given by :

$$\mathbf{K} = \mathbf{k} \; / \, \rho_f \, g$$

It should be mentionned that Biot (1956) in his original work has introduced an inertial coupling term between the two phases in the Darcy equation :

$$-\mathbf{grad}\, p + \rho_f \mathbf{g} = \mathbf{K}^{-1} \partial_t \mathbf{u}_{rf} + \rho_f (\partial_{tt} \mathbf{u}_s + \partial_{tt} \mathbf{u}_{rf} / n) + \rho_a \partial_{tt} \mathbf{u}_{rf} \quad \rho_a \geq 0$$

Here the vanishing of ρ_a is equivalent to saying that while imposing an acceleration, a turtuosity parameter may be introduced :

$$\Xi = 1 + n\, \rho_a / \rho_f \qquad \Xi \geq 1$$

$$-\mathbf{grad}\, p + \rho_f \mathbf{g} = \mathbf{K}^{-1} \partial_t \mathbf{u}_{rf} + \rho_f (\partial_{tt} \mathbf{u}_s + \Xi\, \partial_{tt} \mathbf{u}_{rf} / n)$$

This parameter depends on the porosity and the geometry of the pores. Here this inertia coupling term will be neglected mainly because of the lack of any quantitative estimation.

The mass conservation of each phase must also be supplied and has the following expression for the solid:

$$\mathrm{div}\, ((1\text{-}n)\rho_s \, \partial_t \mathbf{u}_s) + \partial_t ((1\text{-}n)\, \rho_s) = 0$$

By developping the above quantities it can be written as:

$$(\mathrm{Tr}\, \partial_t \sigma) / 3 K_s + (1-n)\, \mathrm{div}\, \partial_t \mathbf{u}_s - \partial_t n = 0$$

In a similar way the mass conservation of the fluid phase is given by :

$$n \quad \partial_t p / K_f + \quad n\, \mathrm{div}\, \partial_t \mathbf{u}_f \quad + \partial_t n = 0$$

By adding the two last equations and by invoking the decompostion of the total stress the mass conservation for the two phases may finally be written :

$$\mathrm{div}\, \partial_t \mathbf{u}_{rf} + \alpha\, \mathrm{div}\, \partial_t \mathbf{u}_s = - \partial_t p / Q$$

whith Q as :

$$1/Q = n \, / K_f + (\alpha - n) \, / K_s$$

In soil mechanics Q is mainly depending on the compressibilty of the water especially in the case of partially saturated soil and the following simplified assumptions may be used :

$$\alpha = 1$$

$Q = K_f / n$

In summary the three sets of equations which describe the (u_s-u_{rf}) model are given by :

$$\mathbf{Div}\ \sigma" + \alpha\ Q\ \mathbf{grad}\ (\ div\ (u_{rf} + \alpha\ u_s\ )) + \rho\ \mathbf{g}\ =\ \rho\ \partial_{tt}u_s + \rho_f\ \partial_{tt}u_{rf}$$

$$Q\ \mathbf{grad}\ (\ div\ (u_{rf} + \alpha\ u_s\ )) + \rho_f\mathbf{g} = K^{-1}\ \partial_t u_{rf} + \rho_f\ (\ \partial_{tt}u_s + \Xi\ \partial_{tt}u_{rf} / n)$$

$$p = -\ Q\ (div\ u_{rf} + \alpha\ div\ u_s\ )$$

which in the case of non linear elastoplastic materials must be supplemented by the constitutive equations :

$$\begin{cases} f\ (\sigma",\alpha) \leq 0 \\ \partial_t\sigma" = D^{ep}\ (\ \sigma",\alpha\ ,\varepsilon\ (\ \partial_t u_s)) : \varepsilon\ (\ \partial_t u_s) \\ \partial_t\alpha = L\ (\ \sigma",\alpha\ ,\varepsilon\ (\ \partial_t u_s)) \end{cases}$$

where f stands for the yield surface, D^{ep} for the elastoplastic matrix, L for the flow rule of the internal parameters (α). Some hints will be given at the end of this paper on the type of constitutive behaviour that is currently used in our work but no detailed description will be given as it is considered to be out of the scope of this paper.

The boundary conditions will obviously be more complex than for one phase material and must be decomposed into boundary conditions relative to the bulk material and to the fluid flow. The boundary Γ of the domain Ω is thus decomposed into two first parts :

$$\Gamma\ =\ \Gamma_\sigma \cup \Gamma_{us}$$

On Γ_{us} a displacement boundary condition is selected :

$$u_s\ =\ 0,\quad x \in\ \Gamma_{us}$$

On Γ_σ a traction boundary condition is applied :

$$\sigma\,n\ =\ T,\quad x \in\ \Gamma_\sigma$$

Then a second partition of the boundary is introduced to deal with the fluid flow :

$$\Gamma\ =\ \Gamma_p \cup \Gamma_{urf}$$

On Γ_{urf} a no drainage type boundary condition is applied :

$$u_{rf} = u_{rf}*, \ x \in \Gamma_{urf}$$

and on Γ_p a pressure boundary condition is considered :

$$p \ = \ p*, \ x \in \Gamma_p$$

Now when the intial conditions are supplied all the equations necessary to the description of the (u_s-u_{rf}) model are discussed. So that the (u_s-u_{rf}) model can be summarized as :

P(u_s-u_{rf}):

Find the following fields : u_s (x,t) , u_{rf} (x,t) , (σ'' (x,t) ,α (x,t)), satisfying the conservation of momentum, Darcy's law and conservation of mass together with the boundary conditions and the initial conditions : u_s (x,0), u_{rf} (x,0), $\partial_t u_s$ (x,0),$\partial_t u_{rf}$ (x,0),

(σ''(x,0), α (x,0)) .

The weak form of these equations especially suited for the numerical approximation by finite elements will be discussed in a forthcoming paragraph. The description of a simplified version of the model is presented first.

3.2 The simplified (u_s- p) model

The preceeding model will be shown later to be interesting for higher frequencies. However for the frequency range suited in earthquake engineering it seems that a simplified version would be more cost effective. In this model the relative acceleration of the fluid is neglected so that the momentum conservation and Darcy's law may be written :

$$\text{Div } \sigma'' \ - \ \alpha \text{ grad } p + \rho \ g \ = \ \rho \ \partial_{tt} u_s$$

$$\text{-grad } p + \rho_f \ g = \ K^{-1} \ \partial_t u_{rf} + \rho_f \ \partial_{tt} u_s$$

The main consequence of this assumption is that now by using the conservation of mass :

$$\text{div } \partial_t u_{rf} + \alpha \text{ div } \partial_t u_s \ = \ - \ n \ \partial_t p / K_f \ - \ (\alpha - n) \ \partial_t p / K_s$$

the relative displacement of the fluid may be eliminated :

$$\alpha \text{ div } \partial_t u_s - \text{div } (\ \textbf{K grad } (p - \rho_f \ g \cdot x)) \ + \text{div } (\ \textbf{K } \rho_f \ \partial_{tt} u_s) + \partial_t p / Q = 0$$

so that the displacement and the pore pressure are the only dependant unknowns. Corresponding to this simplified model the following boundary conditions are employed. The first partition of the boundary Γ is considered:

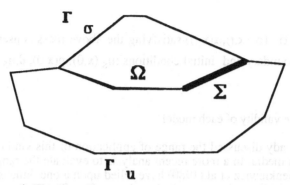

Figure 1: The domain and the mechanical boundary conditions

$$\Gamma = \Gamma_\sigma \cup \Gamma_{us}$$

On Γ_{us} the displacement is imposed while the total traction boundary condition is retained on Γ_σ. For the fluid flow boundary conditions another partition is considered :

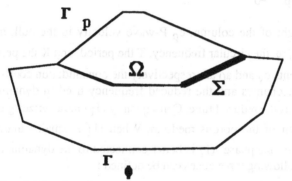

Fifure 2: The domain and the hydraulic boundary conditions

$$\Gamma = \Gamma_p \cup \Gamma_\varphi$$

On Γ_p there is a pore pressure boundary condition :

$$p = p^* \quad , X \in \Gamma_p$$

while on Γ_φ there is a normal velocity condition, weaker than in the full (u_s-u_{rf}) model:

$$\partial_t u_{rf}.n = \varphi \quad , X \in \Gamma_\varphi$$

In summary the simplified model is characterized by the following problem:

P(u_s-p) :

Find u_s (x,t), p (x,t) (σ"(x,t),α(x,t)) satisfying the above mass conservation, Darcy's law together with boundary and initial conditions :u_s (x,0), p(x,0), $\partial_t u_s$ (x,0), (σ"(x,0), α(x,0)) .

3.3 Comments on the validity of each model

Biot (1956) has already discussed the range of application of this kind of model for the dynamics of porous media. In a more recent analysis to evaluate the range of validity of each formulation Zienkiewicz et al.(1980) have relied upon a one dimensional study of a column where a sinusoidal loading is applied on ther surface. The drainage is allowed only at the head of the column. The fluid is assumed to be inviscid and the solid is elastic so that the analytical solution is easily carried out. Two adimensional parameters Π_1 et Π_2 govern essentially the problem. They are :

$$\Pi_1 = K \rho c_p^2 / (\omega L^2) = k \ [(\lambda+2\mu)+K_f/n] \ T / 2\pi \ (\rho_f g L^2) = c_v T / 2\pi \ L^2$$
$$\Pi_2 = \omega L / c_p = a_0$$

where L is the height of the column, c_p P-wave velocity in the bulk medium ($c_p^2 = [(\lambda+2\mu)+K_f/n] / \rho$), ω the circular frequency, T the period, and K the permeability (= k / $\rho_f g$). The coefficients c_v and a_0 are respectively the consolidation coefficient classically employed in soil mechanics and the reduced frequency used in dynamic soil-structure interaction. Π_1 is thus the reduced time. Consequently Π_1 characterizes globally the one or two phase behaviour of the porous medium. When Π_1 is small (undrained case) the mixture is essentially one phase. Π_2 is better connected to the dynamic behaviour of the medium. Now the following three zones can be defined :

  zone 1 : quasistatic regime
  zone 2 : dynamic regime (relative acceleration of the fluid phase negligible)
  zone 3 : fast dynamic regime (full model)

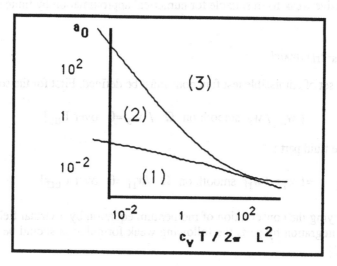

Figure 3: The different ranges of behaviour (after Zienkiewicz et al. (1980))

It may be noticed that the quantity $\Pi_1 \cdot (\Pi_2)^2$ which corresponds to the limit of the $(u_s\text{-}p)$ model is approximately equal to one. The corresponding frequency is given by :

$$f \approx \rho_f \, g / (2\pi \, \rho \, k)$$

which is very close to the expression of the characteristic frequency given by Biot which thus should give an upper limit of validity of the simplified model. For rather permeable soils ($k = 10^{-1}$ to 10^{-2} m/sec) the characteristic frequency may belong to the frequency range dealt with in earthquake engineering. For values of Π_1 larger than 10^2 the problem may be considered as drained as the loading is not fast enough and the pore pressure has some time to dissipate. For faster loading conditions (Π_1 very low et Π_2 large) the soil behaviour may be considered as undrained. One phase computations may be considered inside this domain but at the boundary where drainage is possible a boundary layer will develop to satisfy the corresponding boundary condition.

4. WEAK FORM OF THE TWO MODELS

Many integral signs are involved in the variationnal formulation of the above two problems. This is the reason why the following notations are introduced for either two vector fields (v_1, v_2), or two tensor fields (τ'_1, τ'_2) defined on the domain Ω or its boundary Γ:

$$(\tau_1, \tau_2)_\Omega \;\; = \;\; \int_\Omega \tau_1 : \tau_2 \, d\Omega \;\;\; = \sum_{ij} \int_\Omega \tau_{1ij} \cdot \tau_{2ij} \, d\Omega$$

$$(v_1, v_2)_\Omega \;\; = \;\; \int_\Omega v_1 \cdot v_2 \, d\Omega \;\;\; = \sum_i \int_\Omega v_{1i} \cdot v_{2i} \, d\Omega$$

$$< v_1, v_2 >_\Gamma \;\; = \;\; \int_\Gamma v_1 \cdot v_2 \, d\Gamma \;\;\; = \sum_i \int_\Gamma v_{1i} \cdot v_{2i} \, d\Gamma$$

The respective weak form suitable for numerical approximation by finite elements are now described.

4.1 The (u_s-u_{rf}) model

As usual a set of admissible test functions must be defined. First for the solid part :

$$V_s \quad =\{ \ w_s \ / w_s \ \text{smooth on} \ \Omega \ / \ w_s = 0 \ \text{over} \ \Gamma_{us}\}$$

then for the fluid part :

$$V_{rf} \quad =\{ \ w_{rf} \ / w_{rf} \ \text{smooth on} \ \Omega \ / w_{rf} = 0 \ \text{over} \ \Gamma_{urf}\}$$

By multiplying the conservation of momentum equation by a virtual field $w_s \in V_s$ and by using an integration by parts the following weak formulation should be satisfied by both u_s and u_{rf} :

$$((\rho \ \partial_{tt} u_s + \rho_f \ \partial_{tt} u_{rf}) , w_s)_\Omega + (\ \alpha \ Q \ \text{div} \ (u_{rf} + \alpha \ u_s) \ , \ \text{div} \ w_s \ )_\Omega + (\sigma", \epsilon(w_s))_\Omega$$

$$= (\rho \ g, w_s \ )_\Omega + <T, w_s>_{\Gamma_\sigma}$$

With a similar procedure for the admissible w_{rf} field the following weak form is established :

$$(\rho_f(\partial_{tt} u_s + \Xi \ \partial_{tt} u_{rf}/n), w_{rf})_\Omega + (Q \ \text{div}(u_{rf} + \alpha u_s), \text{div} \ w_{rf})_\Omega$$

$$+ (K^{-1} \ \partial_t u_{rf}, w_{rf})_\Omega = (\rho_f g, w_{rf})_\Omega + <p^*, w_{rf} \ n>_{\Gamma_p}$$

It can be checked that the boundary conditions described in the preceeding section are naturally taken into account with this formulation. The only equations left are the constitutive behaviour which will be written also in a weak form :

$$\begin{cases} (\partial_t \sigma", \theta \ )_\Omega = (\ D^{ep} \ (\ \sigma", \alpha \ , \partial_t \epsilon \ (\ u_s)) : \partial_t \epsilon \ (\ u_s), \theta \ )_\Omega, \ \forall \theta \in \ \Theta \\ \\ (\partial_t \alpha \ , \varpi \ )_\Omega = \ L \ (\ \sigma", \alpha \ , \partial_t \epsilon \ (\ u_s)), \varpi \ )_\Omega \forall \ \varpi \in \ \Pi \end{cases}$$

and the initial conditions. So that the (u_s-u_{rf}) model may be summarized as :

P(u_s-u_{rf}) :

Find $u_s(t) \in V_s$, $u_{rf}(t) \in V_{rf}$ and $(\ \sigma"(t), \alpha(t) \ ) \in \ \Theta \times \Pi$, satisfying the above variational formulation together with initial conditions.

4.2. The (u_s-p) model

In a similar way test functions are introduced for the displacement field for this simplified model :

$$V_s = \{ \ w_s \ / w_s \ \text{smooth on} \ \Omega \ / \ w_s = 0 \ \text{over} \ \Gamma_{us} \}$$

and for the pressure field :

$$Q = \{ \ q \ / q \ \text{smooth on} \ \Omega \ / q = 0 \ \text{over} \ \Gamma_p \}$$

The two corresponding variational formulations are subsequently defined :

$$(\rho \ \partial_{tt} u_s w_s) \Omega + (\sigma'', \varepsilon(w_s)) \Omega - (\alpha \ p, \text{div} \ w_s) \Omega = (\rho \ g, w_s) \Omega + <T, w_s> \Gamma_\sigma$$

$$- (\rho_f \ \mathbf{K} \ \text{div} \ \partial_{tt} u_s, q) \Omega - (\alpha \ \text{div} \ \partial_t u_s, q) \Omega - (\partial_t p \ / \ Q, q) \Omega - (\mathbf{K} \text{grad} \ p, \text{grad} \ q) \Omega$$

$$= < \varphi, q > \Gamma_\varphi - (\mathbf{K} \text{grad} \ (\rho_f \ g \ X), \text{grad} q) \Omega$$

So that the (u_s-p) problem may be defined by :

P(u_s-p) :

Find $u_s(t) \in V_s$, $p(t) \in Q$ and ($\sigma''(t)$, $\alpha(t)$) $\in \Theta \times \Pi$ satisfying the above mentionned initial conditions.

5. FINITE ELEMENT APPROXIMATION

Having established the weak form the finite element approximation of $u_s(t)$, $u_{rf}(t)$ $p(t)$ defined by $u_{sh}(t)$, $u_{rfh}(t)$ et $p_h(t)$ will be computed by restricting the test functions to finite dimensional spaces for the displacement of the solid, the fluid or of the pressure respectively : V_{sh}, V_{rfh} et Q_h and for the stresses and internal variables : $\sigma''(t)$ et $\alpha(t)$ Θ_h et Π_h .

The two problems (u_s-u_{rf}), and (u-p) are now replaced by their corresponding approximations :

P(u_s-u_{rf}))$_h$:

Find $u_{sh}(t) \in V_{sh}$, $u_{rfh}(t) \in V_{rfh}$ and ($\sigma''_h(t)$, $\alpha_h(t)$) $\in \Theta_h \times \Pi_h$ satisfying the weak forms $\forall \ w_{sh} \in V_{sh}$, $\forall \ w_{rfh}(t) \in V_{rfh}$

P(u_s-p)$_h$:

Find $u_{sh}(t) \in V_{sh}$, $p_h(t) \in Q_h$ and $(\sigma''_h(t), \alpha_h(t)) \in \Theta_h \times \Pi_h$ satisfying the weak forms $\forall\, w_{sh} \in V_{sh}$, $\forall\, q_h \in Q_h$.

Some considerations are now given to the different quantities to be computed to arrive at the solution.

5.1 Implementation of $(u_s - u_{rf})_h$

The finite element spaces V_{sh}, V_{rfh} have shape functions : $w_{sJ}(x) \cdot e_j$ and $w_{rfJ}(x).e_j$ where e_j is the unit basis vector of the euclidean space and $w_{sJ}(x)$ and w_{rfJ} are finite element shape functions. In this manner the following expansion may be written :

$$u_{sh} = \sum_{Ii} u_{sIi} \cdot w_{sI}\, e_i , \qquad u_{rfh} = \sum_{Ii} u_{rfIi} \cdot w_{rfI}\, e_i$$

Correspondingly the test functions will be chosen successively equal to each member of the unit basis that is :

$$w_{sh} = w_{sJ} \cdot e_j , \qquad\qquad w_{rfh} = w_{rfJ} \cdot e_j$$

Obviously the shape function for the solid and the fluid may be chosen differently. However only the equality case will be further considered so that :

$$w_J = w_{sJ} = w_{rfJ}$$

Now when the above expansions are plugged into the finite dimensional weak forms, the following non linear system of differential equations is obtained first for the momentum :

$$\sum \partial_{tt} u_{sIi}\, \delta_{ij}\, (\rho\, w_I, w_J) + \sum \partial_{tt} u_{rfIi}\, \delta_{ij}\, (\rho_f\, w_I, w_J)$$
$$+ \alpha\, Q\, \{\, \sum u_{rfIi}\, (\text{div } w_I\, e_i, \text{div } w_J\, e_j) + \alpha \sum u_{sIi}\, (\text{div } w_I\, e_i, \text{div } w_J\, e_j)\,\}$$
$$+ (\sigma'', \varepsilon(\, w_J \cdot e_j\,))_\Omega \quad = \quad (\rho\, g, w_J \cdot e_j\,)_\Omega + <T, w_J \cdot e_j >_{\Gamma\sigma}$$

then for the interphase conservation :

$$\sum \partial_{tt} u_{sIi}\, \delta_{ij}\, (\rho_f\, w_I, w_J) + \sum \partial_{tt} u_{rfIi}\, \delta_{ij}\, (\Xi\, \rho_f/n\; w_I, w_J) +$$
$$Q\, \{\, \sum u_{rfIi}\, (\text{div } w_I\, e_i, \text{div } w_J\, e_j) + \alpha \sum u_{sIi}\, (\text{div } w_I\, e_i, \text{div } w_J\, e_j)\,\}$$
$$+ \sum (K^{-1} \partial_t u_{rfIi}\, \delta_{ij}\, (w_I, w_J)$$
$$= \quad (\rho_f\, g\,, w_J\, e_j\,)_\Omega + <p\; w_J, e_j\, n>_{\Gamma p}$$

5.2 Implementation of $(u_s\text{-}p)h$

In a similar manner developements may be performed for the simplified (u-p) model where this time the solid displacement and the pore pressure are approximated :

$$u_{sh} = \sum u_{sIi} \cdot w_{sI} \, e_i, \qquad p_h = \sum p_I \cdot w_{fI}$$

and by choosing the same test functions for both solid displacement components and pore pressure another nonlinear differential system is obtained :

$$\sum \partial_{tt}u_{sIi} \, \delta_{ij} \, (\rho \, w_I , w_J) + (\sigma'', \epsilon(\, w_J \cdot e_j \,))_\Omega - \alpha \sum p_I \, (w_I , \, \text{div} \, w_J \, e_j \,)$$
$$= (\rho \, g, w_J \cdot e_j)_\Omega + <T, w_J \cdot e_j >_{\Gamma_\sigma}$$

$$- \sum \partial_{tt}u_{sIi} \, (\text{div} \, \rho_f \, Kw_I \, e_i, w_J) - \alpha \sum \partial_t u_{sIi} \, (\text{div} \, w_I \, e_i, w_J)$$
$$- \sum \partial_t p_I \, (1/Qw_I , w_J \,) - \sum p_I \, (K.\text{grad} \, w_I , \, \text{grad} \, w_J \,)$$
$$= < \varphi , w_J >_{\Gamma_\varphi} - \, (K\text{grad} \, (\rho_f \, g \, X), \, \text{grad} \, w_J \,)_\Omega$$

The finite element approximation corresponds to the space discretization. The final full time discretization is discussed in the next chapter.

6. MIXED IMPLICIT/EXPLICIT TIME DISCRETIZATION

In structural dynamics where different components may have different characteristic periods the concept of implicit/explicit partition has been introduced some years ago by Hughes et Liu(1978) especially with respect to finite element partitionning as opposed to nodal partitionning. It just turns out that element partitionning can easily be implemented in a very clear frame. It is why this choice has been made here. Obviously the main reason for developing such schemes is to circumvent the shortcomings of both approaches, mainly the hard stability requirements of explicit schemes difficult to satisfy for fine meshes and stiff sediments and the higher cost of implicit schemes due to the need to solve nonlinear systems of equations at each time step.

The mixed approach will be considered as above for both models $(u_s\text{-}u_{rf})h$ and (u-p)h. Let Ω, Ω_I, and Ω_E be respectively the whole, the implicit and the explicit domains. Let]0,T[be the time interval for the analysis, Δt the time step and n the step number n. Then the following quantities: $u_{sh}(n \, \Delta t)$, $\sigma''_{sh}(n \, \Delta t)$, $u_{rfh}(\, n\Delta t))$, $p_h(\, n\Delta t) \, ...$, will be approximated by the corresponding series : u_{shn}, σ''_{shn}, u_{rfhn}, p_{hn}. Furthermore in order to simplify the notations the index h which stands for the spatial approximation by finite elements will dropped from $u_{sh}(t)$, $\sigma''_{sh}(t)$, $u_{rfh}(\, t))$, $p_h(\, t)$.

6.1. The predictor-corrector Newmark scheme

The scheme is best introduced by formulating the usual Newmark method into two components. Let (β, γ), and the required series assumed to be known at time step n :$(u_{sn}, \sigma'_{sn}, v_{sn}, a_{sn})$, $(u_{rfn}, v_{rfn}, a_{rfn})$. The Newmark predictor is then given by :

$$u_{n+1}{}^* = u_n + \Delta t \, v_n + \Delta t^2 (1 - \beta/2) \, a_n$$
$$v_{n+1}{}^* = v_n + \Delta t \, (1 - \gamma) \, a_n$$
$$\sigma''_{sn+1} {}^* = \sigma''_{sn} + \Delta t \, D^{ep}: \varepsilon \, (u_{n+1}{}^* - u_n)$$

The Newmark corrector is on the other hand given by:

$$u_{n+1} = u_{n+1}{}^* + \Delta t^2 \, \beta \, a_{n+1}$$
$$v_{n+1} = v_{n+1}{}^* + \Delta t \, \gamma \, a_{n+1}$$
$$\sigma''_{sn+1} = \sigma''_{sn} + \Delta t \, D^{ep}: \varepsilon \, (u_{n+1} - u_n)$$

The mixed scheme will be built from these two operators.

6.2 Partitionning of the domain Ω

During the prediction phase of the computation the quantities $\{ u_{sn+1}, \sigma''_{sn+1}, \partial_t u_{sn+1}, \partial^2_{tt} u_{sn+1} \}$ are respectively replaced by : $\{ u_{sn+1}{}^*, \sigma''_{sn+1}{}^*, v_{sn+1}{}^*, a_{sn+1} \}$, with the corresponding operations for the fluid phase. During the correction phase in the implicit zone $\{ u_{sn+1}, \sigma''_{sn+1}, \partial_t u_{sn+1}, \partial^2_{tt} u_{sn+1} \}$ will be replaced by $\{ u_{sn+1}, \sigma''_{sn+1}, v_{sn+1}, a_{sn+1} \}$ In the explicit zone $\{ u_{sn+1}, \sigma''_{sn+1}, \partial_t u_{sn+1}, \partial^2_{tt} u_{sn+1} \}$ will be replaced by $\{ u_{sn+1}{}^*, \sigma''_{sn+1}{}^*, v_{sn+1}{}^*, a_{sn+1} \}$. Thus the correction in the explicit zone affects only the acceleration vector.

6.3. Time discretization of $(u_s\text{-}u_{rf})$

Following Chouvet (1983), Modaressi (1987) in the $(u_s\text{-}u_{rf})$ model the scheme is supplied underneath :

1. At time step (n+1), let : $\{ u_{sn}, \sigma''_{sn}, v_{sn}, a_{sn} \}$, $\{ u_{rfn}, v_{rfn}, a_{rfn} \}$ be known

2. <u>Predictor phase</u> : Compute both in Ω_E and in Ω_I the following estimators :

$$u_{sn+1}{}^* = u_{sn} + \Delta t \, v_{sn} + \Delta t^2 (1 - \beta/2) \, a_{sn}$$
$$v_{sn+1}{}^* = v_{sn} + \Delta t \, (1 - \gamma) \, a_{sn}$$

$$\sigma''_{sn+1}{}^* = \sigma''_{sn} + \Delta t \, D^{ep}: \varepsilon \, (u_{n+1}{}^* - u_n)$$

$$u_{rfn+1}{}^* = u_{rfn} + \Delta t \, v_{rfn} + \Delta t^2 (1 - \beta/2) \, a_{rfn}$$
$$v_{rfn+1}{}^* = v_{rfn} + \Delta t \, (1 - \gamma) \, a_{rfn}$$

Compute from them the accelerations :

$$a_{sn+1} = 1/(\Delta t^2 \beta) \, (u_{sn+1} - u_{sn+1}{}^*)$$
$$a_{rfn+1} = 1/(\Delta t^2 \beta) \, (u_{rfn+1} - u_{rfn+1}{}^*)$$
$$v_{sn+1} = v_{sn+1}{}^* + \gamma/(\Delta t \, \beta) \, (u_{sn+1} - u_{sn+1}{}^*)$$
$$v_{rfn+1} = v_{rfn+1}{}^* + \gamma/(\Delta t \, \beta) \, (u_{rfn+1} - u_{rfn+1}{}^*)$$

3. <u>Corrector phase</u> : Compute $\{ u_{sn+1}, \sigma''_{sn+1}, v_{sn+1}, a_{sn+1}\}$, $\{ u_{rfn+1},$ $v_{rfn+1}, a_{rfn+1}\}$ in Ω_I, Ω_E, such that :

$$(1/\beta\Delta t^2 \, (\rho \, u_{sn+1} + \rho_f \, u_{rfn+1}), w_s)\Omega$$
$$+ (\alpha \, Q \, \text{div} \, (u_{rfn+1} + \alpha \, u_{sn+1}) , \text{div} \, w_s)\Omega_I + (\sigma''_{sn+1} , \varepsilon(w_s))\Omega_I$$
$$= (\rho \, g, w_s)\Omega + <T_{n+1}, w_s> \Gamma_\sigma$$
$$+ (1/\beta\Delta t^2 \, (\rho u_{sn+1}{}^* + \rho_f \, u_{rfn+1}{}^*), w_s)\Omega - (\sigma_{sn+1}{}^*, \varepsilon(w_s))\Omega_E$$

$$(\rho_f /\beta\Delta t^2 \, (u_{sn+1} + \Xi /n \, u_{rfn+1}), w_{rf})\Omega$$
$$+ (Q \, \text{div}(u_{rfn+1} + \alpha u_{sn+1}) , \text{div} \, w_{rf})\Omega_I + (\gamma/\beta\Delta t K^{-1} u_{rfn+1}, w_{rf})\Omega_I$$
$$= (\rho_f \, g, w_{rf})\Omega + <p, w_{rf} \, n> \Gamma_p$$
$$+ (\rho_f /\beta\Delta t^2 \, (u_{sn+1}{}^* + \Xi /n \, u_{rfn+1}{}^*), w_{rf})\Omega - (p_{n+1}, \text{div} \, w_{rf})\Omega_E$$
$$+ (\gamma/\beta\Delta t \, K^{-1} \, u_{rfn+1}{}^*, w_{rf}) \, \Omega_I - (K^{-1} \, v_{rfn+1}{}^*, w_{rf})\Omega$$

Obviously the nonlinearity of the constitutive equation implies the solution of a nonlinear system in the implicit zone. Some comments on its solution are provided underneath.

When either Ω_E or Ω_I is void the scheme reduces to respectively the standard implicit or explicit method. It is easy to check that the so-called dynamic stiffness reduces to the mass matrix in the explicit zone and the standard dynamic stiffness in the implicit zone. In this manner if the mass matrix is chosen lumped in the explicit zone the profile of the dynamic stiffness will be considerably reduced on the overall.

6.4. Time discretization of $(u_s\text{-}p)$

The time discretization of the $(u_s\text{-}p)$ model proceeds along the same lines. Let $\{ u_{sn}, \sigma''_{sn}, v_{sn}, a_{sn} \}$, $\{ p_n, s_n \}$ be known from past computations.

1. Predictor phase : Compute

$u_{sn+1}{}^* = u_{sn} + \Delta t\, v_{sn} + \Delta t^2(1 - \beta/2)\, a_{sn}$

$v_{sn+1}{}^* = v_{sn} + \Delta t\, (1 - \gamma)\, a_{sn}$

$\sigma''_{sn+1}{}^* = \sigma''_{sn} + \Delta t\, D^{ep}{:}\, \varepsilon\,(u_{n+1}{}^* - u_n)$

$p_{n+1}{}^* = (1/2 + 1/\beta)\, p_n + (1/2 - 1/\beta)\, p_n{}^* + \Delta t\, s_n$

$s_{n+1}{}^* = 1/\gamma\, s_n + (1 - 1/\gamma)\, s_n{}^*$

$a_{sn+1} = 1/(\Delta t^2\, \beta)\, (u_{sn+1} - u_{sn+1}{}^*)$

$v_{sn+1} = v_{sn+1}{}^* + \gamma/(\Delta t\, \beta)\, (u_{sn+1} - u_{sn+1}{}^*)$

$s_{n+1} = s_{n+1}{}^* + \gamma/(\Delta t\, \beta)\, (p_{n+1} - p_{n+1}{}^*)$

2. Corrector phase: Compute $\{u_{sn+1}, \sigma''_{sn+1}, v_{sn+1}, a_{sn+1}\}, \{p_n, s_n\}$ from :

$(\rho/\beta\Delta t^2\, u_{sn+1}, w_s)_\Omega - (\alpha\, p_{n+1}, \operatorname{div} w_s)_{\Omega I} + (\sigma''_{sn+1}, \varepsilon(w_s))_{\Omega I}$

$= (\rho\, g, w_s)_\Omega + <T_{n+1}, w_s>_{\Gamma\sigma}$

$+ (\rho/\beta\Delta t^2\, u_{sn+1}{}^*, w_s)_\Omega - (\sigma_{n+1}{}^*, \varepsilon(w_s))_{\Omega E}$

$- (\alpha\, \operatorname{div} u_{sn+1}, q)_{\Omega I} - (1/Q\, p_{n+1}, q)\, \Omega_I - (\beta\Delta t/\gamma\, K \operatorname{grad} p_{n+1}, \operatorname{grad} q)_{\Omega I}$

$- (1/\gamma\Delta t\, K\rho_f\, u_{sn+1}, \operatorname{grad} q)_\Omega$

$= < \beta\Delta t/\gamma\, \varphi_{n+1}, q >_{\Gamma\varphi}$

$- (\beta\Delta t/\gamma\, K \operatorname{grad}(\rho_f\, g\, X), \operatorname{grad} q)_\Omega - (\operatorname{div} u_{sn+1}{}^*, q)_{\Omega I}$

$+ (\beta\Delta t/\gamma\, \operatorname{div} v_{sn+1}{}^*, q)_\Omega - (1/Q\, p_{n+1}{}^*, q)\, \Omega_I + (\beta\Delta t/\gamma\, Q\, \partial_t p_{n+1}{}^*, q)_\Omega$

$- (\beta\Delta t/\gamma\, K \operatorname{grad} p_{n+1}{}^*, \operatorname{grad} q)_{\Omega E} - (1/\gamma\Delta t\, K\rho_f\, u_{sn+1}{}^*, \operatorname{grad} q)_\Omega$

It is to be noticed that the last term is treated explicitely to keep a symetric form to the stiffness matrices. This factor is important for the stability of the mixed scheme. In the numerical experiments that have been conducted with the computer code **GEFDYN** (Aubry et al ., 1985) this destabilizing effect has not been noticed. In the frequency range which has been investigated no significant variation is implied by the omission of this factor but it should be recalled that when it is dropped out of the formulation then the Darcy law is no more of the "generalized" type.

7 .SOLUTION OF THE NONLINEAR SYSTEM

It has been shown for each model (u_s-u_{rf}) or (u_s-p) that the discretized variational formulation leads to the solution of a nonlinear system of equation at each time step as soon as the implicit domain is not void. The algorithm to solve these systems is now presented.

7.1 Solution of (u_s-u_{rf})

Assume that from previous computations the series at time step n $\{u_{sn}, v_{sn}, a_{sn}, \sigma''_{sn}, u_{rfn}, v_{rfn}, a_{rfn}\}$ is known. Then the updated series at time step (n+1) is looked for. In the iterative process m will stand for the iteration number. The following auxiliary elasticity matrix is introduced in the implicit zone (D^*) which will imply the corresponding elastic energy:

$$(D^* : \varepsilon(u_{sn+1}) , \varepsilon(w_s))_{\Omega_I}$$

The underneath notation is used for the increment of displacements either in the solid or in the fluid phase :

$$\Delta u_{n+1}^{m+1} = u_{n+1}^{m+1} - u_{n+1}^m$$
$$\Delta u_{n+1}^* = u_{n+1}^{*m} - u_{n+1}$$

Let b_s (.,.) be the following bilinear form :

$$b_s (\Delta u_{sn+1} , \Delta u_{rfn+1}) =$$
$$1/\beta\Delta t^2 ((\rho\Delta u_{sn+1} + \rho_f\Delta u_{rfn+1}), w_s)_\Omega + (\alpha Q div(\Delta u_{rfn+1} + \alpha\Delta u_{sn+1}), div w_s)_{\Omega_I}$$
$$+ (D^* : \varepsilon (\Delta u_{sn+1}) , \varepsilon(w_s))_{\Omega_I}$$

and b_{rf} (.,.) the corresponding one for the fluid equations :

$$b_{rf}(\Delta u_{sn+1}, \Delta u_{rfn+1}) =$$
$$1/\beta\Delta t^2 (\rho_f(\Delta u_{sn+1} + \Xi/n \Delta u_{rfn+1}), w_{rf})_\Omega$$
$$+ (Q div (\Delta u_{rfn+1} + \alpha \Delta u_{sn+1}), w_{rf})_{\Omega_I} + (\gamma/\beta\Delta t K^{-1} \Delta u_{rfn+1}, w_{rf})_{\Omega_I}$$

and finally let r_{sn+1}^m and r_{rfn+1}^m stand for the right hand sides defined by :

$$r_{sn+1}^m =$$

$$1/\beta\Delta t^2 (\rho\Delta u_{sn+1}^* + \rho_f\Delta u_{rfn+1}^*), w_s)_\Omega$$
$$- (\sigma_{sn+1}^*, \varepsilon(w_s))_{\Omega_E} - (\sigma_{sn+1}^m, \varepsilon(w_s))_{\Omega_I} + (\rho g, w_s)_\Omega + <T_{n+1}, w_s>_{\Gamma_\sigma}$$

$$r_{rfn+1}^m =$$

$$1/\beta\Delta t^2 (\rho_f(\Delta u_{sn+1}^* + \Xi/n \Delta u_{rfn+1}^*), w_{rf})_\Omega$$

$$+ (\gamma/\beta\Delta t \ K^{-1} \ \Delta u_{rfn+1}{}^*, w_{rf})\Omega_I - (K^{-1} \ v_{rfn+1}{}^*, w_{rf})\Omega$$

$$- (p_{n+1}{}^*, div w_{rf})\Omega_E - (p_{n+1}{}^m, div \ w_{rf})\Omega_I + (\rho_f \ g, w_{rf})\Omega + <p^\wedge, w_{rf}n> \Gamma_p$$

Then the nonlinear scheme may be described by the following steps :

◊ Initialization   (m = 0, step n+1) :

$$u_{\phi n+1}{}^0 = u_{\phi n} \qquad\qquad v_{\phi n+1}{}^0 = v_{\phi n} \qquad\qquad a_{\phi n+1}{}^0 = a_{\phi n}$$
$$\phi = s , rf$$

$$\sigma''_{n+1}{}^0 = \sigma''_n \qquad\qquad \alpha_{n+1}{}^0 = \alpha_n$$

◊ iteration m+1 :
Assume $\{u_{sn+1}{}^m, v_{sn+1}{}^m, a_{sn+1}{}^m, \sigma''_{sn+1}{}^m\}$, $\{u_{rfn+1}{}^m, v_{rfn+1}{}^m, a_{rfn+1}{}^m\}$

◊ Outer global iteration: compute $\{ u_{sn+1}{}^{m+1}$ et $u_{rfn+1}{}^{m+1} \}$ from :

$$b_s (\Delta u_{sn+1}{}^{m+1}, \Delta u_{rfn+1}{}^{m+1}) = r_{sn+1}{}^m$$

$$b_{rf} (\Delta u_{sn+1}{}^{m+1}, \Delta u_{rfn+1}{}^{m+1}) = r_{rfn+1}{}^m$$

◊ Inner local iteration : compute$\{ \sigma''_{n+1}{}^{m+1}, \alpha_{n+1}{}^{m+1} \}$ from constitutive eq.

◊ Convergence criterion : stop when simultaneously

$$\| u_{sn+1}{}^{m+1} - u_{sn+1}{}^m \| \le Utol \cdot \|u_{sn+1}{}^{m+1}\|$$
$$\| u_{rfn+1}{}^{m+1} - u_{rfn+1}{}^m\| \le Utol \cdot \| u_{rfn+1}{}^{m+1} \|$$

$$\| r_{sn+1}{}^{m+1} \| \le Rtol \cdot \| r_{sn+1}{}^0 \| \ and \qquad \| r_{rfn+1}{}^{m+1} \| \le Rtol \cdot \| r_{rfn+1}{}^0 \|$$

7.2. Solution of (u-p)

The solution of the nonlinear system for the (u-p) model can be described along the same lines for the following series : $\{u_{sn}, v_{sn}, a_{sn}, \sigma''_{sn}, p_n, s_n\}$

$$\Delta u_{n+1}{}^{m+1} \ = \ u_{n+1}{}^{m+1} - u_{n+1}{}^m$$
$$\Delta p_{n+1}{}^{m+1} \ = \ p_{n+1}{}^{m+1} - p_{n+1}{}^m$$
$$\Delta u_{n+1}{}^* \ = u_{n+1}{}^* - u_{n+1}{}^m$$
$$\Delta p_{n+1}{}^* \ = p_{n+1}{}^* - p_{n+1}{}^m$$

The bilinear forms are now defined respectively by :

$$b_s (\Delta u_{sn+1}, \Delta p_{n+1}) =$$
$$1/ \beta \Delta t^2 (\rho \ \Delta u_{sn+1}, w_s)$$
$$- (\alpha \ p_{n+1}, \text{div } w_s)_{\Omega I} + (D^* : \epsilon(\Delta u_{sn+1}) , \epsilon(w_s))_{\Omega I}$$

$$b_{rf} (\Delta u_{sn+1} , \Delta p_{n+1}) =$$
$$- (\alpha \text{ div } \Delta u_{sn+1}, q)_{\Omega I} - (1/Q \ \Delta p_{n+1}, q) \ _{\Omega I}$$
$$- (\beta \Delta t/\gamma \ \mathbf{K} \text{grad } \Delta p_{n+1}, \text{grad } q)_{\Omega I}$$

and the right hand sides by :

$$r_{sn+1}{}^m =$$
$$1/\beta \Delta t^2 (\rho \Delta u_{sn+1}{}^*, v_s) - (\sigma_{sn+1}{}^*, \epsilon(w_s))_{\Omega E} - (\sigma_{sn+1}{}^m, \epsilon(w_s))_{\Omega I}$$
$$+ (\rho \ g, w_s)_\Omega + <T_{n+1}, w_s>_{\Gamma\sigma}$$

$$r_{rfn+1} =$$
$$- \quad (\text{div } \Delta u_{sn+1}{}^*, q) \ _{\Omega I} + (\beta \Delta t/\gamma \text{ div } v_{sn+1}{}^* , q)_\Omega$$
$$- (1/Q \ \Delta \ p_{n+1}{}^*, q) \ _{\Omega I} + (\beta \Delta t/\gamma \ Q \ s_{n+1}{}^*, q)_\Omega$$
$$- (\beta \Delta t/\gamma \ \mathbf{K} \text{ grad } p_{n+1}{}^*, \text{ grad } q)_{\Omega E} - (\beta \Delta t/\gamma \ \mathbf{K} \text{ grad } p_{n+1}{}^m, \text{ grad } q)_{\Omega I}$$
$$+ < \beta \Delta t/\gamma \ \varphi_{n+1}, q>_{\Gamma\varphi} - (\beta \Delta t/\gamma \ \mathbf{K} \text{ grad}(\rho_f \ g \ X), \text{grad} q)_\Omega$$
$$- (1/\Delta t \ \gamma \ \mathbf{K} \ \rho_f \ \Delta \ u_{sn+1}{}^*, \text{ grad } q)_\Omega$$

The iterative process consists in almost the same phases and will not be described furthermore here (see H. Modaressi (1987)).

7.3 Local integration of the constitutive behaviour at the Gauss points

Due to the complexity of the cyclic behaviour of the soil the computation of the effective stresses at time step (n+1) must be carefully designed. By using the above global scheme we must compute locally at each integration station the new effective stresses σ'_{sn+1} with the history quantitites at time step n known and the estimated displacement $u_{n+1}{}^{m+1}$ coming from the iteration number (m+1). From this quantity the rate of deformation $\partial_t \epsilon$ may be estimated :

$$\partial_t \epsilon \cong [\ \epsilon \ (u_{n+1}{}^{m+1}) - \ \epsilon \ (u_n)] / \Delta t$$

and then the incremental behaviour must be solved for the time interval $t_n \leq t \leq t_{n+1}$:

$$\partial_t \sigma'(t) \quad = \quad \mathbf{D}^{ep} \left(\sigma'(t), \alpha(t) , \partial_t \varepsilon \right) : \partial_t \varepsilon$$

$$\partial_t \alpha (t) \quad = \quad \mathbf{L} \left(\sigma'(t), \alpha(t) , \partial_t \varepsilon \right)$$

with the initial values :

$$\sigma'(t_n) \quad = \quad \sigma'_n$$

$$\alpha(t_n) \quad = \quad \alpha_n$$

This differential system is solved by using subtime steps Δt^* with an implicit Euler scheme with radial return and substep number k :

$$\varepsilon^* = \varepsilon \, \Delta t^* \qquad \sigma'_k = \sigma'(k \, \Delta t^*)$$

$$\sigma'_k - \sigma'_{k-1} \quad = \quad \mathbf{D}^{ep} \left(\sigma'_k, \alpha_k , \varepsilon^* \right) : \varepsilon^*$$

$$\alpha_k - \alpha_{k-1} \quad = \quad \mathbf{L} \left(\sigma'_k, \alpha_k , \varepsilon^* \right)$$

$$f \left(\sigma'_k, \alpha_k \right) \leq 0$$

The influence of an accurate integration of the local behaviour is important for the global convergence speed. The substep must be adaptively estimated because in some areas it may be chosen safely equal to the global time step Δt while in areas where the strain path is more complex many subtime steps may be required. The projection of the fictive elastic stress increment with respect to the normal to the last activated yield surface must be such that the value of this yield surface is small enough to avoid a drift error which usually has an important destabilization effect on the overall computations.

7. ELASTOPLASTIC SOIL BEHAVIOUR

Only a few remarks are addressed towards a unified approach mainly by using the level of strain as an indicator to the type of behavior because it has a major influence on cyclic soil properties and in dynamic analysis the strain range may be extremely wide. Experimental results show that when the cyclic shear strain amplitude is less than 10^{-6} the soil response is rather reversible and non linear elasticity due to the effects of the confinement may describe rather well the stress-strain curve. Between 10^{-6} and 10^{-4} irreversible deformations begin to appear but the cycles are rather stable. Then an approach like the linear equivalent moduli with hysteretic damping may be of some value. For larger shear strain amplitudes the loops get modified at each cycle due either to the densification of the material under drained conditions or to the pore pressure increase under undrained conditions.

The prediction of the pore pressure increase and of permanent deformations can only be done by using an incremental constitutive equation in the framework of elastoplasticity theory with due account of the evolution of the internal variables like the density of the material. In such an approach stress increments are related to strain increments the relationship being driven by a yield function such as the following :

$$f^m \, (\, \sigma', r_n, e, n \,) \; = \; \| \, \tau_n \, \| \; + \; \sigma'_{nn} \; r_n \; F \, (\, p, e \,) \; tg \, \phi$$

where σ', r_n, e, ϕ, n stand respectively for the effective stresses, the degree of mobilization of the shear friction mechanism, e the void ratio, ϕ the friction angle of the material and n the unit normal to a plane in the case of directional properties. The function $F \, (\, p, e \,)$ is introduced to take into account the so-called critical state of the material. Essentially F tends to unity when the current void ratio goes to its critical void ratio which is a fundamental property of the solid skeleton at a given confining pressure.

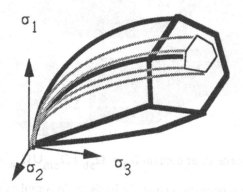

Figure 4: Cyclic and monotonous yield surfaces

As shown in the above picture the generalization to cyclic loading can be performed by using cyclic loading functions with a memory of the last loading reversal. Several soils have been described with such a constitutive equation including sands and clays and a good agreement with observed experimental investigations is generally obtained. From the point of view of the coupling between the solid skeleton and the pore fluid the two following fundamental aspects of the model should be noted :

  - effective stresses are used in the above equations which is unavoidable if basic soil behaviour is to be described,
  - a dilatancy rule is incorporated in the plastic strains which is also fundamental to predict correctly the volumetric strains.

As mentionned at the very beginning of this section space limitations do not allow to discuss more precisely these topics and the reader is referenced to Aubry et al.(1982).

8. REMARKS ON ABSORBING BOUNDARY CONDITIONS

In the same spirit as in the preceeding paragraph some indications are presented on the modelling of diffraction problems which occur in earthquake engineering where the loading is not explicitely given. The seismic event is defined by an incident displacement field u_i which is assumed to be known beforehand and which is compatible with the outer domain Ω_s' but of course not coherent with the heterogeneity which are present in the inner domain filled with a porous material modelled either by the (u_s-u_{rf}) or (u-p) equations. The inner domain is assumed to be bordered by an ealstic domain where a simplified version of wave propagation is assumed.

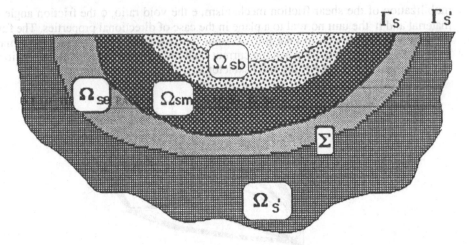

Figure 6 : The composite inner domain $\Omega_s = \Omega_{sb} \cup \Omega_{sm} \cup \Omega_{se}$ and the outer domain Ω_s'

This approximate wave propagation equation leads in the simplest case to the following relation between the displacement at the boundary and the traction vector for outwards propagationg waves :

$$t_{s'} = A_0(\partial_t u_s)$$

Obviously this relation is of the dashpot type. Nevertheless for higher order approximations more complex equations come out. The radiated field is defined by:

$$u'_r = u_{s'} - u_i \quad \text{satisfying the Sommerfeld radiation condition}$$

The preceeding relation must be applied to the radiated field only :

$$t_{s'} (u'_r) = A_0(\partial_t u'_r)$$

But the stress vector reciprocity relation on Σ gives :

$$t_s (\partial_t u_s) = - t_{s'} (u'_i) - A_0(\partial_t u_s) + A_0(\partial_t u'_i)$$

Now the principle of virtual work in the inner domain becomes :

$$(\rho\, \partial_{tt} u_s\, ,\, w\,)_{\Omega sm} + (\sigma_s\, ,\, \varepsilon(w)\,)_{\Omega sm} - <t_s,\, w>_\Sigma = (\,\rho\, g\, ,\, w\,)_{\Omega sm}$$

which by taking into account the impedance relation leads to :

$$(\rho\,\partial_{tt}u_s\,,w\,)_{\Omega sm} + (\,\sigma_s\,,\epsilon(w)\,)_{\Omega sm} + <A_0(\partial_t u_s)\,,w>_\Sigma$$
$$= (\,\rho\,g\,,w)_{\Omega sm}\ + <\cdot\ t_{s'}(u'_i) + A_0(\partial_t u'_i)\,,w>_\Sigma$$

The preceeding equation shows clearly how the seismic loading can be described with at the same time a damping matrix and a right hand side where both the distribution of the incident displacement and the incident traction vectors are to be known.

Applications and numerical results for the analysis of seismic site effects are presented in (Aubry and Modaressi, 1988)

REFERENCES

1. Aubry D., Hujeux J.C., Lassoudière F. & Meimon Y. (1982) A double memory model with multiple mechanisms for cyclic soil behaviour, Int. Symp. Num. Mod. Geomech., Balkema.3-13.
2. Aubry D., Chouvet D., Modaressi A. & Modaressi H. (1985) GEFDYN_5 : Logiciel danalyse du comportement statique et dynamique des sols par éléments finis avec prise en compte du couplage sol-eau-air. Rapport Scientifique Ecole Centrale de Paris.
3. Aubry D., Modaressi H. (1988) Numerical modelling of the dynamics of saturated anelastic soils, Seismic hazards in Mediterranean regions (Ed. J. Bonnin, M. Cara, A. Cisternas) Kluwer, Dordrecht, p 151-172.
4. Auriault J.L., Sanchez Palencia E. (1977) Etude du comportement macroscopique d'un milieu poreux saturé déformable. J. Mécanique, vol 16, 4, 573-603.
5. Bear J., Pinder G.F. (1983) Porous medium deformation in multiphase flow. J. Geotech. Eng., 109, 5, 736-737.
6. Chouvet D. (1983) Calcul elastoplastique d'interaction dynamique sols-Structure ; Application aux ouvrages de soutenement. Doctoral thesis. Ecole Centrale de Paris.
7. Hughes TJR, Liu IK (1977) Implicit explicit finite elements in transient analysis : stability theory. J. Applied Mechanics. Vol 45, 371-375
8. Modaressi H. (1987) Modélisation numérique de la propagation des ondes dans les milieux poreux anélastiques, Doctoral Dissertation , Ecole Centrale de Paris.
9. Park K.C. (1983) Stabilization of partitionned solution procedures for pore-fluid interaction analysis Int. J. Num. Eng., vol 19, 1669-1673.
10. Zienkiewicz O.C. & Shiomi T. (1984) Dynamic behaviour of saturated porous media ; The generalized Biot formulation and its numerical solution. Int. Jour. Num. and Anal. Meth. Geomech. vol 8,71-96.

IMPLEMENTATION OF MODERN CONSTITUTIVE LAWS AND ANALYSIS OF FIELD PROBLEMS

I.M. Smith
University of Manchester, Manchester, UK

DEVELOPMENT OF MODERN CONSTITUTIVE LAWS AND
ANALYSIS OF FIELD PROBLEMS

ABSTRACT

Nonlinear material models are being used increasingly in the analysis of complex
geotechnical engineering works. Such analyses have to answer, if possible, two basic
questions. Firstly, what is the nature of the ultimate or limit state of the works and,
secondly, what are the likely deformations under normal loading conditions.

The type of calculation to be carried out is likely to differ from project to project.
In some cases it will be necessary to use more complicated material models, whereas
in others, relatively simple nonlinear material models will suffice.

1. INTRODUCTION

1.1 Constitutive Laws

Despite many years of research, it is still common geotechnical engineering practice to use different constitutive laws for the limit state and for the serviceability state of a construction. In the former case a rigid–plasticity law is still typical and in the latter a linear elasticity law. This is in spite of the observation that the factors of safety which apply in geotechnical engineering are often much lower than those which apply in, for example, structural engineering. Therefore, the departure from linearity in the serviceability state can be quite pronounced.

In these lectures it is assumed that the minimum complexity which will be permitted in analyses involves a linear elastic–perfectly plastic constitutive law. In principle, therefore, both the limit state and the serviceability state can be computed using the same model. This will be demonstrated for a number of field problems.

For certain other problems, however, usually those involving excess water pressures in the soil, adequate limit state solutions will not result from the use of "simple" models such as the linear elastic–perfectly plastic group. In these problems, we must resort to more "complicated" constitutive models for soil, which describe in some detail the stress–strain–volume change characteristics of various skeletal materials in various initial states. When drainage is absent, or partial, then volume change (dilatancy) characteristics will give rise to excess porewater pressures (positive or negative) which will dominate the performance of the geotechnical construction. The lectures are arranged in four parts. The first deals with a description of the constitutive laws – both "simple" and "complicated" – which will be used. In the other three parts, different field problems are addressed. "Simple" models are applied to problems of deep excavations, and then to problems of reinforced earth. In the final part, a "complicated" model is applied to "deep" and "shallow" foundation problems.

2. CONSTITUTIVE LAWS

2.1 The simplest acceptable law

Given that we seek to compute both the serviceability and the limit states in a single calculation, the simplest acceptable stress–strain law will include a measure of stiffness, that is the rate of increase of strain with stress at low load levels – the serviceability state – and a measure of strength, that is the stress condition which cannot be exceeded anywhere in the soil mass at high stress levels – the ultimate state. Figure 1 shows typical data from a triaxial test conducted on a loose sand at a given confining pressure both in the drained (no excess porepressure) and the undrained states. When data for various confining pressures are assembled, limiting stress conditions for drained sands are found to be reasonably well approximated by the Mohr–Coulomb surface shown in Figure 2. A simple constitutive law for drained sand might therefore assume linear elastic behaviour for any stress point inside the Mohr–Coulomb surface and perfectly plastic behaviour for any stress point situated on the Mohr–Coulomb surface. When such a law is fitted to test data, as shown in Figure 1, "reasonable" fits can be obtained both to the observed deviator stress versus axial strain curve, and to the volumetric strain versus axial strain curve as shown.

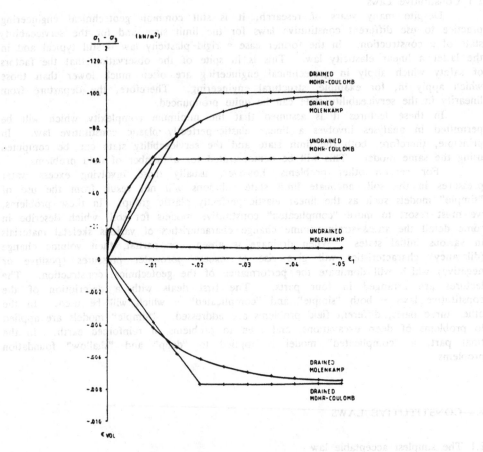

1.   Drained and undrained triaxial compression tests -
       'simple' and 'complicated' models.

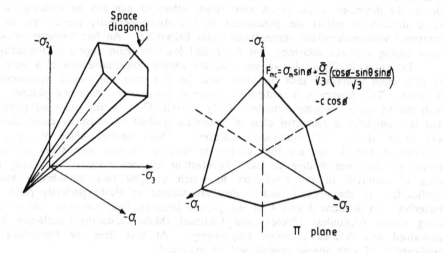

2.   Mohr-Coulomb failure criterion.

The bilinear approximation to the real curve does of course involve compromise but, given that we seek to fit all states between the serviceability and the limit, further sophistication is of doubtful value. If, it goes without saying, a design is so safe that the strains are tiny and the limit state never remotely approached, a far stiffer elastic modulus would be appropriate.

Data from drained triaxial tests on a denser sand are shown in Figure 3, together with simple bilinear curve fits using a constant elastic stiffness. Although "reasonable" fits to the deviatoric stress versus axial strain behaviour can still be achieved, it is now more difficult to adjust the parameters in the elastic–perfectly plastic law to fit the measured volumetric–strain versus axial strain behaviour. An improvement would result from adding a plastic dilatancy, but the model loses its main appeal of simplicity.

Returning to Figure 1, we see that the simple constitutive law is a very bad fit to data from undrained tests on loose sand (in this case the overall volumetric strain is zero). This feature is amplified in Figure 4 where more data are plotted, together with the fit using the simple model. Only material C is in any sense adequately fitted and it exemplifies a restricted class of materials typified by normally consolidated clays and sands at "critical" confining pressures. These neither dilate strongly, as do materials A and B, nor are they subject to skeleton collapse with super–generation of excess porepressures leading to static liquefaction as does material D. They are the class of materials in soil mechanics to which a $\varphi_u = 0$ limit equilibrium analysis is applicable. In the present context they are treated as elastic–perfectly plastic Tresca materials. In Sections 3 and 4 of the present lectures, field problems will be analysed using simple frictionless (Tresca) and frictional (Mohr–Coulomb) constitutive laws for undrained and drained problems respectively. At that time the limitations to the applicability of such simple models will be discussed.

2.2 "Complicated" constitutive laws

In the previous sub–section, we saw that "simple" constitutive laws could only partially capture the real stress–strain–volume change characteristics of soils. In particular, linearly elastic–perfectly plastic models represented volume change characteristics of the skeleton badly and because of the effective stress principle, this meant in general poor modelling of the undrained state. Although dilation could be modified in the plastic region, the additional complexity seemed hardly justified in most cases.

Instead, it is preferable to use one of the many constitutive laws specially developed to describe the general stress–strain–volume change behaviour observed in tests on soil. Boundary value problems in the field can then be analysed, to see whether the main interesting aspects of field behaviour have been captured.

At a recent symposium [1], over 30 such "complicated" constitutive models were presented, and used in a predictive role to forecast the results of laboratory tests on different sands. In general these predictions were rather good and while some were definitely superior to others, amongst the leading group there were at least 10 perfectly adequate models. What matters now is the incorporation of any of these models into computer programs for the solution of boundary value problems.

In Reference [1], the main distinction was between models employing elasto–plasticity with yield surfaces, and models which did not require the notion of a yield surface for their definition. The ratio of the former to the latter was about 2:1. The model favoured by the writer is elasto–plastic with two yield surfaces – one of the so–called "double–hardening" group. Figure 5 shows the failure surface which

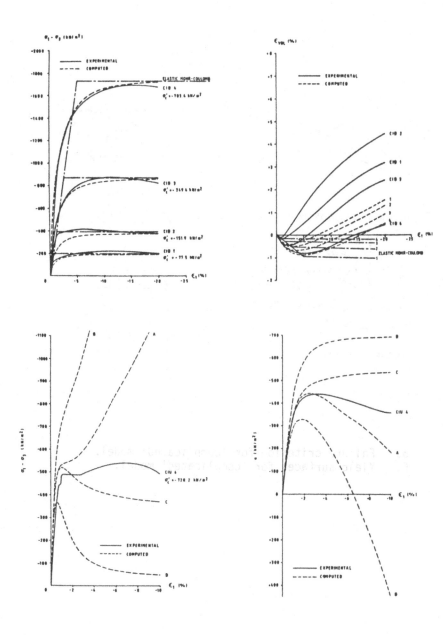

3. Drained triaxial compression tests - 'simple' and 'complicated' models.
4. Undrained triaxial compression tests - 'complicated' model.

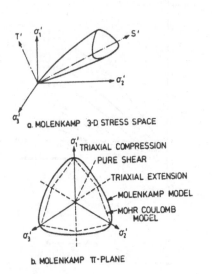

a. MOLENKAMP 3-D STRESS SPACE

b. MOLENKAMP π-PLANE

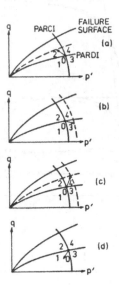

5. Failure criterion for 'complicated' model.
6. Yield surfaces for 'complicated' model.

can be compared with the "simple" one of Figure 2. The additional complications involve curvature of the failure surface both in the "deviatoric plane" and in the "π-plane". In the former case this enables simulation of reduced soil shear strengths (T) at high mean normal pressure (S') and in the latter case the simulation of higher soil shear strengths in plane strain and in triaxial extension compared with triaxial compression. However, the curved failure surfaces can be reduced to the "simple" straight Mohr–Coulomb ones of Figure 2 as a special case.

Figure 6 shows the most important additional complexity in the "complicated" model. Whereas in Figure 2, all stress points inside the Mohr–Coulomb envelope represented elastic soil, in Figure 6 there are two curved yield surfaces inside the failure surface. From some initial state, for example with both yield surfaces collapsed to lines and points on the S' axis ($K_0=1$) or on a line T/S'=constant, yielding and plastic flow can begin from the initiation of loading. Thus there need be no elastic zone at all, although elastic behaviour results on subsequent unloading with respect to either yield surface. Figure 6 shows a stress probe involving loading with respect to the deviatoric surface and unloading with respect to the isotropic surface. By positioning the yield surfaces beyond the current stress point, an initial elastic region is created, thus modelling "overconsolidated", "cemented" and other such initial states.

A useful feature of the presence of two yield surfaces is that it simplifies the specification of a test programme to isolate the parameters which define the constitutive model. The minimum of tests required comprises drained isotropic compression with unloading, drained triaxial compression at 3 confining pressures with unloading and undrained triaxial compression.

The characteristics of the model have been extensively described elsewhere in References [2] to [12]. Returning to Figure 3, the much improved modelling of both stiffness and volumetric strains by the "complicated" model are evident. In the final section of this paper, its use in the solution of foundation problems will be described. Before that, we return to analysis of field situations using the "simple" constitutive laws.

3. ANALYSIS OF EXCAVATIONS IN UNDRAINED CLAY

3.1 Introduction

In Section 2 we saw that one particularly appropriate application of "simple" soil models was to the analyses of undrained, normally consolidated, clays. These materials have a shear strength which depends only upon previous history and not on subsequent loading (the so-called "$\varphi_u=0$" effect). Since deformation is undrained, there is essentially no volume change which can be achieved in the elastic domain by setting Poisson's ratio to nearly 0.5 and, in the plastic range, by adopting the associated flow rule to the Tresca yield (failure) criterion. The only parameters needed for the calibration of the constitutive model are the undrained strength, C_u, and a modulus of elasticity, E or G for example. For preliminary analyses, estimates of the ratio E/C_u are available, even in the absence of more detailed measurements.

3.2 The Limit State

At the outset, we emphasised that possibly the most important attribute of any deformation, e.g. finite element, analysis is its ability to capture well "collapse" or "limiting" conditions. For vertical cuts in clay the geometry of a walled excavation is illustrated in Figure 7. Also illustrated in the Figure (Reference [13]) are various limit analysis solutions for braced excavations. For example, for the case shown in Figure 7(ii), the depth at which base heave failure should occur has been estimated [14] to be

$$D_f = \frac{N_c C_u - p}{C_u} \qquad (D' > 0.7B, \ D = D_f - H)$$
$$\gamma - \frac{C_u}{0.7B}$$

where p represents any overburden pressure and the value of N_c is approximately 5.

At the other extreme, a vertical cut could be completely open, i.e. without any wall, in which case estimates of the limiting depth are:

$$D_f = \frac{4C_u}{\gamma}$$

assuming a plane failure surface orientated at 45^0 passing through the wall, and:

$$D_f = \frac{3.85C_u}{\gamma}$$

assuming a circular wall slip surface.

When these problems are analysed, using even a crude mesh containing less than 100 eight-noded isoparametric finite elements, the results in Figure 8 are obtained for the open excavation. Collapse occurs at $D_f \simeq 4C_u/\gamma$ and inspection of the displacement vectors close to failure, Figure 9, reveals that the finite element solution appears to favour the 45^0 plane mechanism in this case.

Very different displacement mechanisms are shown for walled excavations in Figure 10. The "unsupported" wall is free to move, in both directions, whereas the "fixed" wall is restrained from any movement in the horizontal direction. When the excavation process is analysed, Figure 11 shows the collapse of the excavation by base heave. The wall fixed against movement in the horizontal direction experiences failure at depths close to those previously quoted from Reference [14]. For an "unsupported" wall, failure depths are intermediate between the "fixed" wall and "completely open" excavations respectively. There are also secondary effects of Poisson's ratio and of the ratio of total horizontal to total vertical stress, termed K_0 in the Figure.

3.3 The Serviceability State

The progress of deformations towards failure are shown in Figure 12 for the unsupported wall and in Figure 13 for the wall fixed against movement in the horizontal direction.

Detailed pictures of ground and wall movements for the two extreme cases are shown in Figures 14 and 15. These illustrate how the unsupported wall fails eventually by a combination of base heave and of the wall "kicking" into the excavation. The fixed wall, of course, fails completely by a base heave mechanism.

Turning to earth pressures, Figure 16 shows how the earth pressures on the wall, both "active" and "passive", vary with excavation depth. No tension has been allowed to develop in the soil or between pile and soil. In the Figure, α is the proportion of C_u which is theoretically mobilised as wall adhesion.

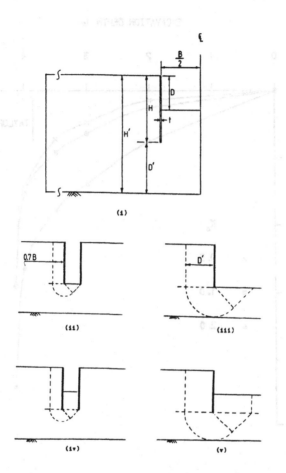

7.   Mechanisms of base-heave failure for deep excavations.

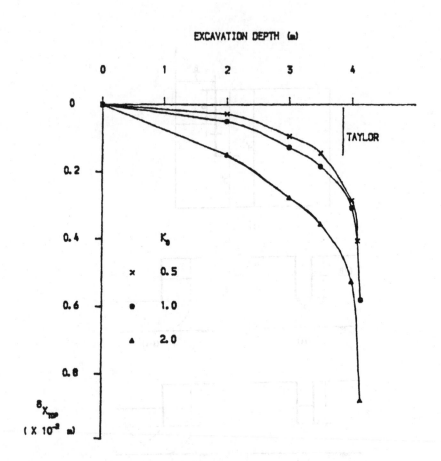

8.   Lateral deflection versus excavation depth for
     different initial stress conditions (plane strain).

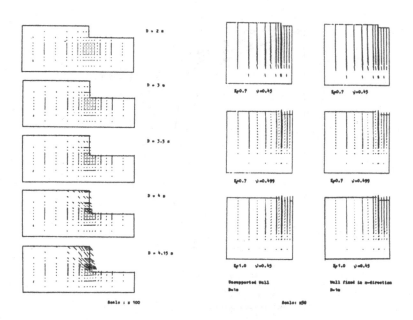

9. Displacement vector plots at various excavation depths (plane strain, $K_0 = 1$).
10. Displacement vector plots. Effect of K_0 and ν for different support conditions.

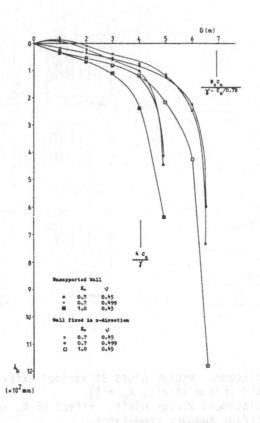

11. Base heave against excavation depth. Effect of K_o
and ν for different support conditions.

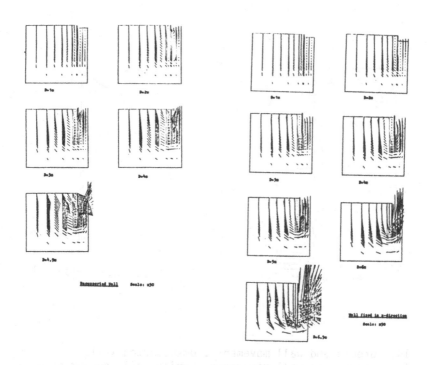

12. Displacement vector plots with excavation depth -
    unsupported wall.
13. Displacement vector plots with depth - horizontally
    supported wall.

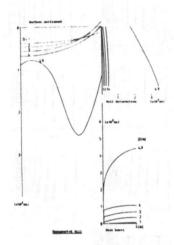 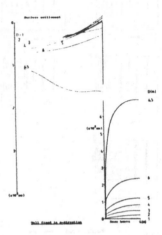

14. Ground and wall movements, unsupported wall.
15. Ground and wall movements, horizontally supported wall.

For an unsupported wall, bending moments vary with depth during excavation, as illustrated in Figure 17. The foregoing results indicate that a finite element analysis involving a "simple" constitutive relationship, is capable of reproducing all the salient features of excavations in undrained, normally consolidated clays. The same program, based upon Reference [15], has been successfully used to simulate a field situation [13]. In the latter case, the problem is considerably more complex, due to the construction process of inserting struts as the excavation proceeds. One then has the difficulty of estimating the strut stiffness, which may vary a great deal from the nominal one due to the method of installation [16]. Nevertheless, all of these factors can be simulated, at least to give upper and lower bounds to potential performance.

The approach is only accurate as long as the "undrained" assumption is valid. In practice, it is also possible for tension cracks to form, and for free water to exist in any such crack. In the "long term" it would be possible to analyse the wall as standing in a frictional (c', φ') soil, subject to the Mohr–Coulomb failure criterion. It could still happen that the wall would experience a critical stress condition in the interim, partially drained, state. Such an anlysis is considerably more difficult, and would best be approached using a "complicated" soil model.

4. ANALYSIS OF REINFORCED EARTH RETAINING WALLS

4.1 Introduction

In the previous section we saw that a "simple" constitutive relationship was adequate for the analysis of excavations in saturated, normally consolidated, undrained clay. The most important aspect of analysing field problems did not lie in the details of the constitutive relationship, but rather in modelling the sequence of construction. Totally different results would be obtained were a wall to be analysed as "unsupported" or "horizontally fixed" and by implication for all the intermediate possible strutting configurations. It would be completely inappropriate to conduct experiments, say using a centrifuge, in which it was not possible to follow the construction sequence quite closely – insertion of struts in some specific order, quality of the strutting installation, and so on.

In the present section we deal with another problem which is dominated by construction sequence, namely that of building reinforced earth retaining walls. These would mainly involve the use of cohesionless material for the fill, and exist in a drained state (no excess porewater pressures) and so for a first analysis, one could think of adopting a "simple" constitutive relationship of the elastic–Mohr Coulomb plastic variety. As we saw in Section 2, this will not give a very good picture of dilatancy in the fill surrounding the reinforcement. Nor will it cope well with compaction stresses and their residual once the compaction equipment has been removed. However, it will allow an insight into how a particular reinforcement system "works", depending upon the nature of the reinforcement (strip, sheet, etc.), its flexibility, and on the density of its spacing. It may be noted that the computation of the limit state will be of considerable interest since there are several potential modes of collapse for any wall configuration and these modes may be activated and then deactivated as deformation proceeds, making any simple application of partial factors of safety a subjective matter.

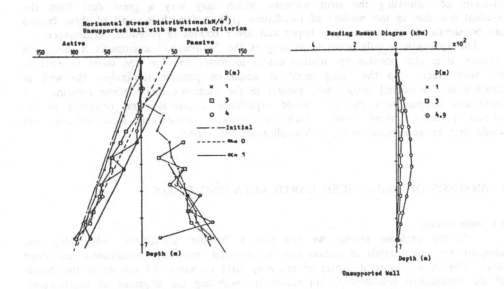

Horizontal Stress Distributions(kN/m²)
Unsupported Wall with No Tension Criterion

Active   Passive

Bending Moment Diagram (kNm)

Unsupported Wall

16. Active and passive earth pressures during excavation
(upsupported wall)
17. Bending moments during excavation (unsupported wall)

4.2 Two–dimensional versus three–dimensional analyses

For sheet reinforcement, a situation of near to plane strain will exist and so a two–dimensional analysis will be essentially valid. For strip reinforcement, the demands on computer storage and power are such that, although the system is essentially three–dimensional, the temptation to try to reduce to two dimensions has been very strong. This has led to various approximate analyses, for example [17].

In the present lectures, this simplification is avoided, and sheet–reinforced systems are analysed in two dimensions, while strip–reinforced systems are analysed, as they really are, in three dimensions.

4.3 Sheet–reinforced walls : limit state

Analysis of the reinforced soil problem is dominated by representation of the "interface" region, that is a region of narrow but indeterminate thickness (possibly 20 soil grain diameters upwards) immediately adjacent to the reinforcement. While many special "interface elements" have been suggested, mesh design is made much easier if the same elements can be used for soil, interface and reinforcement. Only the material properties need then differ.

Figure 18 shows such a mesh of 8–noded, two–dimensional isoparametric elements (in plane strain), which are arranged to simulate a pull–out test, on sheet reinforcement embedded in a frictional soil (c'=0, φ'=30^0). The load–deflection curves are shown in Figure 19 where it can be seen that clear ultimate pullout loads are achieved, consistent with the interface failure criterion $\tau_{sn}/\sigma_n=\sin\varphi'$, assuming also that the initial K_0 remains constant at 1.0 [13]. A further important aspect of mesh design is that the reinforcement must be able to separate or "de–bond" from the soil at its embedded end. We must recall that in our "simple" soil constitutive model, no allowance has been made for the presence of dilatant volumetric strains, which will almost invariably exist. However, as a first step, load–deflection curves such as those in Figure 19 can be fitted to the results of experiments. Note that in the Figure, two different calculation methods have been used and yield more or less identical results [15].

4.4 Sheet–reinforced walls : serviceability state

Having established accurate limiting conditions for the behaviour of individual strips, we may, using the finite element method, build up a reinforced wall as shown in Figure 20. There are many variables in the construction, concerning the stiffness of the wall facing, the stiffness of the sheet reinforcement, and its spacing. Some typical results for wall displacements, sheet tensile stresses, earth pressures and lines of maximum tension in the sheets are shown in Figures 21 and 22, which model a stiff wall with extensible sheeting, and a flexible wall with inextensible sheeting respectively. The pictures show how wall and reinforcement respond to the construction sequence. Progress of yield towards the limiting state can be assessed for the two different walls in Figures 23 and 24 (stiff/extensible and flexible/inextensible respectively). In the former case, an "active" zone forms in the vicinity of the wall whereas in the latter, a "block" failure is seen to form behind the reinforced section. In the present analyses, unyielding sheeting has been modelled, but this aspect can of course be handled in the analyses as well [13].

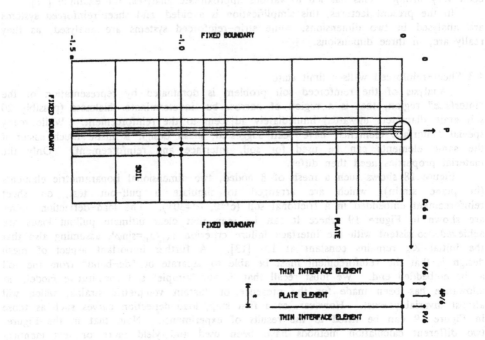

18. Mesh for vertical pull-out test (2-d).

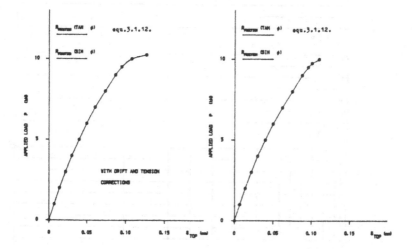

19. Pullout load-deflection curves.

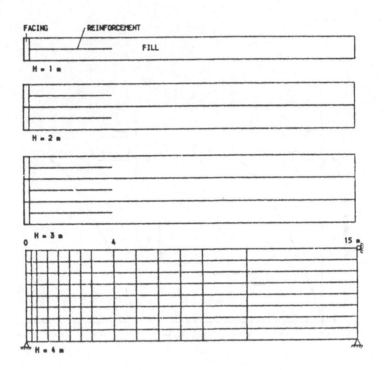

20.   Wall construction meshes (2-d).

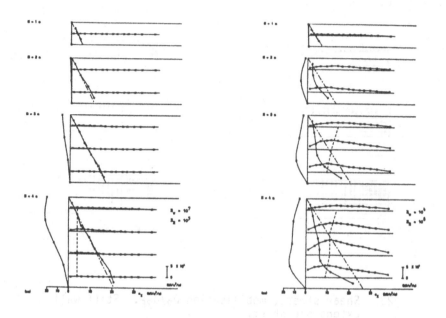

21. Wall deflections, earth pressures, strip tensions,
    line of maximum tension.   Stiff wall, extensible sheet.
22. Wall deflections, earth pressures, strip tensions,
    line of maximum tension. Soft wall, inextensible sheet.

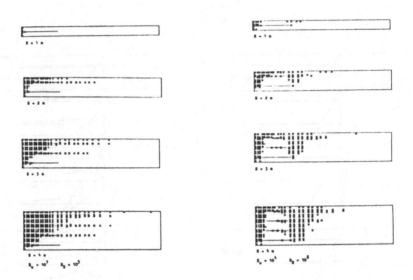

23. Shear stress, mobilisation J_2/J_{2F}. Stiff wall,
     extensible sheet.
24. Shear stress mobilisation J_2/J_{2F}. Soft wall,
     inextensible sheet.

4.5 Strip-reinforced walls : limit state

In the previous sub-section dealing with sheet-reinforced walls, the interfaces were assigned the same frictional properties as the surrounding soil. The same assumption has been made when dealing with strip reinforcement. Figure 25 shows the mesh arrangement, consisting this time of 8-noded brick elements. Debonding at the embedded end of the strip is again a vital component of the modelling process.

Simulation of pullout load-deflection behaviour for a simple strip is shown in Figure 26. As mentioned previously, the assumption of elastic-perfectly Mohr Coulomb plastic soil (c'=0, φ'=30°) means that real dilatancy in the soil is not modelled. In practice, such dilatancy could be introduced and, allied with varying the interface thickness, would be used to simulate the results of actual pullout tests.

4.6 Strip-reinforced walls : serviceability state

Figure 27 shows a typical stage construction of a wall using three-dimensional elements for soil, strips and interfaces (both horizontal and vertical). Various patterns, symmetric and staggered, can be used for the reinforcement in elevation, with various strip spacings. Typical results for the deformations of a wall system are shown in Figures 28 to 30 which show the consequences of differing lengths of reinforcement and different arrangements in elevation, in terms of the horizontal spacing between reinforcement. In these cases the facing of the "wall" is essentially absent.

Typical stress distributions in terms of mobilised shear stress contours are shown in Figures 31 and 32 which are for horizontally closely spaced and remotely spaced reinforcement respectively. For closely spaced reinforcement the bunching of stress contours between the reinforcement illustrates the interactive strengthening which takes place. One can imagine this effect being accentuated by dilatancy. For the remotely spaced reinforcement, Figure 32, there is no equivalent interaction between strips at full height.

5. ANALYSIS OF FOUNDATIONS

5.1 Introduction

In this final section, two sets of analyses involving the "complicated" soil model described in Section 2 are reported. The first concerns a vertically loaded foundation bearing on a saturated subsoil with a compressible skeleton, while the second concerns a horizontally loaded foundation where the bearing material is a saturated relatively loose quartz sand.

5.2 Calibration of model for foundation problem 1

The calibration process is described in some detail in reference [10]. Drained isotropic tests with unloading and drained triaxial compression tests with unloading serve to isolate the isotropic and deviatoric components of the model respectively. Since the problem at hand was at least partly undrained, fits to undrained test data are essential to refine the model calibration. Since the laboratory samples were obtained from 250m below sea level, in a fairly variable subsoil, Figures 33 and 34

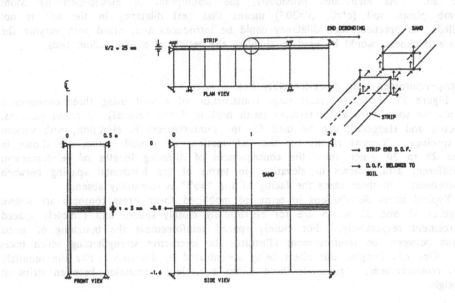

25. Mesh for horizontal pull-out test (3-d).

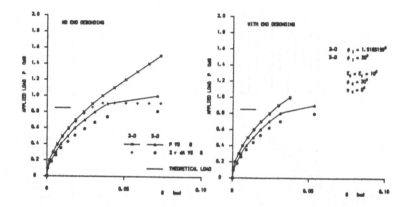

26. 2-d and 3-d pullout test comparisons.

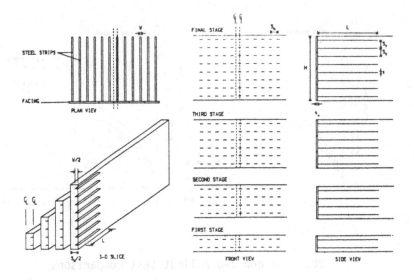

27. Wall construction sequence (3-d).

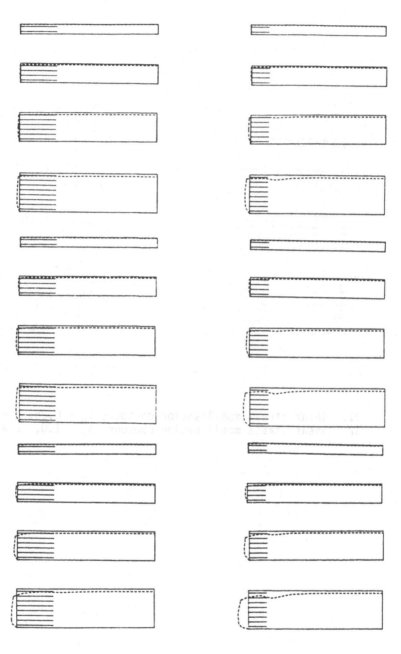

28. Deformed shape $S_y = 0.5$, $S_H = 0.25$, $L = 4,2$.
29. Deformed shape $S_y = 0.5$, $S_H = 0.5$, $L = 4,2$.
30. Deformed shape $S_y = 0.5$, $S_H = 1.0$, $L = 4,2$.

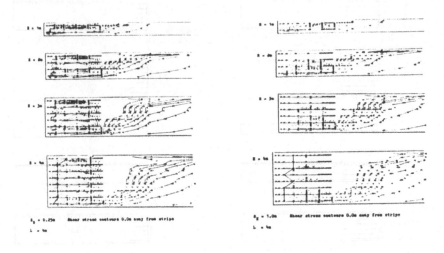

31.   Shear stress mobilisation contours S_H = 0.25, L = 4.
32.   Shear stress mobilisation contours S_H = 1.0, L = 4.

illustrate the kind of variability in results which is unavoidable in practical circumstances. Figure 33 contains data from saturated isotropically consolidated undrained tests and Figure 34 contains data from saturated anisotropically consolidated undrained tests. It is clear that the model only fits the data in a general sense, but what was of paramount importance for a conservative design was that excess porewater pressures in undrained tests should not be underestimated so that undrained strengths were not overestimated.

5.3 Analyses of limit and serviceability states

Figure 35 shows a typical mesh of 8-noded axisymmetric isoparametric elements which were used in the analyses, while Figure 36 shows typical load–settlement results for a circular foundation 4.5m in diameter. In Figure 35, five different material types were identified to form a layered configuration. Figure 36 illustrates the different limit states achieved for undrained "u" and drained plus undrained "d+u" situations. For the purely drained situation "d", no limit state was achieved and allowable deflections would govern the design.

The question of whether the field situation would be drained or undrained is addressed in Figure 32 which shows results from analyses conducted at different loading rates. It can be seen that only two orders of magnitude of the permeability parameter k separate the fully drained from the essentially undrained states. Once the field problem had been identified as potentially undrained, there was no escape from using a "complicated" soil model in the analyses.

5.4 Calibration of model for foundation problem 2

In this case also, concern centred on the possibility that relatively loose material, in this case sand, would be loaded under essentially undrained conditions. The model calibration details are given in reference [9] and Figure 38 shows the fit between two calibrations "A" and "B" and the laboratory data in the form of four saturated isotropically consolidated undrained tests, "C1U1" to "C1U4". Since the quartz sand was reconstituted in the laboratory before testing, the variability in data is far less than for problem 1, where "undisturbed" samples were used. Ultimately, four calibrations were done, with a non-dilating category "C" and a supercritical category "D" being added to the original "A" and "B" of Figure 38. The full range is illustrated in Figure 39.

5.5 Analyses of limit and serviceability states

A typical mesh of 8-noded isoparametric elements for the analysis of the horizontally loaded foundation is shown in Figure 40. Load–displacement plots are shown in Figure 41 for the four calibrations in both the drained and undrained states. For the more dilatent material, A/B, no limit load was achieved but for the non-dilatent materials, C and D, low undrained limit loads were computed. Again the question is – drained or undrained? Figure 42 shows load–displacement results for varying degrees of drainage. Again, as in problem 1, roughly two orders of magnitude of change of loading rate (or of change in permeability k at constant loading rate) differentiate the essentially undrained state from the purely drained state. If undrained, or partially undrained rates of loading are anticipated, the "complicated" type of model must be employed.

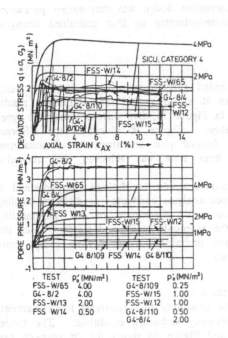

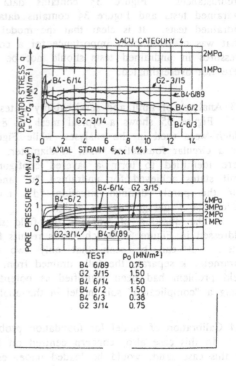

33. Test data and model fits - saturated isotropically
    consolidated undrained tests.
34. Test data and model fits - saturated anistropically
    consolidated undrained tests.

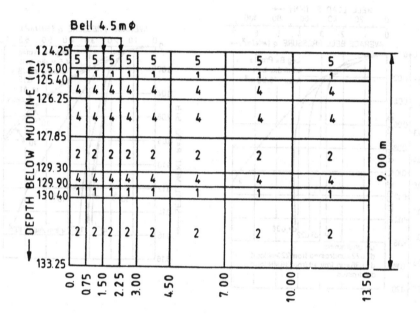

35. Axisymmetric FE mesh.

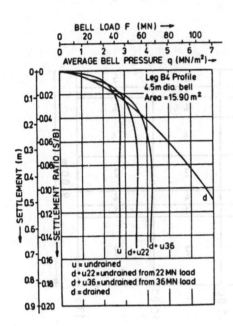

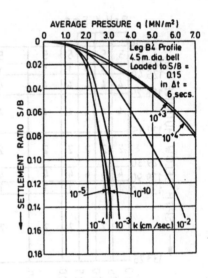

36. Load - displacement curves.
37. Effect of drainage on load-displacement behaviour.

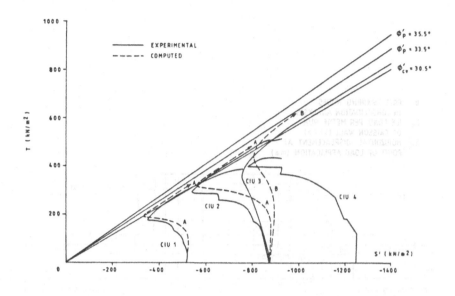

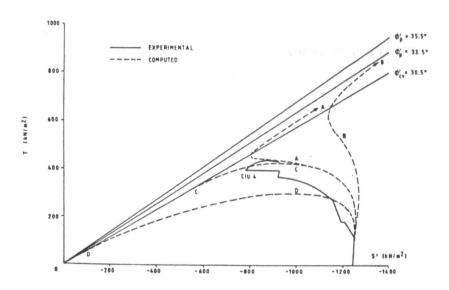

38.   Undrained stress paths, measured and fitted.
39.   Undrained stress paths, model categories A to D.

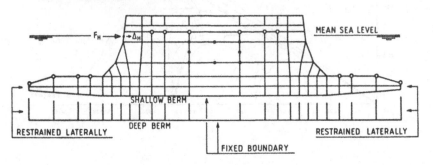

○ FREE DRAINING NODES
   IN CONSOLIDATION ANALYSIS
F_H ICE LOAD PER METRE RUN
   OF CAISSON WALL (kN/m)
Δ_H HORIZONTAL DISPLACEMENT AT
   POINT OF LOAD APPLICATION (mm)

40.   Plane FE mesh.

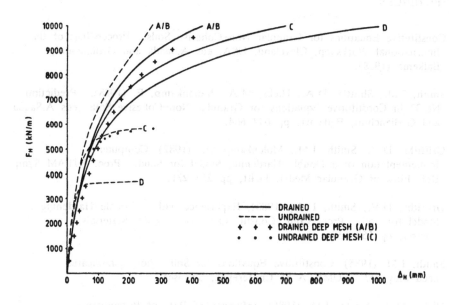

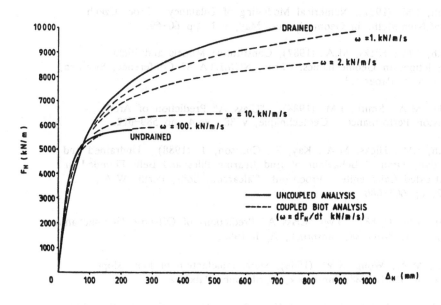

41. Load - displacement curves.
42. Effect of drainage on load-displacement behaviour.

6. REFERENCES

[1] Constitutive Equations for Granular Non–Cohesive Soils,   Proceedings of the
     International Workshop, Cleveland, USA, eds A.Saada and G.Bianchini,
     Balkema (1988).

[2] Smith, I.M., Shuttle, D.A., Hicks, M.A., Molenkamp, F. (1988),   Prediction
     No.32 in Constitutive Equations for Granular Non–Cohesive Soils. eds A.Saada
     and G.Bianchini, Balkema, pp 647–664.

[3] Griffiths, D.V., Smith, I.M., Molenkamp, F., (1982), Computer
     Implementation of a Double–Hardening Model for Sand.   Proc. IUTAM Symp.
     Def. Flow of Granular Media, Delft, pp 213–221.

[4] Griffiths, D.V., Smith, I.M., (1983), Experience with a Double–Hardening
     Model for Soil.   Proc. Conf. Constit. Laws for Engrng. Materials,
     Tucson, pp 553–559.

[5] Smith, I.M. (1985), Constitutive Equations for Soil : how complicated
     need they be?   NUMETA 85 Conference, Swansea.

[6] Hicks, M.A., Smith, I.M. (1986), Influence of Rate of Porepressure
     Generation on the Stress–Strain Behaviour of Soils.   I.J.Num.Methods in
     Engineering, v 22, No.3, pp 597–621.

[7] Smith, I.M. (1987), Numerical Modelling of Dilatancy.   Proc. Czech.
     Conf.Num.Meth. in Geomechanics, May, v 1, pp 60–69.

[8] Smith, I.M., Hicks, M.A. (1987), Constitutive Models and Field
     Predictions in Geomechanics.   Proc. NUMETA 87 Conference, Swansea,
     July, v 2, Paper C3.

[9] Hicks, M.A., Smith, I.M. (1988),   "Class A" Prediction of Arctic
     Caisson Performance.   Geotechnique, v 38, No.4, pp 589–612.

[10] Smith, I.M., Hicks, M.A., Kay, S., Cuckson, J. (1988),   Undrained and
     Partially Drained Behaviour of End Bearing Piles and Bells Founded in
     Untreated Calcarenite.   Proc.Conf. Calcareous Soils, Perth, W.A.,
     v 2, pp 663–680.

[11] Smith, I.M. (1988),   Two "Class A" Predictions of Offshore Geomechanics.
     Proc. ICONMIG 88, Innsbruck, April 1988.

[12] Hicks, M.A., Wong, S.W. (1988), Static liquefaction of loose slopes.
     Proc. ICONMIG 88, Innsbruck, April, pub. Balkema.

[13] Ho, D.K.H. (1989), Analysis of Geotechnical Constructions by the Finite
     Element Method, PhD Thesis, University of Manchester.

[14] Terzaghi, K., (1943), Theoretical Soil Mechanics;
     John Wiley, New York, pp510.

[15] Smith I.M., Griffiths, D.V. (1988), Programming the Finite Element Method;
     John Wiley, Chichester, UK, pp 469.

[16] Peck, R.B. (1969), Deep Excavations and Tunnelling in Soft Ground,
     Proc. 7th Int.Conf. Soil Mech, Mexico, State-of-the-Art Volume, pp225-290.

[17] Naylor, D.J., Richards, H. (1978), Slipping Strip Analysis of Reinforced
     Earth.   Int.J.Num.Analytical Meth.Geomech., v2, pp

NUMERICAL MODELLING OF TUNNELS

G. Swoboda
University of Innsbruck, Innsbruck, Austria

ABSTRACT

Until a few years ago tunnel construction was based exclusively on experience. Numerical methods, however, constitute a very valuable supplement. Shown here are the approximations necessary for 2D analysis. The most important load cases, such as dead load or water pressure, are also illustrated. The damage tensor theory needed for realistic simulation of jointed rock is also presented. In future 3D analysis will take on increasing significance, for which reason the pertinent models are also dealt with.

1 Introduction

This chapter is intended as a guide to those engineers responsible for the design of tunnels and aims to give them an overview of the state of the art of practical numerical models for tunneling. For many years the design of tunnels was based solely on experience and sometimes on very simple analytical considerations. This does not mean that a tunnel cannot be designed without the use of a numerical model. The fact is that such computer models are very helpful tools in understanding the mechanical background of the excavation process.

Every tunnel design begins by selecting from engineering drawings the most critical cross section for construction of the tunnel. This can be a function of the excavated area, the distance to other tunnels, the height of overburden, the influence of buildings on the surface and the geological conditions.

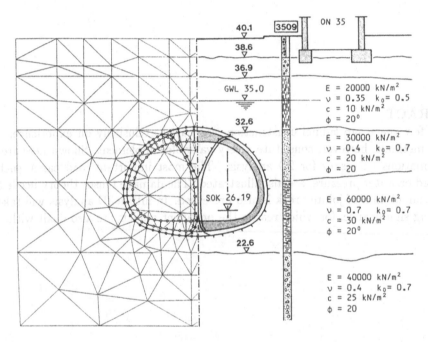

Figure 1: Numerical model for a subway tunnel

Fig. 1 shows a typical numerical model for a subway tunnel. Modern mesh generation allows such a numerical model to be modified in a wide range due to geometric and geological modifications between different cross sections.

The question of safety is so important for subway tunnels that hardly any design or construction work can be performed without numerical model. Numerical study of stress redistribution was of great importance for the progress of modern cross sections and new excavation procedures. Fig. 2 shows different complex subway cross sections.

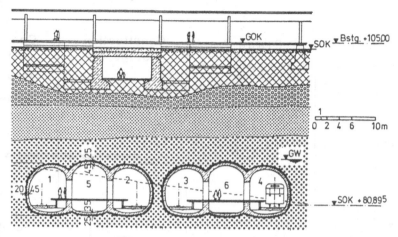

Figure 2: Complex cross section for subway tunnels

2 Two-dimensional models

Numerical models in tunneling started with a full-load two-dimensional model (Fig. 3) that was based on a rigid supporting shield driven in undisturbed primary state of stress. Thus protected, the primary support (shotcrete lining) was erected without stress. Only then was the imaginary supporting shield removed, with static pressure being completely redistributed to the shotcrete. This type of model is very conservative and stays on the safe side. It was not able to be used for tunnels with great overburden.

For this reason, the 3D support effect, as simulated by the strain relief in front of the face, was taken into account.

Three models are widely used to simulate the three-dimensional effect. These are the

- stiffness reduction model
- load reduction model
- modified stiffness reduction model

The stiffness reduction model uses a support core with a modified modulus of elasticity. For excavation, the full-stress field in front of the tunnel face is applied to this core. Modification of the original modulus of elasticity E to the modulus of the support core is done with

$$E_S = \alpha_a E \tag{1}$$

When the α value is infinite, $\alpha = \infty$, we find the original initial stress field in front of the face after stress relief. The stiffness factor α is calibrated by displacement measurement based on the relation:

$$\frac{u_A}{u_B} = \frac{\bar{u}_A}{\bar{u}_B} \tag{2}$$

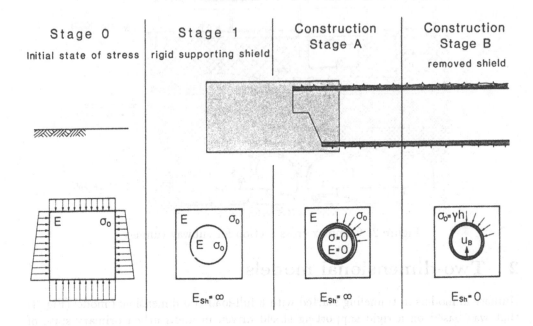

Figure 3: Full load model

which means that the relation of the displacement on the front of the face u_A to the final displacement u_B must be the same in the field $(\bar{u}_A, \bar{u}_B)$ as in the numerical mode (u_A, u_B).

In the load reduction method, mainly used in Germany, the entire load from the initial stresses is applied. At the same time, however, a support load βN is applied and the effective load on $(1 - \beta)N$ is reduced. In order to prevent uplift of the system, the load factor has to be assumed variably over the circumference. The load factor can be within $0 < \beta < 1$.

The modified stiffness reduction method attempts to create a synthesis of these two methods. Here the load factor β is applied supplementary to the stiffness reduction factor, with the result that the factor α_b is within $0 < \alpha_b < 1$.

3 Loading conditions

The final forces in the primary support system (shotcrete lining) are based on the following influences:

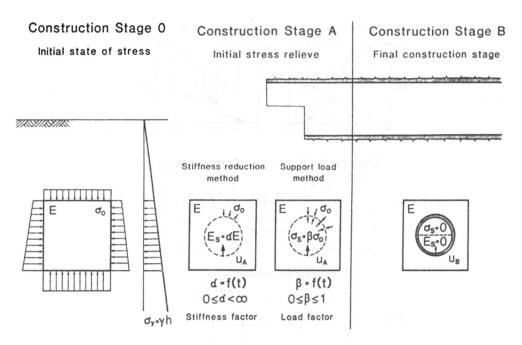

Figure 4: Two-dimensional simulation of the three-dimensional effect

3.1   Stress redistribution due to excavation

Tunnel excavation is simulated in a numerical model using the following steps:

- The elements are eliminated in the excavation area. This can be achieved by reducing the elasticity modulus to zero or by deleting the elements from the list of elements.
- At the same time, the primary stresses σ_0 present in the element have to be eliminated by integration over the element.

This integration is performed element by element according to [7] with

$$\{f_e\} = \int_V [B]^T \{\sigma_0\} \cdot dV \tag{3}$$

The nodal forces $\{f\}$ thus determined for an element are equilibrium loads. The total vector $\{F\}$ is the sum of all the elements E involved in the excavation.

$$\{F\} = \sum_{e=1}^{E} \{f_e\} \tag{4}$$

Since only the nodes are located on the excavation boundary, this total vector has to be divided into two vectors

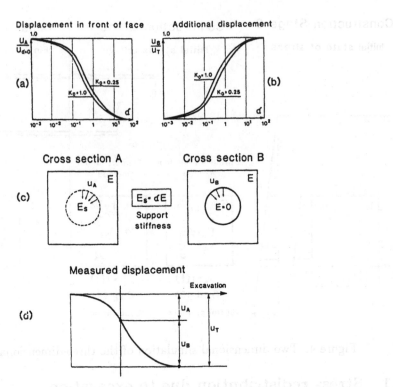

Figure 5: Relation between the displacement in front of the face and final displacement

$${F} = \begin{Bmatrix} F_a \\ F_i \end{Bmatrix}$$

$${F_i} = {0}$$

(5)

whereby $\{F_a\}$ represents the loads on the excavation boundary and $\{F_i\}$ those in the interior, which, however, are not to be considered. Only now is the equilibrium load vector $\{F\}$ transformed into a load vector $\{F_a\}$, whereby the sum corresponds to the dead load of the excavated material.

The dead load of the excavated material acts in the global y direction and gives rise to the following problem. The size of the displacement from this load is dependent on the size of the chosen FEM mesh. However, this is a problem that also occurs when attempting an analytical solution.

In order to examine the question of loads in a continuum, a surface load is considered in place of the internal dead load because it makes the problem easier to illustrate. In the two-dimensional halfspace (Fig. 7) the displacement of a certain point $P(x, y)$ can be determined according to Girkmann [8]

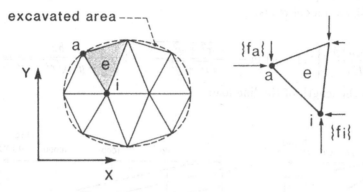

Figure 6: Simulation of excavation

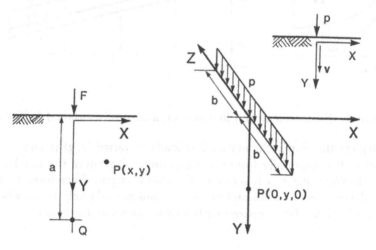

Figure 7: Line load in two- and three-dimensional halfspace

$$v = \frac{(1+\mu)\,F}{E\,\pi}\,\frac{F}{h} \cdot \left[(1-\mu)\ln\frac{x^2+y^2}{a^2} - \frac{y^2}{x^2+y^2}\right] \tag{6}$$

whereby h represents the thickness and a the distance to a reference point Q, where the vertical displacement is $v = 0$. If the distance is infinite, the displacement V is infinite. This means that we have a problem identical to the FEM model, namely displacement dependent on a reference point.

Since this problem only occurs in the loading "dead load of the excavated material" and, due to the type of driving employed, the loading only affects a short distance, it should be examined whether similar problems occur in three-dimensional halfspace. Since finite displacements occur as a result of concentrated loads after Boussinesq, this question was already investigated by Girkmann [9]. The displacement of a line load p, which corresponds to the concentrated load F in the change from the two- to three-dimensional model reads

$$v = \frac{(1+\mu)\,p}{2\,E\,\pi} \cdot \left[4\,(1-\mu)\,\ln\frac{b+\sqrt{x^2+y^2+b^2}}{\sqrt{x^2+y^2}} + 2\frac{b\,x^2}{(x^2+y^2)\sqrt{x^2+y^2+b^2}}\right] \qquad (7)$$

whereby b is the length of the line load.

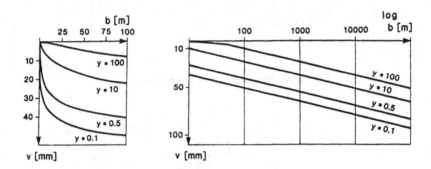

Figure 8: Displacement of a line load on a three-dimensional halfspace

Fig. 8 depicts the displacements for line loads of varied lengths and for points at various depths. It is especially from the logarithmic diagram of the line load that we see that the displacement is a function of the load's length. Therefore, it can be said that the displacements of an infinitely long line load are infinitely large, whereby only an infinite line load can be compared with a two-dimensional example.

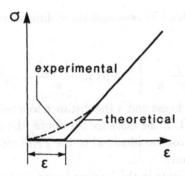

Figure 9: Nonlinearity in the origin of the stress-strain curve

The reason why it is so difficult to obtain an unmistakable solution is that the nonlinearity in the origin (Fig. 9) is not taken into consideration in the constitutive law in general use today. For points whose initial strain of the theoretical stress-strain diagram is not exceeded, integration of the analysis has to be interrupted if the strains are less than the initial strain of the material ε_0. In order for the numerical model to take account of the dead load, since this is a very short line load and thus the strains very quickly drop below the initial strains, the mesh size has to be reduced accordingly. The program is easily revised to meet this situation, in that the elasticity modulus is set at infinite for all mesh elements near the boundary where the strain ε_0 is not reached. As indicated in Fig. 10, this brings about a modification in the support conditions.

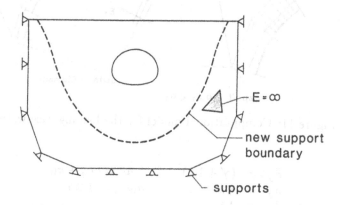

Figure 10: Modification of the boundary condition

3.2   Water pressure

The calculation of water load presents a special problem in the construction of shallow tunnels in loose material. With earth pressure it can be assumed that the earth pressure is reduced by the spatial carrying effect. A similar assumption is in no way possible for water load, however since it will always exert its full load on the concrete arch.

When calculating this load case it must be decided whether the soil is damp, zone a in Fig. 11, or water-saturated, zones b and c. The soil stresses in the first case are

$$\begin{aligned} \bar{\sigma}_y &= \gamma\,h \\ \bar{\sigma}_x &= \lambda\,\bar{\sigma}_y \end{aligned} \tag{8}$$

whereby γ represents the specific weight of the damp soil and λ the coefficient of earth pressure at rest. The stresses $\bar{\sigma}_x$ and $\bar{\sigma}_y$ are identical to the effective stresses $\bar{\sigma}_g$ in the soil's grain. Thus, the occurring expansion can be determined by removing the soil stress from the following elastic relationship:

$$\{\varepsilon\} = [D]\{\sigma\} = [D]\left\{ \begin{array}{c} \bar{\sigma}_x \\ \bar{\sigma}_y \end{array} \right\} \tag{9}$$

For zone b, which is under water pressure, the soil stress is:

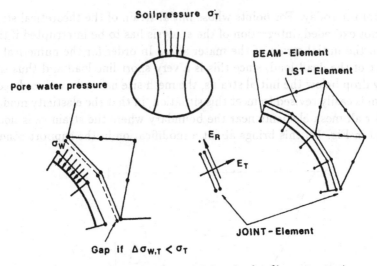

Figure 11: Computational model for the loading case water

$$
\begin{array}{ll}
\bar{\bar{\sigma}}_y = (\gamma' + 1.0) & h = \sigma_g + \sigma_W \\
\bar{\bar{\sigma}}_g = \gamma' h & \sigma_W = 1.0\,h \\
\bar{\bar{\sigma}}_x = \lambda \bar{\bar{\sigma}}_y &
\end{array}
\tag{10}
$$

whereby γ' is the specific weight under uplift. The soil stresses are divided here into the effective soil stresses $\bar{\bar{\sigma}}_g$ and the pore-water pressure σ_W.

If the water level drops due to construction, the geomechanical changes in stresses have two causes:

(i)   **Change in effective soil stresses**

Removal of the pore-water pressure in zone b causes an increase in the effective soil stress from $\bar{\bar{\sigma}}_g$ to $\bar{\sigma}_g$. For the related expansion it must be assumed that the expansions related to $\bar{\bar{\sigma}}_g$ have already taken place. This means the newly occurred expansions in zone b are

$$
\{\varepsilon\} = [D]\{\Delta\sigma_g\} = [D] \left\{ \begin{array}{c} \bar{\sigma}_{g,x} - \bar{\bar{\sigma}}_{g,x} \\ \bar{\sigma}_{g,y} - \bar{\bar{\sigma}}_{g,y} \end{array} \right\}
\tag{11}
$$

(ii) **Pore-water pressure on a joint face**

If the drop in the water level causes water to accumulate along a joint face (Fig. 12, Stage II) for example sheet piling, the full pore-water pressure σ_W will become effective here.

The redisposition of forces described in (i) and (ii) produce a new primary stress state in the soil. This is the decisive stress for the earth pressure during ecxavation, meaning that pressure on the concrete arch is σ_T. If the water pressure becomes effective after completion of the tunnel, the reversible processes described in (i) and (ii) will take place. This means the effective soil stress will be reduced and the pore-water pressure removed from the joint face.

(iii) **Pore-water pressure on the concrete arch**

As shown in Fig. 12, the pore-water pressure σ_W will act on the joint face of the tunnel arch identical to load (ii). In this way the tangential earth pressure changes by the amount of $\Delta\sigma_{W,T}$. If the tensile stress $\Delta\sigma_{W,T}$ becomes larger than the original earth pressure, tensile stresses would occur between the concrete arch and the soil, which however, cannot be the case. This is solved by the structure of the calculation model shown in Fig. 11. A nonlinear JOINT element is inserted between the concrete arch (BEAM elements) and the soil (LST elements).

When the tensile failure condition

$$\sigma_T - \Delta\sigma_{W,T} > 0 \qquad \rightarrow \qquad E_T = 0 \tag{12}$$

is reached the tangential modulus of elasticity equals 0. In this way the gap between the soil and the concrete arch can be simulated.

4   Jointed rock

Until now numerical models have not been applied to jointed rock because it is much more difficult to predict the orientation of a joint system than any soil parameter.

But in tunnels with pilot tunnel excavation systems the joint system is well known and can be considered in the numerical model.

Goodman's [15] works first presented an element that describes the rock jointing. This element can only be used to describe individual faults and is also numerically very instable. Improvements for this element have been proposed by many authors. In recent years, models have been developed to break down the discontinuum into blocks. In the rigid body model (RBM) by Cundall [19], the rock mass is broken down into rigid blocks. This is a model that does not take consideration of the elastic deformations and is certainly not suitable for calculating shallow tunnel structures. In the further developments by Asai, the elastic properties are simulated by springs in the nodes. In

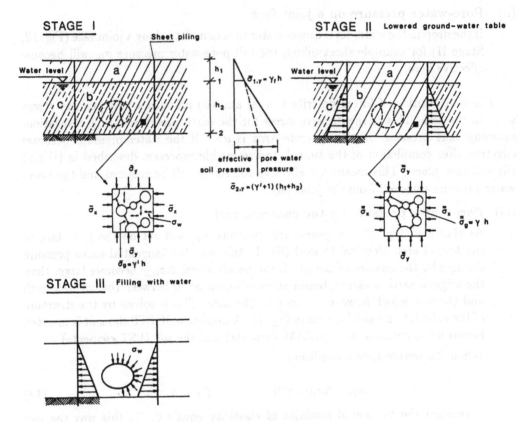

Figure 12: Loading case water

places where the block's displacement share is very large, a usable formulation of a model around the discontinuum definitely has to be described.

The approach taken here, namely that of the "decoupled finite element method", uses finite elements as elastic blocks. According to the faults, these are coupled by constraint elements [21]. These elements can give consideration to the opening and closing of the joint.

4.1 Decoupled finite element theory

4.1.1 Theory

When a continuum consists of two separate systems A and B (Fig. 13), the equation system (Fig. 14) breaks down into two independent blocks.

$$[K]\{a\} - \{F\} = 0 \tag{13}$$

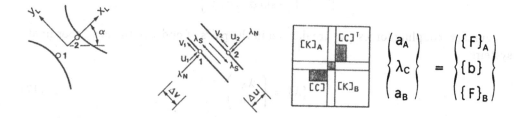

Figure 13: Coupled system

with

$$\begin{bmatrix} K_A & 0 \\ 0 & K_B \end{bmatrix} \left\{ \begin{matrix} a_A \\ a_B \end{matrix} \right\} - \left\{ \begin{matrix} F_A \\ F_B \end{matrix} \right\} = 0 \qquad (14)$$

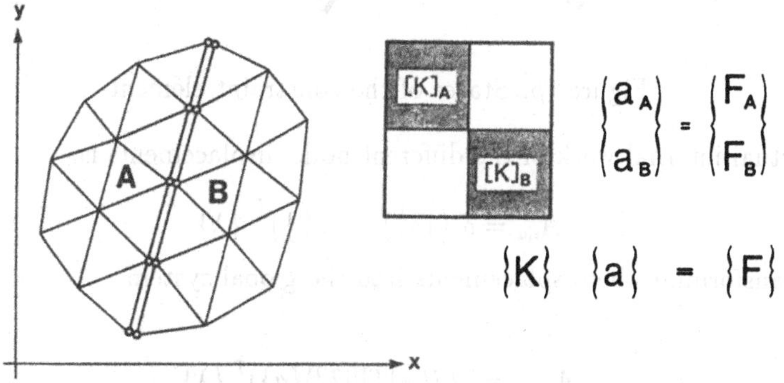

Figure 14: Decoupled constraint element

Coupling of the region A with the region B is done by means of two-node constraint elements (Fig. 14). For this purpose, pairs of nodes with separate local displacements are defined along the joint

$$\{a_1^L\} = \left\{ \begin{matrix} u_1 \\ v_1 \end{matrix} \right\} \qquad \{a_2^L\} = \left\{ \begin{matrix} u_2 \\ v_2 \end{matrix} \right\} \qquad (15)$$

or in a global system

$$\{a_1\} = [T]\{a_1^L\} \qquad \{a_2\} = [T]\{a_2^L\} \qquad (16)$$

with

$$[T] = \begin{bmatrix} \cos\phi & \sin\phi \\ -\sin\phi & \cos\phi \end{bmatrix}$$

Further, coupling forces $\{\lambda\}$ are defined in the pairs of nodes in the local coordinate system

$$\{\lambda\} = \left\{ \begin{array}{c} \lambda_N \\ \lambda_S \end{array} \right\} \tag{17}$$

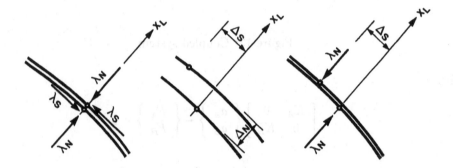

Figure 15: States of the constraint element

The virtual internal work of the different node displacements is:

$$A_{i,a} = \delta\left(\{a_2^L\} - \{a_1^L\}\right)^T \{\lambda\} \tag{18}$$

or, after transforming the displacements into the global system

$$
\begin{aligned}
A_{i,a} &= \delta\left((-[T]|[T])\{a\}\right)^T \{\lambda\} \\
&= \delta([C]\{a\})^T\{\lambda\} \\
&= \delta\{a\}^T[C]^T\{\lambda\}
\end{aligned}
\tag{19}
$$

The corresponding external work is

$$A_{e,a} = \delta\{a\}^T\{F\} \tag{20}$$

whereby $\{F\}$ is the nodal forces from the external load. Beyond this, the additional internal forces also perform virtual work

$$A_{i,\lambda} = \delta\{\lambda\}^T \{[C]\{a\}\} \tag{21}$$

The corresponding external work reads

$$A_{e,\lambda} = \delta\{\lambda\}^T\{a_r\} \tag{22}$$

whereby $\{a_r\}$ is the initial displacements of the nodes.

By equating internal and external work the following results:

$$\begin{aligned} A_i &= A_e \\ A_{i,a} + A_{i,\lambda} &= A_{e,a} + A_{e,\lambda} \end{aligned} \tag{23}$$

and by inserting (19), (20), (21) and (22)

$$\delta\left\{\begin{matrix} a \\ \lambda \end{matrix}\right\}^T \begin{bmatrix} 0 & [C]^T \\ [C] & 0 \end{bmatrix} \left\{\begin{matrix} a \\ \lambda \end{matrix}\right\} = \delta\left\{\begin{matrix} a \\ \lambda \end{matrix}\right\}^T \left\{\begin{matrix} F \\ a_r \end{matrix}\right\} \tag{24}$$

the stiffness matrix of the constraint joint element is received

$$\begin{bmatrix} 0 & [C]^T \\ [C] & 0 \end{bmatrix} \left\{\begin{matrix} a \\ \lambda \end{matrix}\right\} = \left\{\begin{matrix} F \\ a_r \end{matrix}\right\} \tag{25}$$

From this it is seen that this element has no elastic stiffness; only additional constraints are introduced, by means of which coupling can be forced.

In order to be able to orient the local system of coordinates, two additional nodes, 3 and 4, are introduced, as shown in Fig. 15. These are also used to formulate the failure criterion for decoupling in stresses and not in forces.

The calculation is performed incrementally in load steps, whereby the fracture criteria are examined. This means that the criteria will be examined for each of these conditions (K) starting with the condition of the last iteration $(K-1)$. This gives the following forces and displacements:

joint displacement

$$\{\Delta\}^K = \{\Delta\}^{K-1} + \{\Delta\}$$

coupling forces

$$\{\lambda\}^K = \{\lambda\}^{K-1} + \{\lambda\} \tag{26}$$

loading

$$\{F\}^K = \{F\}^{K-1} + \{F\}$$

whereby the added values of the pertinent load steps are without indices.

The friction law developed by Coulomb is used as the failure criterion, which gives the maximum coupling forces:

shear

$$F_t^K = (c - \sigma_N \tan\varphi)\, L\, d\frac{\lambda_S^K}{|\lambda_S|^K} \tag{27}$$

tension

$$F_z^K = \sigma_z L d \tag{28}$$

In these equations, c is the cohesion, φ the angle of friction, σ_z the tension failure stresses, L is the influence of the element and d the thickness of the element. The decision matrix in Table 1 gives the stiffness matrix described in [8] for the conditions *fix state*, *free state* and *slip state*. The conditions are illustrated in Fig. 15.

Table 1: Decision Table

Iteration n n - 1	Fix	Slip	Free
Fix	$\lambda_N^K \leq F_z^K$ $\lambda_S^K \leq F_t^K$	$\lambda_N^K \leq F_z^K$ $\lambda_S^K > F_t^K$	$\lambda_N^K > F_z^K$
Slip	$\lambda_N^K \leq F_z^K$ $\Delta_s F_t^K < 0$	$\lambda_N^K \leq F_z^K$ $\Delta_s F_t^K \geq 0$	$\lambda_N^K > F_z^K$
Free	$\Delta_N^K < 0$		$\Delta_N^K \geq 0$

4.1.2 Mesh generation of jointed rock

Jointed rock is described in the finite element mesh as rock along given possible fault lines, where constraint elements make it possible for the failure mechanisms described above to take place. In the framework of interactive mesh generation by means of a digitizer, the NETDIG program [22] includes the possibility of introducing individual fault lines after completion of the mesh topology. For this purpose, the node numbers on both sides of the fault line are reassigned, whereby one node retains its original number and the second, newly introduced node receives the highest node number available. The reference nodes in Fig. 15 have to be included in the generation.

Special problems arise in this algorithm when two fault lines intersect. In this case, appropriate constraint element combinations have to be generated, as shown in Fig. 16, in order to permit all elements to move along the given lines. A total of up to eight intersecting lines can be generated in the NETDIG program system.

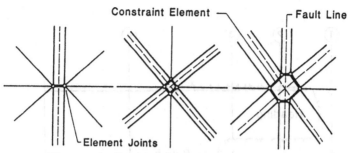

Figure 16: Crossing fault lines of jointed rock

4.1.3   Calculation of a tunnel in discontinuum

According to the New Austrian Tunneling Method (NATM) [1], driving is performed in partial excavation, with rapid securement of the cavity using shotcrete. This effects an activation of the surrounding rock and economic calculation of the cavity's securing measures. After excavation of the top heading, the primary stresses in the undisturbed soil or rock are redistributed transverse to the direction of driving and also via the face by arch action. This redistribution of loads causes displacements in the roof, which in shallow tunnels can extend to the surface in the form of settlement. The area behind the face also undergoes displacement that is caused on the one hand by the arch action following excavation, and on the other hand by the free surface of the face, which does away with the support action in the tunnel's longitudinal direction. Instead of plane displacement, plane stress is approximated.

Due to the extensive calculations entailed, the work is generally calculated on a plane model, whereby the driving is approximated by individual sections transverse to the tunnel axis. Literature makes reference to three simulation procedures that take consideration of the influences in the tunnel's longitudinal direction. For example, there is the possibility of variation of the modulus of elasticity of the shotcrete arch, or calculation according to the partial-load or support-load method [23], [24].

In the work at hand, simulation is performed using the stiffness reduction method, that has been seen to be the simplest possibility. This method takes account of the displacements caused by strain relief by considering the future excavation area as a support core. The stresses here are applied on the excavation's periphery as nodal forces. The support core, which has a certain defined stiffness, acts as an elastic bed for the surrounding elements. If the support core has a theoretically infinitely large stiffness, preliminary relief of strain will preclude any displacements. Thus, the state of stress from the previous loads remains unchanged. Realistic values for the stiffness of the support core, that were confirmed in comparisons with calculations using in-situ measurements, are in a range of 0.2 to 3 times the rock's module of elasticity.

Fig. 17 shows the excavation procedure as well as the analyzed cross section of the individual construction steps.

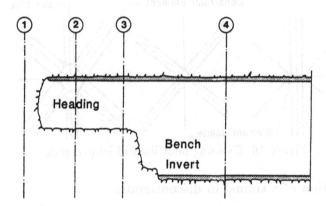

Figure 17: Excavation sequence

Tunnel driving is subsequently examined in a discontinuous rock mass with unin-terrupted fault system, as shown in Fig. 18. In Fig. 19, construction step 3, a chimney-like cave-in can be seen, a failure mechanism often observed in such jointed rock. The load applied by the failure body presses particularly on the foot of the top heading's shotcrete lining, causing the roof to heave slightly in construction step 4.

4.2   Damage tensor theory

Many rock types can be described as a homogeneous body with a number of cracks located in one plane. The planes have a regular distance and can be overloaded by other planes with different orientations. The damage tensor theory of rock is part of the system of a continuum theory. The basis is the geological investigation of the fault system of the rock.

4.2.1   The damage tensor - a linear representation

Let us assume that we have a three-dimensional element with one plane of cracks (Fig.20). In the plane of cracks are the axes x_1 and x_2 of the cartesian coordinate system x_1, x_2 and x_3. The x_3 axis coincides with the direction of $\{n\}$, the orientation of the damaged rock. The share of damage or the remaining rock bridges are described by the damage factor ω [33].

If we assume a general plane in the undamaged rock ABC with the area S, the normal vector of this plane is

$$\{S\} = S\{\nu\} = S_1\{e_1\} + S_2\{e_2\} + S_3\{e_3\} \tag{29}$$

$$\{e_3\} = \{n\}$$

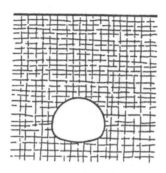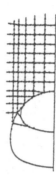

Figure 18: Tunnel in rock with horizontal and vertical jointing, arrangement of constraint elements.

In this relation, $\{\nu\}$ is the normal vector and $\{e_i\}$ $(i = 1, \cdots, 3)$ are the canonical base vectors of the chosen coordinate system.

The effect of damage can be considered by reducing the effective area of OAB, in the plane of cracks, by the factor $(1 - \omega)$, while the area of the planes OBC and OAC remains unchanged.

Thus the reduced effective area will be described by OA^*B^*. Geometrically, we change from the undamaged plane ABC to the damaged plane $A^*B^*C^*$ by reducing sides OA and OB by the factor $\sqrt{1 - \omega}$ while side OC is enlarged by the factor $1/\sqrt{1 - \omega}$. The area of the thus obtained triangle $A^*B^*C^*$ is equal to S^*, and its normal vector is $\{\nu^*\}$.

The normal vector of the damaged plane is

$$\{S^*\} = S^*\{\nu^*\} = S_1\{e_1\} + S_2\{e_2\} + (1 - \omega) \cdot S_3\{e_3\} \tag{30}$$

and with the help of equation (29) we find

$$\{S^*\} = S\{\nu\} - \omega S \cdot [\{e_3\} \otimes \{e_3\}] \cdot \{\nu\}$$

whereby $\otimes$ is the tensor product:

$$[A] = [\{n\} \otimes \{n\}] \quad \Rightarrow \quad A_{ij} = n_i \cdot n_j$$

If a tensor $[\Omega]$, the so-called damage tensor, is defined as

$$[\Omega] = \omega \cdot [\{n\} \otimes \{n\}]$$

then $\{S^*\}$ is

$$\{S^*\} = S \cdot [I - \Omega] \cdot \{\nu\} = [I - \Omega] \cdot \{S\} \tag{31}$$

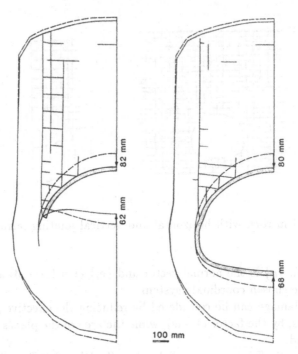

Figure 19: Fault system and displacements in construction steps 3 and 4

4.2.2   The damage tensor in rock mechanics

Let us consider a cube with a side length L and a volume V (Fig. 12) that has been cut out of the rock. Through this cube we cut several parallel planes at intervals of l, namely the mean distance of the individual joint planes. Next, we imagine the cracks in the rock as being in these planes. The rock between them is homogeneous and isotropic. Since $\frac{L}{l}$ represents the number of planes, the largest possible total area of all cracks in the considered joint pattern is calculated as follows:

$$S = L^2 \cdot \frac{L}{l} = \frac{V}{l} \tag{32}$$

The k-te crack of this joint pattern ("i") has the area $a_{k,i}$ and the normal vector $\{n_i\}$. Its related area is thus

$$\omega_{k,i} = \frac{a_{k,i}}{S_i} \qquad k = 1, \ldots, N \tag{33}$$

Now the damage tensor of this special crack can be calculated:

$$[\Omega_{k,i}] = \omega_{k,i} \left[\{n_i\} \otimes \{n_i\} \right] \tag{34}$$

The following is thus true for the entire joint pattern:

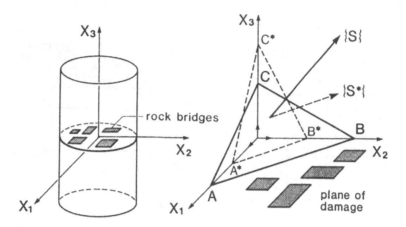

Figure 20: A continuum theory of creep and creep damage

$$[\Omega_i] = \sum_{k=1}^{N}[\Omega_{k,i}] = \frac{l_i}{V}[\{n_i\} \otimes \{n_i\}] \sum_{k=1}^{N} a_{k,i} \tag{35}$$

If the cube is cut through by several joint patterns, $[\Omega_{total}]$ is obtained as the sum of the individual damage tensors:

$$[\Omega_{total}] = \sum_{i=1}^{M}[\Omega_i] = \frac{1}{V}\sum_{i=1}^{M} l_i [\{n_i\} \otimes \{n_i\}] \sum_{k=1}^{N} a_{k,i} \tag{36}$$

Based on the damage factor of each set of cracks the damage tensor can also be found by the following relation

$$[\Omega_{total}] = \sum_{k=1}^{N} \omega_k [\{n_k\} \otimes \{n_k\}]$$

where N is the number of sets of cracks.

4.2.3 Introduction of net stresses

Let us consider a 3D element V containing several cracks parallel to the y axis. The separation factor is ω $(0 \leq \omega < 1)$. A normal stress of σ_x is at work on the element. For reasons of equilibrium, a larger stress must be present inside the element (Point 1) since, due to the cracks, the effective resisting area is smaller than the area on which the load is imposed. We term these strains net strain σ_x^*. They are expressed as

$$\sigma_x^*(1 - \omega) = \sigma_x \tag{37}$$

or

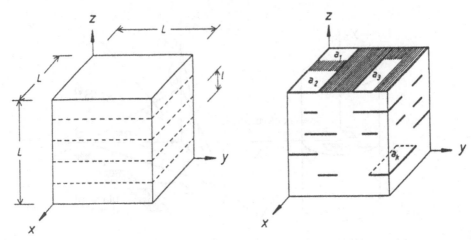

Figure 21: Joint system and joint planes

$$\sigma_x^* = \sigma_x(1 - \omega)^{-1} \tag{38}$$

When these net stresses are applied to an isotropic, homogeneous element, the element is deformed under the given load, as the below element with cracks shows.

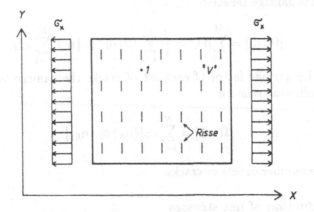

Figure 22: Damaged rock

4.2.4 Mathematical formulation

Once the stresses on a homogeneous and isotropic system have been calculated, the net stresses $[\sigma^*]$ are determined. If $[\sigma]$ is the Cauchy stress tensor and $\{P\}$ the resulting vector of force on the tetrahedron surface ABC in Fig. 18, the following holds true:

$$[\sigma] = \begin{bmatrix} \sigma_x & \tau_{yx} & \tau_{zx} \\ \tau_{xy} & \sigma_y & \tau_{zy} \\ \tau_{xz} & \tau_{yz} & \sigma_z \end{bmatrix}$$

$$\{p\} = [\sigma] \cdot \{\nu\} \tag{39}$$

$$\{P\} = S \cdot \{p\} = S \cdot [\sigma] \cdot \{\nu\} = [\sigma] \cdot \{S\}$$

$$\{P\} \doteq [\sigma^*] \cdot \{S^*\} \quad \Rightarrow \quad [\sigma^*] \cdot \{S^*\} = [\sigma] \cdot \{S\}$$

$$\Rightarrow \quad [\sigma^*] = [\sigma] \cdot [I - \Omega]^{-1} \tag{40}$$

Since the effective area of the element is reduced by the cracks in the rock, these net stresses must be larger than the stresses in a homogeneous system. The linear representation $[I - \Omega]^{-1}$ describes these mechanical effects of the joints in the rock.

Now, however, it must be determined whether the stresses acting normally on the cracks are positive or negative. Tensile stresses must be absorbed exclusively by the effective area. Thus the net stresses are precisely calculated according to the above-given formula. The compressive stresses, in contrast, can also be transferred through the joint since some of the cracks are closed. The effective area is thus larger than that described by the joint tensor. The same is also true for the shear stresses acting parallel to the joint. The prevailing state of stress thus effects the size of the effective area of any element.

If a matrix $[T]$ is defined, whose columns are the eigenvector $\{\phi_i\}$ of the damage tensor $[\Omega]$ and if a diagonal matrix $[\Omega']$ contains the eigenvalues ω_i, namely

$$[T] = [\{\phi_1\} \{\phi_2\} \{\phi_3\}]; \qquad |\phi_i| = 1 \qquad i = 1, \dots, 3$$

$$[\Omega'] = \begin{bmatrix} \omega_1 & & \\ & \omega_2 & \\ & & \omega_3 \end{bmatrix}$$

the following holds true:

$$[\Omega'] = [T]^T [\Omega] [T]$$

Using $[T]$ the Cauchy stress tensor $[\sigma]$ describing the stresses in the global system of coordinates can be transformed to the direction of the damage tensor's principal axes:

$$[\sigma'] = [T]^T [\sigma] [T] \tag{41}$$

In order to be able to express the above cases in mathematical terms, $[\sigma']$ must be broken down into a tensor $[\sigma'_n]$ that contains the normal stresses and a tensor $[\sigma'_t]$ with only shear stresses:

$$[\sigma'] = [\sigma'_n] + [\sigma'_t] \tag{42}$$

In its explicit form this relation reads as follows:

$$
\begin{bmatrix}
\sigma'_1 & \tau'_{21} & \tau'_{31} \\
\tau'_{12} & \sigma'_2 & \tau'_{32} \\
\tau'_{13} & \tau'_{23} & \sigma'_3
\end{bmatrix}
=
\begin{bmatrix}
\sigma'_1 & 0 & 0 \\
0 & \sigma'_2 & 0 \\
0 & 0 & \sigma'_3
\end{bmatrix}
+
\begin{bmatrix}
0 & \tau'_{21} & \tau'_{31} \\
\tau'_{12} & 0 & \tau'_{32} \\
\tau'_{13} & \tau'_{23} & 0
\end{bmatrix}
$$

If the crack area were completely plane, no shear force could be transferred across it. In this case the effective area would have to be calculated with $[\mathbf{I} - \mathbf{\Omega'}]$. Since, however, its surface is very uneven, a rough effect occurs that permits part of the shear stresses to be absorbed. For this reason the effective area is modified as follows:

$$[\mathbf{I} - \mathbf{\Omega'}] \mapsto [\mathbf{I} - C_t \mathbf{\Omega'}]; \qquad 0 \leq C_t \leq 1$$

As already stated, this is correspondingly true for compressive stresses:

$$[\mathbf{I} - \mathbf{\Omega'}] \mapsto [\mathbf{I} - C_n \mathbf{\Omega'}] \quad in\ case\ \sigma < 0; \quad 0 \leq C_n \leq 1$$

Contrarily, no modification is necessary for tensile stresses. The net stresses transformed to the principal axis of the damage tensor are thus calculated as follows:

$$[\sigma^{*\prime}] = [\sigma'_t] \cdot [\mathbf{I} - C_t \mathbf{\Omega'}]^{-1} + [\sigma'_n] \cdot [[\mathbf{H}]\langle\sigma'_n\rangle\,[\mathbf{I} - \mathbf{\Omega'}]^{-1} + [\mathbf{H}]\langle-\sigma'_n\rangle\,[\mathbf{I} - C_n \mathbf{\Omega'}]^{-1}] \tag{43}$$

whereby $[\mathbf{H}]\langle\cdot\rangle$ is an operator that permits us to distinguish between positive and negative normal stresses:

$$\mathbf{H}_{ij}\langle\sigma'_n\rangle = \begin{cases} 1 & in\ case\ \ \sigma'_{ij} > 0 \\ 0 & in\ case\ \ \sigma'_{ij} \leq 0 \end{cases}$$

Now the stresses $[\sigma^{*\prime}]$ must be transformed to the original global system of coordinates:

$$[\sigma^*] = [\mathbf{T}][\sigma^{*\prime}][\mathbf{T}]^T$$

If $[\sigma]$ is the originally calculated Cauchy stresses, the differential stress tensor $[\psi]$ is defined as follows:

$$[\psi] = [\sigma^*] - [\sigma]$$

4.2.5   Analysis in a finite element program

The equation to be solved in the framework of the principle of virtual work generally reads as follows:

$$\int_v \{\sigma\}^T \cdot \delta\{\varepsilon\} \, dv = \int_s \{t^0\}^T \cdot \delta\{u\} \, ds_t + \int_v \{f\}^T \cdot \delta\{u\} \, dv \qquad (44)$$

or

$$[K]\{u\} = \{F\}$$

After inserting the above relation, the equation reads:

$$\int_v \{\sigma^*\}^T \cdot \delta\{\varepsilon\} \, dv = \int_s \{t^0\}^T \cdot \delta\{u\} \, ds_t + \int_v \{f\}^T \cdot \delta\{u\} \, dv + \int_v \{\psi\}^T \cdot \delta\{\varepsilon\} \, dv \qquad (45)$$

or

$$[K]\{u\} = \{F\} + \{F^*\}$$

It is noteworthy that the stiffness matrix $[K]$ only takes the characteristic values for isotropic and homogeneous rock. All mechanical effects of the cracks are covered by the term $\{F^*\}$.

$$\{F^*\}^e = \int_v [B]^T \{\psi\}^e \, dv \qquad (46)$$

When one single element is isolated from the entire system, $\{F^*\}^e$ is the joint load vector that causes the element to deform according to the differential stresses $[\psi]$, or that causes these differential stresses to occur in the element.

4.2.6   Practical procedure

(1) The equation $[K]\{u'\} = \{F\}$ is solved and the stresses of an isotropic sysstem are calculated from the joint displacements $\{u'\}$.

(2) The net stresses $[\sigma^*]$ are calculated from the Cauchy stress tensor $[\sigma]$ (under consideration of the mentioned influences of the particular state of stress) with the help of the damage tensor $[\Omega]$.

(3) The differential stress tensor $[\psi] = [\sigma^*] - [\sigma]$ is calculated followed by the joint load vector $\{F^*\}$.

(4) Once the system in question is statically determined, the final joint displacements $\{u\}$ of the jointed systems are obtained by solving the equation

$$[K]\{u\} = \{F\} + \{F^*\}$$

(5) In the other case (see Point 4), solving the equation will only provide an approximation for the desired joint displacements. Since the introduction of jointed elements
to the system (namely in the form of an additional joint load vector $\{F^*\}$) is equivalent to changing the stiffness matrix, the stresses in the system are differently
distributed than in the isotropic calculation (see Point 1) that was the prereq-
uisite for calculation of the joint load vector. For this reason, the desired joint
displacement must be calculated by iterative means. The stress tensor $[\tilde{\sigma}]$, which
corresponds to a load of $\{F\} + \{F^*\}$, is calculated according to Point 4. Since the
joint load vector $\{F^*\}$, however, is not present in the actual system, its part of the
total stress $[\tilde{\sigma}]$ must not be considered in the new calculation of the net stresses.
For this reason, the differential stress tensor $[\psi_{old}]$ (from which $\{F^*\}$ was originally
calculated) must be subtracted from $[\tilde{\sigma}]$. This permits the new net stresses, the
new differential stresses and thus the new joint load vector to be calculated as
follows:

$$
\begin{aligned}
{[\sigma]} &= [\tilde{\sigma}] - [\psi_{old}] \\
{[\sigma^*]} &= f([\sigma], [\Omega]) \\
{[\psi_{new}]} &= [\sigma^*] - [\sigma] \\
{\{F^*\}_{new}} &= f(\{\psi_{new}\})
\end{aligned}
$$

(6) The strain condition present can now be given on the element for every point with
the help of the $[B]$ matrix:

$$
\{\varepsilon\}^e_{(x,y,z)} = [B]^e_{(x,y,z)} \{u\}
$$

This strain condition is independent of whether or not cracks exist in the element
in question because the strain condition depends solely on the joint displacement and
the chosen shape function.

Once the matrix of elasticity $[D]^e$ is known, the stresses at every point are also
known:

$$
\{\sigma\}^e = \{\sigma\}^e_{(x,y,z)} = [D]^e \{\varepsilon\}^e = [D]^e [B]^e \{u\}
$$

As far as the normal stresses are concerned, these must occur in this form in the
rock bridges of the jointed rock. The shear stresses, however, are still those of the
isotropic system, but the desired net shear stresses can be calculated from these very
simply.

If $[\sigma]$ is the stress tensor equivalent to $\{\sigma\}^e$, then

$$
\begin{aligned}
{[\sigma']} &= [T]^T [\sigma][T] \\
{[\sigma']} &= [\sigma'_n] + [\sigma'_t] \\
{[\sigma^{*\prime}]} &= [\sigma'_t] \cdot [I - C_t \Omega']^{-1} + [\sigma'_n] \\
{[\sigma^*]} &= [T][\sigma^{*\prime}][T]^T
\end{aligned}
$$

Unfortunately, this procedure does not always bring the desired results. Some
systems have been seen to have a critical separation factor of the joints, above which
the solution no longer converges.

4.2.7 Calculation of the damage tensor

The sphere diagram is generally accepted for geomechanical representation of the fault system, namely the sum of all the 3D data of a network of joint surfaces. For this reason, calculation of the damage tensor proceeds from the assumption that the fault system is given in this form.

On first inspection the network of joints appears to be very complicated. When it is statistically analyzed by a suitable procedure, however, it is seen to contain main directions. Individual joints can be ascribed to joint patterns with various orientations. A joint pattern is understood as a number of joint planes that mainly have the same (or a similar) position in space.

Minen any several joint patterns there are (usually) two or three that are more pronounced in the frequency of their individual components and thus statistically significant as compared to other joint patterns or individual joints. These are known as main joint patterns and are mathematically defined by the damage tensor.

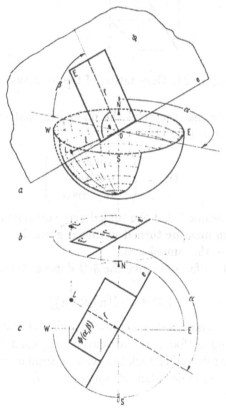

Figure 23: Sphere diagram [1]

In the sphere diagram (or in its representation) every joint is clearly defined by its penetration point diagram (Fig. 21). If we imagine the plane as being pushed to the

center of the sphere and if we take the normal vector to this plane, the desired point
is defined as the one where the surface of the sphere is penetrated. This point can be
more simply obtained by having the plane approach the sphere only until it touches
the sphere. This point is the desired penetration point.

In order to calculate the damage tensor, we need the penetration point diagram of
a maximum of three main joint patterns, whereby the particular damage tensor is used
to determine these. The tensor that describes the entire network of joints is the sum
of these.

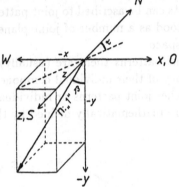

Figure 24: Orientation of the joint system

First, we need the normal vector of the main joint pattern in question, which is
easily obtained as shown in Fig. 22:

$$\{n\} = \left\{ \begin{array}{c} -\sin\beta\sin\alpha \\ -\cos\beta \\ \sin\beta\cos\alpha \end{array} \right\} \tag{47}$$

It is automatically assumed that the tunnel axis runs north-south. If this is not the
case, the sphere diagram must be turned so that α gives the "strike" of the main joint
pattern with reference to the tunnel axis.

$\{n\}$ and the separation factor ω give the 3-D damage tensor $[\Omega]$:

$$[\Omega] = \omega \cdot [\{n\} \otimes \{n\}] \tag{48}$$

If the rock is under a plane strain condition, it should also be described as a plane
problem since this is simpler than a 3D calculation. $[\Omega]$ must therefore be converted to
a 2D damage tensor that describes rock behavior as similar as possible to the original
tensor (in a 3D calculation under plane strain condition).
For a unit stress condition

$$[\sigma] = \left[\begin{array}{ccc} 1 & & \\ & 1 & \\ & & 0 \end{array} \right]$$

the pertinent net stresses $[\sigma^*]$ are calculated analogous to Chapter 4.2.4.:

$$[\sigma^*] = [\sigma] \cdot [\mathbf{I} - \Omega]^{-1} \tag{49}$$

$$[\sigma^*] = \begin{bmatrix} \sigma_{11}^* & \sigma_{12}^* & \sigma_{13}^* \\ \sigma_{21}^* & \sigma_{22}^* & \sigma_{23}^* \\ \sigma_{31}^* & \sigma_{32}^* & \sigma_{33}^* \end{bmatrix}$$

From the 3D stress tensor $[\sigma^*]$, those components are isolated that exclusively cause plane displacements in the area in question:

$$[\sigma^*]_{2\times2} = \begin{bmatrix} \sigma_{11}^* & \sigma_{12}^* \\ \sigma_{21}^* & \sigma_{22}^* \end{bmatrix}$$

The desired plane damage tensor is then given as follows:

$$[\Omega]_{2\times2} = [\mathbf{I}]_{2\times2} - [\sigma^*]_{2\times2}^{-1}$$

A check of the formula's accuracy is simple: in the 2D tensor the unit strain condition is given as $[\mathbf{I}]_{2\times2}$. The net stresses $[\tilde{\sigma}^*]$ are thus given as:

$$[\tilde{\sigma}^*] = [\tilde{\sigma}] \cdot [\mathbf{I} - \Omega]^{-1} = [\mathbf{I}] \cdot [\mathbf{I} - \Omega]^{-1} = [\mathbf{I} - \Omega]^{-1} \tag{50}$$

By inserting $[\Omega]_{2\times2}$, the desired net stresses are obtained for the chosen plane strain condition:

$$[\tilde{\sigma}^*] = [\sigma^*]_{2\times2}$$

4.2.8 Comparison of the results of this work with conventional methods

It is customary for the rock with its joints to be simulated by a material with an isotropic material behavior. For this reason, the results of such calculations are employed here in order to get a feel for the mechanical effect of the damage tensor.

The compressive test described in Fig. 23 is an example of this. The joint planes are inclined 45° away from the x axis and have a mean separation factor of 0.3. The damage tensor thus reads:

$$\{n\} = \left\{ \begin{array}{c} \cos 45° \\ \sin 45° \end{array} \right\}$$

$$[\Omega] = \omega \cdot [\{n\} \otimes \{n\}] = \begin{bmatrix} 0.15 & 0.15 \\ 0.15 & 0.15 \end{bmatrix}$$

The finite element network has the following appeareance:

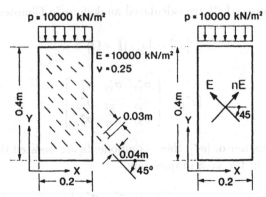

Figure 25: Comparison between damage theory and anisotropy

Fig. 25 shows the results of a damage tensor analysis for various combinations of C_t and C_n. Since the net shear stresses do not cause any additional displacements, the factor C_t has no influence on the size of the displacements. In this example these are solely dependent on C_n and $[\Omega]$. The maximum values of the displacements are summarized in a table at the end of this work.

The transverse isotropic model roughly describes the displacement behavior of the jointed rock when using the factor $(1 - C_n\omega)$ for n (the relation of the moduli of elasticity). The stresses calculated are, however, incorrect because the calculations were based entirely on different prerequisites.

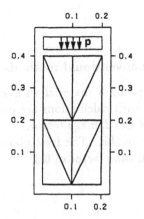

Figure 26: Numerical model

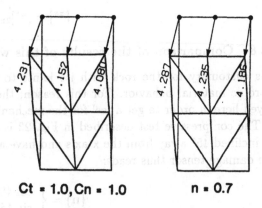

Figure 27: Displacement for different damage constants [25]

5 Three-dimensional numerical models

Particularly in tunnels with large displacements, however, numerous aspects of the construction works cannot be resolved without examining the static influence of the driving face. The influence of time, in particular, cannot be realistically accounted for in a 2D model. As shown in Fig. 28, complex interaction between driving velocity v, rheological displacement of the ground s and the time-dependent change in the elasticity modulus of the shotcrete combined with its viscoelasticity must be polued in order to define the stresses on the tunnel face. Since these processes take place directly behind the face, the use of 3D models is an absolute necessity.

5.1 Numerical modelling of the shotcrete lining

A sandwich shell element (Fig. 29) is used to simulate the shotcrete lining. Its multilayer system permits very good simulation of shotcrete application.

The isoparametric sandwich shell with quadratic shape function (ISSQ) [30] is based on Ahmad's concept for degenerated isoparametric elements [29], whereby the element geometry is described by the mid-surface. According to Fig. 30, every node of the shell can be determined by means of the mid-surface's coordinates, the shell thickness t and the v_3 vector, a vector normal to mid-surface:

$$\left\{ \begin{array}{c} x \\ y \\ z \end{array} \right\} = \left\{ \begin{array}{c} x_0 \\ y_0 \\ z_0 \end{array} \right\} + v_3 \frac{t}{2} \zeta \tag{51}$$

with

$$\left\{ \begin{array}{c} x_0 \\ y_0 \\ z_0 \end{array} \right\} = \frac{1}{2} \left\{ \left\{ \begin{array}{c} x_{top} \\ y_{top} \\ z_{top} \end{array} \right\} + \left\{ \begin{array}{c} x_{bot} \\ y_{bot} \\ z_{bot} \end{array} \right\} \right\} \qquad v_3 = \frac{1}{t} \left\{ \left\{ \begin{array}{c} x_{top} \\ y_{top} \\ z_{top} \end{array} \right\} - \left\{ \begin{array}{c} x_{bot} \\ y_{bot} \\ z_{bot} \end{array} \right\} \right\}$$

Interpolation of the coordinates from the element coordinates is performed with the help of the shape function

$$\left\{\begin{array}{c} x(\xi,\eta,\zeta) \\ y(\xi,\eta,\zeta) \\ z(\xi,\eta,\zeta) \end{array}\right\} = \sum_{i=1}^{n} N_i(\xi,\eta) \left\{\begin{array}{c} x_{0,i} + \frac{1}{2}\zeta t_i v_{3,x_i} \\ y_{0,i} + \frac{1}{2}\zeta t_i v_{3,y_i} \\ z_{0,i} + \frac{1}{2}\zeta t_i v_{3,z_i} \end{array}\right\}$$

$$= [N]^T \left\{\begin{array}{c} \{x_0\} + \frac{1}{2}\zeta\{t\} \otimes \{V_{3x}\} \\ \{y_0\} + \frac{1}{2}\zeta\{t\} \otimes \{V_{3y}\} \\ \{z_0\} + \frac{1}{2}\zeta\{t\} \otimes \{V_{3z}\} \end{array}\right\} \tag{52}$$

whereby

$$\{a\} \otimes \{b\} = \{a_1 b_1 \quad a_2 b_2 \quad a_3 b_3 \quad \cdots\}^T$$

Interpolation of the displacements is performed in the same way as for the coordinates

$$\left\{\begin{array}{c} u \\ v \\ w \end{array}\right\} = \left\{\begin{array}{c} u_0 \\ v_0 \\ w_0 \end{array}\right\} + \frac{t}{2}\zeta \left[\left\{\begin{array}{c} v_1 x \\ v_1 y \\ v_1 z \end{array}\right\} \alpha - \left\{\begin{array}{c} v_2 x \\ v_2 y \\ v_2 z \end{array}\right\} \beta\right] \tag{53}$$

whereby α and β represent the rotations around the local axes x_L and y_L, and thus the displacement from the element node displacements

$$\left\{\begin{array}{c} u(\xi,\eta,\zeta) \\ v(\xi,\eta,\zeta) \\ w(\xi,\eta,\zeta) \end{array}\right\} = \sum_{i=1}^{n} N_i(\xi,\eta) \left\{\begin{array}{c} u_{0,i} + \frac{1}{2}\zeta t_i(v_{1,x_i}\alpha - v_{2,x_i}\beta_i) \\ v_{0,i} + \frac{1}{2}\zeta t_i(v_{1,y_i}\alpha - v_{2,y_i}\beta_i) \\ w_{0,i} + \frac{1}{2}\zeta t_i(v_{1,z_i}\alpha - v_{2,z_i}\beta_i) \end{array}\right\}$$

$$= [N]^T \left\{\begin{array}{c} \{u_0\} + \frac{1}{2}\zeta\{t\} \otimes (\{v_{1,x}\} \otimes \{\alpha\} - \{v_{2,x}\} \otimes \{\beta\}) \\ \{v_0\} + \frac{1}{2}\zeta\{t\} \otimes (\{v_{1,y}\} \otimes \{\alpha\} - \{v_{2,y}\} \otimes \{\beta\}) \\ \{w_0\} + \frac{1}{2}\zeta\{t\} \otimes (\{v_{1,z}\} \otimes \{\alpha\} - \{v_{2,z}\} \otimes \{\beta\}) \end{array}\right\} \tag{54}$$

The strains in the element are as follows

$$\{\epsilon_L\} = \left\{\begin{array}{c} \epsilon_{x_L} \\ \epsilon_{y_L} \\ \gamma_{x_L y_L} \\ \gamma_{x_L z_L} \\ \gamma_{y_L z_L} \end{array}\right\} = \left\{\begin{array}{c} u'_{0,x_L} \\ v'_{0,y_L} \\ u'_{0,y_L} + v'_{0,x_L} \\ w'_{0,x_L} + \alpha \\ w'_{0,y_L} - \beta \end{array}\right\} + z_L \left\{\begin{array}{c} \alpha_{x_L} \\ -\beta_{y_L} \\ \alpha_{y_L} - \beta_{x_L} \\ 0 \\ 0 \end{array}\right\} \tag{55}$$

$$u'_{0,x_L} = \frac{\delta u_0}{\delta x_L} \qquad u'_{0,y_L} = \frac{\delta u_0}{\delta y_L}$$

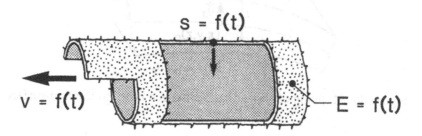

Figure 28: Influences of time: driving velocity, viscoplasticity of rock, viscoelasticity of shotcrete

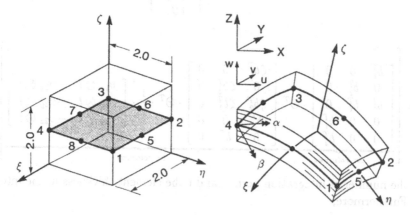

Figure 29: Sandwich shell element for simulation of the shotcrete lining

The displacement function used belongs to the family of Lagrange Polynomes and is of C_1 continuity

$$
\begin{aligned}
N_i &= \tfrac{1}{4}\zeta\zeta_i(1 + \zeta\zeta_i)\,\eta\eta_i\,(1 + \eta\eta_i) & i &= 1, 2, 3, 4 \\
N_i &= \tfrac{1}{2}\zeta\zeta_i(1 + \zeta\zeta_i)(1 - \eta^2) & i &= 6, 8 \\
N_i &= \tfrac{1}{2}\eta\eta_i(1 + \eta\eta_i)(1 - \zeta^2) & i &= 5, 7 \\
N_i &= (1 - \zeta^2)(1 - \eta^2) & i &= 9
\end{aligned}
\tag{56}
$$

Using (56) and (55), the strains can be found by differentiation:

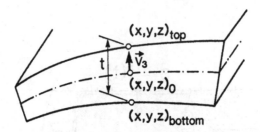

Figure 30: Mapping of three-dimensional geometry to the mid-surface

$$\{\epsilon_L\} = [B] \begin{Bmatrix} \{u_0\} \\ \{v_0\} \\ \{w_0\} \\ \{\alpha\} \\ \{\beta\} \end{Bmatrix} \tag{57}$$

$$[B] = \underbrace{\begin{bmatrix} b_1^T & 0 & 0 \\ 0 & b_2^T & 0 \\ b_2^T & b_1^T & 0 \\ 0 & 0 & b_1^T \\ 0 & 0 & b_2^T \end{bmatrix} \cdot \Theta^T \quad \begin{bmatrix} \zeta b_1^T & 0 & 0 \\ 0 & \zeta b_2^T & 0 \\ \zeta b_2^T & \zeta b_1^T & 0 \\ b_3^T & 0 & 0 \\ 0 & b_3^T & 0 \end{bmatrix} \cdot \Theta^T \cdot \frac{1}{2} \cdot \begin{bmatrix} v_{1x} \otimes t & -v_{2x} \otimes t \\ v_{1y} \otimes t & -v_{2y} \otimes t \\ v_{1z} \otimes t & -v_{2z} \otimes t \end{bmatrix}}_{\substack{3n \text{ columns} \qquad\qquad\qquad\qquad 2n \text{ columns}}} \tag{58}$$

n is the number of integration points, and t the element thickness in the integration points. Furthermore,

$$\begin{aligned} \{b_1\}^T &= [N]_{\eta,\zeta}^T f_{11} + [N]_{\eta,\eta}^T f_{12} \\ \{b_2\}^T &= [N]_{\eta,\zeta}^T f_{21} + [N]_{\eta,\eta}^T f_{22} \\ \{b_3\}^T &= [N]_{\eta}^T f_{33} \end{aligned} \tag{59}$$

The coefficient f_{ij} results from

$$[F] = \begin{bmatrix} f_{11} & f_{12} & f_{13} \\ f_{21} & f_{22} & f_{23} \\ f_{31} & f_{32} & f_{33} \end{bmatrix} = [Q]^T [J]^{-1} \tag{60}$$

whereby $[\theta]$ is the direction cosine matrix and $[J]^{-1}$ the inverted Jacobian matrix. The stiffness matrix of the multi-layered shell elements is thus found as follows

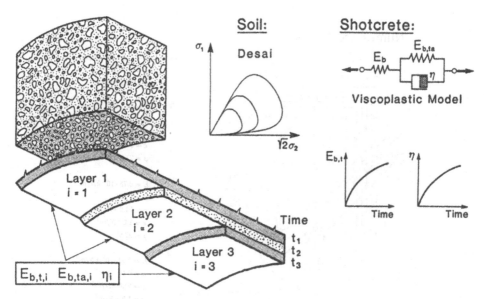

Figure 31: Multilayered shotcrete - soil system

$$[k] = \sum_{k=1}^{l} \int_{-1}^{1} \int_{-1}^{1} \int_{-1}^{1} [B]^T [Q]^{(k)} [B] \frac{s_k}{t} \det |J| \, d\xi \, d\eta \, d\zeta_k \tag{61}$$

The matrix $[Q]^{(k)}$ is the stress-strain equation of the individual layers. Integration is now only performed with the shell thickness s_k and is subsequently added for all layers l.

In order to account for the viscoelasticity of the shotcrete [31], the element load vector $\{F_{vel}\}$ must be determined from the strain increment $\{\Delta\epsilon_{vel}\}$

$$\{\Delta\epsilon_{vel}\} = (1 - e^{-\frac{E_{b,ta}}{\eta} \cdot \Delta t})(\frac{3}{2} \frac{\{\sigma_{tot}^D\}}{E_{b,ta}} - \{\epsilon_{vel}(t_0)\}) \tag{62}$$

This is performed with the following known equation

$$\{\Delta F_{vel}\} = \begin{Bmatrix} F_x \\ F_y \\ F_z \\ M_x \\ M_y \end{Bmatrix} = \left\{ \begin{matrix} \{F\} \\ \{M\} \end{matrix} \right\} = \int_V [B]^T [D]\{\Delta\epsilon vel\} \, dVol \tag{63}$$

whereby η represents the viscosity and $E_{b,ta}$ the time-dependent modulus of elasticity of the shotcrete.

The following equations for the viscoelastic additional forces can be set up for the existing element consisting of l layers:

Numerical Model

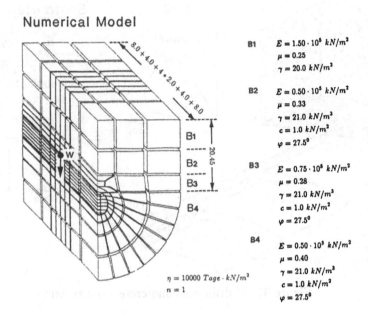

$\eta = 10000\ Tage \cdot kN/m^2$
$n = 1$

B1
$E = 1.50 \cdot 10^5\ kN/m^2$
$\mu = 0.25$
$\gamma = 20.0\ kN/m^3$

B2
$E = 0.50 \cdot 10^5\ kN/m^2$
$\mu = 0.33$
$\gamma = 21.0\ kN/m^3$
$c = 1.0\ kN/m^2$
$\varphi = 27.5^0$

B3
$E = 0.75 \cdot 10^5\ kN/m^2$
$\mu = 0.28$
$\gamma = 21.0\ kN/m^3$
$c = 1.0\ kN/m^2$
$\varphi = 27.5^0$

B4
$E = 0.50 \cdot 10^5\ kN/m^2$
$\mu = 0.40$
$\gamma = 21.0\ kN/m^3$
$c = 1.0\ kN/m^2$
$\varphi = 27.5^0$

Figure 32: Finite element model

$$\{F\} = \sum_{k=1}^{l} \int_{-1}^{1} \int_{-1}^{1} \int_{-1}^{1} [\Theta] \begin{bmatrix} b_1 & 0 & b_2 & 0 & 0 \\ 0 & b_2 & b_1 & 0 & 0 \\ 0 & 0 & 0 & b_1 & b_2 \end{bmatrix} [\Theta]^{(k)} \{\Delta\epsilon_{vel}^k\} \frac{s_k}{t} det|J|\, d\xi\, d\eta\, d\zeta_k \quad (64a)$$

$$\{M\} = \sum_{k=1}^{l} \int_{-1}^{1} \int_{-1}^{1} \int_{-1}^{1} \frac{1}{2} \begin{bmatrix} \{v_{1,x} \otimes \{t\} & \{v_{1,y} \otimes \{t\} & \{v_{1,z} \otimes \{t\} \\ -\{v_{2,x} \otimes \{t\} & -\{v_{2,y} \otimes \{t\} & -\{v_{2,z} \otimes \{t\} \end{bmatrix} \cdot$$
$$\cdot [\Theta] \begin{bmatrix} \zeta b_1 & 0 & \zeta b_2 & b_3 & 0 \\ 0 & \zeta b_2 & \zeta b_1 & 0 & b_3 \\ 0 & 0 & 0 & 0 & 0 \end{bmatrix} [\Theta]^{(k)} \{\Delta\epsilon_{vel}^k\} \frac{s_k}{t} det|J|\, d\xi\, d\eta\, d\zeta_k \quad (64b)$$

The load $\{\Delta F_{vel}\}$ that simulates the viscoelastic displacements of the shotcrete is redefined in every time step and added to the static loads.

Fig. 31 shows the application of the multi-layered model. The shotcrete lining was applied in three layers in this case, whereby each layer has different viscoelastic properties $E_{b,ta}$ and $\eta_{b,ta}$ according to the time of application t_0, t_1 or t_2.

The time-dependent displacements of the soil were treated with the flow rule after Desai [32]. The two regions were coupled by shifting the solid element and the shell element u, v, w. The rotations α and β were not coupled with the soil region.

5.2   Simulation of tunnel excavation

The method used to simulate excavation was not the simplified method, according to which the excavation is simulated in a numerical model by displacing the model without changing the system. This method is not suitable for simulating very jointed rock or for analyzing special fault zones.

Working from a plain strain condition, the excavation is simulated step-by-step by eliminating the elements. The assumption is that the elements concerned in one round are those from n to m. This means that the modulus of elasticity is assumed to be zero for these elements (65) and the stresses in the eliminated elements integrated after (66) and applied as extreme loads $\{F_{ex}\}$

$$E_i \approx 10^{-20} \qquad i = n, m \qquad\qquad (65)$$

$$\{F_{ex}\} = \sum_{i=n}^{m} \int_V [B]^T \{\sigma_0\}\, dV \qquad\qquad (66)$$

The application was performed on a one-track tunnel profile (Fig. 32) for which an appropriate time-displacement curve was available. The soil was loose, mainly quartenary deposits. The driving velocity was 0.666 m/day. The viscosity of the soil was modified as a parameter. The values $\eta = 10000, 25000$ and 50000 $days \cdot KN/m_2$ were calculated. The exponent for the viscosity rule was $n = 1.0$.

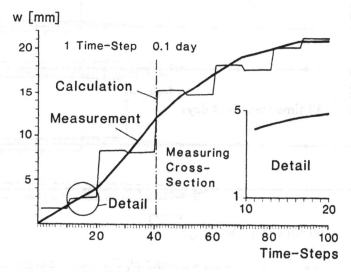

Figure 33: Comparison of measurements and viscoplastic calculations. Roof displacement.

The following characteristic values were chosen for the shotcrete:

Time	E_b [KN/m_2]	$E_{b,ta}$ [KN/m_2]	η [$days \cdot KN/m_2$]
$0 - 2days$	$15 \cdot 10^6$	$6 \cdot 10^6$	$6 \cdot 10^8$
$> 2days$	$25 \cdot 10^6$	$10 \cdot 10^6$	$10 \cdot 10^8$

Fig. 33 shows a comparison of the displacements calculated and those measured in the roof. The reason why the displacements in the numerical model have a pronounced step function is that in the excavation, for example of the top heading, this process of several hours takes place suddenly in the numerical model instead of slowly, step by step under actual site conditions. Furthermore, the measurement curves are smoothed since the measurements are not performed more often than twice a day. From the longitudinal displacement line for the roof and invert, as shown in Fig. 34, it can be seen that the displacements resulting from excavation start just in front of the face. Thus, the redistribution of stresses takes place in a very small region before and after the face.

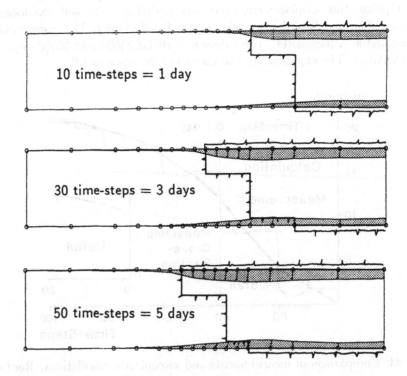

Figure 34: Excavation sequence and longitudinal roof and invert displacement

For various displacement velocities of the soil, and thus various viscosities, a comparison was made of the displacements in the roof (Fig. 35) and the longitudinal normal forces N_y (Fig. 36). Particularly in the case of normal forces, the displacement velocities can sometimes cause very different forces to occur. The longitudinal support effect is especially pronounced in the base of the crown. These forces result from the cantilever effect of the lining in the top heading.

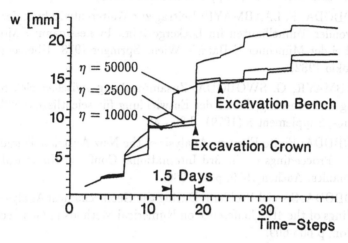

Figure 35: Parameter study of roof displacement. Influence of the soil viscosity

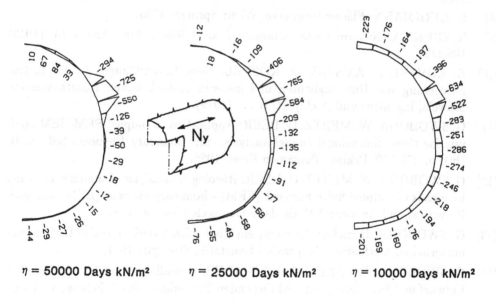

Figure 36: Longitudinal normal forces in the measuring cross section

References

[1]   L. MÜLLER, Salzburg: Der Felsbau, III. Bd., Tunnelbau, F.Enke 1987

[2]   M. BAUDENDISTEL: Zur Bemessung von Tunnelauskleidungen in wenig festem Gebirge. Rock Mechanics, Suppl. 2, 279-312 (1973)

[3]   G. SWOBODA, F. LAABMAYR: Beitrag zur Weiterentwicklung der Berechnung flachliegender Tunnelbauten im Lockergestein. In Lessmann: "Moderner Tunnelbau bei der Münchner U-Bahn". Wien: Springer 1978. Übersetzung: Peking 1980, Tokio 1982.

[4]   F. LAABMAYR, G. SWOBODA: Zusammenhang zwischen elektronischer Berechnung und Messung; Stand der Entwicklung für seichtliegende Tunnel. Rock Mechanics, Supplement 8 (1979), S 29 - 42.

[5]   G. SWOBODA: Finite Element Analysis of the New Austrian Tunneling Method (NATM). Proceedings of the 3rd International Conf. on Numerical Methods in Geomechanics, Aachen 1979, p 581-586.

[6]   G. SWOBODA: Special Problems During the Geomechanical Analysis of Tunnels. Proceedings of the 4th Conference on Numerical Methods in Geomechanics 1982, Edmonton, p 605-609.

[7]   O.C. ZIENKIEWICZ: The Finite Element Method. London: McGraw Hill 1982.

[8]   K. GIRKMANN: Flächentragwerke, Wien: Springer 1983.

[9]   K. GIRKMANN: Zum Halbraumproblem von Michell. Ing. Archiv 14, (1943) 106-112.

[10]   K. KOVARI, C. AMSTAD, J. KÖPPEL: Neue Entwicklungen in der Instrumentierung von Untertagbauten und anderen geotechnischen Konstruktionen, Schweiz, Ingenieur und Architekt, H 41. (1979).

[11]   G. SWOBODA, W. MERTZ, G. BEER: Application of coupled FEM-BEM analysis for three-dimensional tunnel analysis. in: Boundary Elements (ed. Q.H. Du), p. 537-550, Peking: Pergamon Press, 1986.

[12]   G. SWOBODA, W. MERTZ, G. BEER: Rheological analysis of tunnel excavation by means of coupled finite Element (FEM) - boundary element (BEM) analysis. Numerical and Analytical Methods in Geomechanics. (in press)

[13]   G. SWOBODA: Boundary Element, ein Bindeglied zwischen analy- tischen und numerischen Methoden. Baupraxis - Baustatik, Stuttgart 1987.

[14]   G. SWOBODA, F. LAABMAYR, I. MADER: Grundlagen und Entwicklung bei Entwurf und Berechnung im seichtliegenden Tunnelbau. Teil 2, Felsbau 4, (1986) S.184-187.

[15] R.E.GOODMAN, R.L.TAYLOR and T.BREKKE: A model for the mechanics of jointed rock. Journal of Soil Mechanics and Foundation Division, ASCE 94, p. 637-658. (1968).

[16] C.S. DESAI, M.M. ZAMAN, J.G. LIGHTNER, H.J. SIRIWARDANE: Thin-layer element for interfaces and joints. Internal Journal for Numerical and Analytical Methods in Geomechanics 8, p. 19-43 (1984).

[17] W. WITTKE. Static analysis for underground openings in jointed rock. in Ch.S. Desai, J.T. Christian: Numerical Methods in Geotechnical Engineering. McGraw Hill, New York 1977.

[18] O.C. ZIENKIEWICZ, C. DULLAGE: Analysis of non-linear problems in rock mechanics with particular reference to jointed rock systems. Proceedings of the 2nd International Congress on Rock Mechanics Sec., 8-14 (1970).

[19] T. MAINI, P. CUNDALL, J. MARTI, P. BERESFORD, N. LAST, M. ASGIAN: Computer modelling of jointed rock mass. Technical Reprint N-78-4. U.S. Army Engineers, Vicksburg, 1978.

[20] T. ASAI, M. NISHIMURA, T. SAITO, M. TERADA: Effects of rock bolting in discontinuous rock mass. 5th International Conference on Numerical Methods in Geomechanics, Nagoya, p. 1273-1280 (1985).

[21] M.G. KATONA: A simple contact-friction interface element with application to buried culverts. Internal Journal for Numerical and Analytical Methods in Geomechanics, 7, p. 371-384 (1983).

[22] NETDIG: Digitalisierung von Finite-Element-Netzen V 2.1 Manual University of Innsbruck 1985.

[23] M. BAUDENDISTEL: Zum Entwurf von Tunneln mit großem Ausbruchsquerschnitt. Rock Mechanics, Supplement 8, p. 75-100 (1979).

[24] K. SCHIKORA, T. FINK: Berechnungsmethoden moderner bergmännischer Bauweisen beim U-Bahnbau. Bauingenieur, 57, p. 193-198 (1982).

[25] T. KYOYA, Y. ICHIKAWA, T. KAWAMOTO: A damage mechanics theory for discontinous rock mass. 5th Int. Conf. on Num. Meth. in Geom., Nagoya, 1985, p.469-480

[26] G. SWOBODA, G. BEER: Städtischer Tunnelbau - Rechenmodelle und Resultatinterpretation als Grundlage für Planung und Bauausführung. Finite Elemente in der Baupraxis, München, 1984.

[27] G. BEER, G. SWOBODA: On the efficient Analysis of Shallow Tunnels. Comp. and Geomech. 1985, 1, pp. 15-31.

[28] G. SWOBODA: FINAL, Finite Element Analyse linearer und nichtlinearer Strukturen, Programmversion 6.1. Universität Innsbruck, 1988.

[29] S. AHMAD, B. IRONS, O.C. ZIENKIEWICZ: Analysis of Thick and Thin Shell
      Structures by Curved Finite Elements. Inter. J. Num. Meth. Eng., 1970, 2, pp.
      419 - 451.

[30] A. SCHMID: Beitrag zur Berechnung geschichteter Schalentragwerke mittels der
      Methode der Finiten Elemente. Dissertation, Innsbruck, 1988.

[31] W. MERTZ: Numerische Modelle drei-dimensionaler plastischer, visko-plastischer
      und visko-elastischer Spannungen und Verformungen in der Geomechanik unter
      besonderer Berücksichtigung seichtliegender Tunnel. Dissertation, Innsbruck,
      1988.

[32] C.S. DESAI: Somasundaram S., Frantziskonis, G., A hierarchical approach for
      constitutive modelling of geological materials. Inter. Jrnl. Num. and Anal.
      Meth. in Geom., 1986, 11.

[33] F. PACHER: Kennziffern des Flächengefüges. Geol. und Bauw. 24, 223ff, 1959.

SOME APPLICATIONS OF MATHEMATICAL PROGRAMMING
IN GEOMECHANICS

G. Gioda
Università di Udine, Udine, Italy
and
Politecnico di Milano, Milano, Italy

ABSTRACT

Some applications are discussed of mathematical programming algorithms to the solution of problems of geotechnical engineering concerning: a) elasto-plastic analysis, b) limit analysis, c) the determination of the average mechanical properties of soil or rock masses through the back analysis of in situ measurements and d) free surface seepage analyses in steady state regime. A common feature of the solution approaches here adopted for the mentioned problems is that they are based on the minimization, under equality or inequality constraints, of suitably defined functions. The main aspects of the various formulations are presented and some illustrative examples are discussed for each group of problems.

1. INTRODUCTION

In various branches of engineering the solution of relevant problems is amenable to the minimization of a function constrained by equations and inequalities which represents, in mathematical terms, a mathematical programming problem.

In some fields, like plasticity and structural mechanics, mathematical programming had a growing role in the recent years as a promising method for the analysis of engineering problems [1,2]. This is due the fact that mathematical programming: a) provides computer oriented solution procedures which are particularly suited for implementation in programs for automatic elaboration and b) provides a unified theoretical framework for the analysis of discrete problems and, consequently, turns out to be convenient for approaches based on discrete solution techniques like the finite element method.

In other fields of engineering, like geomechanics, mathematical programming had a less prominent role, and its applications have been mainly restricted to rigid-plastic or limit analysis [3]. However, it can be observed that the above mentioned reasons, supporting its use, hold also in geotechnical engineering and that in this field the range of application of mathematical programming is perhaps even larger than in structural engineering.

Here the formulations based on mathematical programming algorithms of some geotechnical engineering problems are presented with reference to elasto-plastic and limit analyses, to the so called back analysis (or calibration of material parameters on the basis of field measurements) and to unconfined seepage problems.

This work does not aim at providing an exhaustive overview of the applications of mathematical programming in geomechanics. Merely an attempt is made to show that these approaches are potentially applicable to engineering problems also in the field of geomechanics and that in some cases they could present advantages with respect to more traditional solution procedures.

2. ELASTO-PLASTIC ANALYSIS

The procedure for elasto-plastic stress analysis here illustrated is based on a quadratic programming algorithms [4,5,6] and subdivides the loading process of the geotechnical medium into a sequence of steps. The solution of each step rests on the following hypotheses:

- the yield condition and the plastic potential are linear functions of the stress components. In geometrical terms this is equivalent to assume a polyhedral yield surface in the stress space. Note that it is always possible to reduce a continuous, non-linear yield condition to this kind of surface through a suitable piecewise linearization, with a degree of approximation depending only on the chosen number of linearizing planes. In addition, in the case of hardening behaviour, the yield condition is also a linear function of the plastic multipliers [7], which represent a measure of the plastic strain increments.

- The stress-strain relationship is reversible within the generic loading step. This assumption is in contrast with the irreversible nature of the plastic behaviour. However, elasto-plastic laws can still be described even under this hypothesis if the plastic strains at the beginning of each loading step are updated to the total plastic strains obtained at the end of the previous step.

- The increase of the plastic deformation within each loading step is holonomic, i.e. elastic unloading from a plastic stress state is prevented. If unloading occurs, plastic strains are completely recovered. The influence of possible errors related to this hypothesis can be reduced to the desired degree of approximation by increasing the number of load increments.

Consider now the i-th loading step at the beginning of which all relevant quantities, denoted by index i-1, are known. The displacements, stresses and strains at the end of the step are

$$\underline{u}_i = \underline{u}_{i-1} + \Delta \underline{u}_i , \tag{1a}$$

$$\underline{\varepsilon}_i = \underline{\varepsilon}_{i-1} + \Delta \underline{\varepsilon}_i , \tag{1b}$$

$$\underline{\sigma}_i = \underline{\sigma}_{i-1} + \Delta \underline{\sigma}_i , \tag{1c}$$

where the increments $\Delta \underline{u}_i$, $\Delta \underline{\varepsilon}_i$ and $\Delta \underline{\sigma}_i$ depend on the external load increment $\Delta \underline{p}_i$. As customary in the standard finite element approach for stress analysis, the displacements are defined at the nodes of the mesh while stresses and strains are defined at the integration points within each element.

The stress and strain increments can be expressed in the following form

$$\Delta\underline{\sigma}_i = \Delta\underline{\sigma}_i^e + \Delta\underline{\sigma}_i^s \quad , \quad \Delta\underline{\varepsilon}_i = \Delta\underline{\varepsilon}_i^e + \Delta\underline{\varepsilon}_i^p \quad , \tag{2,3}$$

where

$$\Delta\underline{\sigma}_i^s = \underline{Z}\,\Delta\underline{\varepsilon}_i^p \quad , \quad \Delta\underline{\varepsilon}_i^e = \underline{D}^{-1}\,\Delta\underline{\sigma}_i \quad . \tag{4a,b}$$

In the above equations $\Delta\underline{\varepsilon}_i^e$ and $\Delta\underline{\varepsilon}_i^p$ are the elastic and plastic (non reversible) strain increments, $\Delta\underline{\sigma}_i^e$ is the elastic response in terms of stresses to the load increment Δp_i (obtained through the linear elastic analysis of the finite element system), $\underline{D}$ is the elastic constitutive matrix and $\Delta\underline{\sigma}_i^s$ are self-equilibrated stresses related to the plastic strains through matrix $\underline{Z}$. In particular, the columns of matrix $\underline{Z}$ represent the vectors of the self-equilibrated stresses induced in the finite elements by unit strains imposed at the integration points.

In order to express the constitutive material law let use the following notation with reference to the a general piecewise linear yield condition with hardening and non-associated flow rule (cf. fig.1):

(i) r is the number of planes of the piecewise linear yield surface;

(ii) $\underline{N}$ is the matrix having as j-th column the gradient of the j-th yield plane, i.e. the outward unit vector normal to it;

(iii) $\underline{\Phi}_i$ is the vector of the yield functions of each yield plane

$$\underline{\Phi}_i = \underline{\Phi}_{i-1} + \Delta\underline{\Phi}_i \quad , \tag{5}$$

and $\underline{\Phi}_{i-1}$ is the vector of the yield functions at the beginning of the load step

$$\underline{\Phi}_{i-1} = \underline{N}\,\underline{\sigma}_{i-1} - \underline{k}_o \quad . \tag{6}$$

$\underline{k}_o$ is the vector collecting the initial radial distances of each yield plane from the origin. The current elastic range in the stress space is represented by the domain $\underline{\Phi}_i < \underline{0}$.

(iv) $\underline{V}$ is the matrix of the gradients of the r plastic potentials having as columns the r outward unit vectors which give the direction of the plastic strain vector for each yield plane.

(v) $\underline{H}$ is the matrix of the hardening coefficients that governs the translation of the yield planes due to plastic deformations.

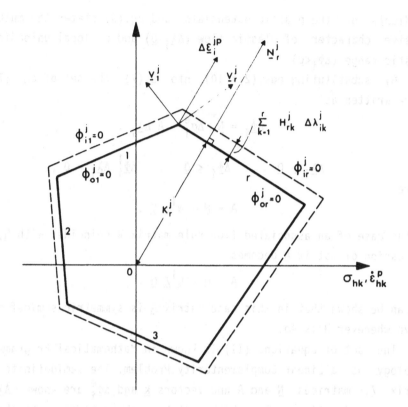

Fig. 1 Notation for the piecewise linear yield surface (i-th loading
        step and j-th finite element).

The constitutive equations for a stress state belonging to the yield condition is expressed by the following relationships

$$\Delta\underline{\Phi}_i = \underline{N}^T \Delta\underline{\sigma}_i - \underline{H} \Delta\underline{\lambda}_i \leq \underline{0} ,$$ (7)

$$\Delta\underline{\lambda}_i \geq \underline{0} \quad , \quad \Delta\underline{\Phi}_i^T \Delta\underline{\lambda}_i = 0 ,$$ (8,9)

$$\Delta\underline{\varepsilon}_i^p = \underline{V} \Delta\underline{\lambda}_i ,$$ (10)

where $\Delta\underline{\Phi}_i$ and $\Delta\underline{\lambda}_i$ are, respectively, the vector of the r non-positive increments of the yield functions and the vector of the non-negative plas-
tic multipliers (eq.8). Eq.(7) defines the geometry of the piecewise linear yield surface. Eq. (10) defines the plastic flow rule, relating the
increment of the plastic strains to the plastic multipliers through the

gradients or the plastic potentials, and eq.(9) states the mutually ex-
clusive character of plastic flow ($\Delta\lambda_i > \underline{0}$) and of local unloading in the
elastic range ($\Delta\underline{\Phi}_i < \underline{0}$).

By substituting eqs.(2,4,10) into eq.(7), the set of eqs.(7-10) can
be re-written as

$$\Delta\underline{\Phi}_i = \underline{N}^T \Delta\underline{\sigma}_i^e - \underline{A} \Delta\underline{\lambda}_i ,$$

$$\Delta\underline{\lambda}_i \geq \underline{0} \quad , \quad \Delta\underline{\Phi}_i \leq \underline{0} \quad , \quad \Delta\underline{\Phi}_i^T \Delta\underline{\lambda}_i = 0 , \tag{11}$$

where

$$\underline{A} = \underline{H} - \underline{N}^T \underline{Z} \underline{V} . \tag{12}$$

In the case of an associated flow rule matrix $\underline{V}$ coincides with $\underline{N}$, and the
expression of matrix $\underline{A}$ becomes

$$\underline{A} = \underline{H} - \underline{N}^T \underline{Z} \underline{N} . \tag{13}$$

It can be shown that in this case matrix $\underline{A}$ is symmetric semidefinite positive whenever $\underline{H}$ is so.

The set of equations (11) is known in Mathematical Programming ter-
minology as a Linear Complementarity Problem. The semidefinite negative
matrix $\underline{Z}$, matrices $\underline{N}$ and $\underline{A}$ and vectors $\underline{k}$ and $\Delta\underline{\sigma}_i^e$ are known. $\Delta\underline{\lambda}_i$ is the
unknown non-negative vector which must be such as to generate through the
first of (11) a non-positive vector $\Delta\underline{\Phi}_i$ orthogonal to it. Having determined the solution vector $\Delta\underline{\lambda}_i$, the increment of stresses and strains can be
calculated through eqs.(10) and (2-4).

Even though algorithms for the solution of the Linear Complementarity Problem (11) are available, for instance a solution procedure applicable in the presence of a positive semidefinite matrix $\underline{A}$ has been presented in [8], from the computational view point it is convenient to re-
write the problem in terms of the minimization of a quadratic function.
This, in fact, would allow to use for solution well known minimization
algorithms already implemented in available computer codes.

Such a step is possible in the case of an associated flow rule since
the Linear Complementarity Problem (11) with positive semidefinite matrix
$\underline{A}$ can be interpreted as Kuhn-Tucker condition [9] of the following Quadratic Programming problem

$$\max_{\Delta\underline{\lambda}_i} \{ -\frac{1}{2} \Delta\underline{\lambda}_i^T (\underline{H} - \underline{N}^T \underline{Z} \underline{N}) \Delta\underline{\lambda}_i + \Delta\underline{\lambda}_i^T (\underline{N}^T \Delta\underline{\sigma}_i^e + \underline{\Phi}_{i-1}) \} \quad , \quad \Delta\underline{\lambda}_i \geq \underline{0} . \tag{14}$$

The case of elastic-perfectly plastic behaviour is derived by assuming $\underline{H}=\underline{0}$. The Quadratic Programming problem then reads

$$\max_{\Delta\underline{\lambda}_i} \{\tfrac{1}{2} \Delta\underline{\lambda}_i^T (\underline{N}^T \underline{Z} \underline{N}) \Delta\underline{\lambda}_i + \Delta\underline{\lambda}_i^T (\underline{N}^T \Delta\underline{\sigma}_i^e + \underline{\Phi}_{i-1})\} \quad , \quad \Delta\underline{\lambda}_i \geq \underline{0} \ . \tag{15}$$

An important aspect of the solution of problems (14) or (15) is their large number of variables, which is given by the number of finite elements in the mesh multiplied by the number of integration points for each element multiplied by the number of planes linearizing the yield condition. The number of variables, i.e. of the entries of vector $\Delta\underline{\lambda}_i$, can be drastically reduced by ignoring the yield planes that are unlikely to be activated during the loading step and that can be chosen on the basis of an estimate of the stress increment obtained from the elastic response $\Delta\underline{\sigma}_i^e$. Then during solution the excluded yield planes should be checked and some of them can be introduced again in the calculation if the variation of the stress state is such as to activate them.

Among various algorithms for function minimization, the procedures of the gradient type turn out to be particularly suitable for the solution of the Quadratic Programming problem. They find the optimal solution through a climbing procedure along a piecewise linear path starting from any given feasible initial point. From the k-th intermediate point $\Delta\underline{\lambda}_i^k$ the climbing direction is generally chosen on the basis of the gradient $\underline{g}$ of the objective function at that point, namely, referring to problem (14), of

$$\underline{g}(\Delta\underline{\lambda}_i^k) = -(\underline{H} - \underline{N}^T \underline{Z} \underline{N}) \Delta\underline{\lambda}_i^k + \underline{N}^T \Delta\underline{\sigma}_i^e + \underline{\Phi}_{i-1} , \tag{16}$$

which represents the direction of the maximum rate of increment of the objective function.

Considering eqs.(4a) and (10), the gradient (16) can be re-written in the following form

$$\underline{g}(\Delta\underline{\lambda}_i^k) = \underline{N}^T \Delta\underline{\sigma}_i^s + [\underline{N}^T \Delta\underline{\sigma}_i^e - \underline{H} \Delta\underline{\lambda}_i^k + \underline{\Phi}_{i-1}] , \tag{17}$$

from which it appears that, apart from the term in square brackets, the gradient is related to the distribution of self equilibrated stresses $\Delta\underline{\sigma}_i^s$ generated by imposing to the finite element system the plastic strains

$$\Delta\underline{\varepsilon}_i^{pk} = \underline{N} \Delta\underline{\lambda}_i^k \ . \tag{18}$$

As to the term in square brackets, it can be easily evaluated on the basis of quantities which are known from the beginning of the step ($\underline{N}$, $\Delta\underline{\sigma}_i^e$, $\underline{H}$ and $\underline{\Phi}_{i-1}$) and of the vector $\Delta\underline{\lambda}_i^k$ that represents the point at which the gradient has to be evaluated. Consequently, each step of the iterative minimization process requires the evaluation of the stress distribution due to given strains imposed on the system and matrix $\underline{A}$ (cf. eq.13) has never to be determined, with a consequent reduction of the computational time.

As an example, the Quadratic Programming approach has been applied to the evaluation of the load-settlement curve of a shallow strip foundation using the finite element grid depicted in fig.2. The results of calculations are shown in fig.3 and compared with those obtained with a standard iterative technique for elasto-plastic stress analysis based on Newton-Raphson method.

3. LIMIT ANALYSIS

The finite element approach for the evaluation of collapse loads described here [10] is based on the static theorem of limit analysis and requires the subdivision of the rigid-plastic medium into triangular, linear stress elements within which the stresses fulfill the equilibrium equations on infinitesimal volume. This choice leads to numerically evaluated collapse loads which are always smaller than the real ones, with an approximation that depends on the shape of the adopted mesh and on the number of elements. Note that, the use of a "kinematic" approach to limit analysis, and consequently of standard "displacement" elements, would have led to collapse loads always larger than the actual one. The problem is studied in plane strain conditions in the x-y plane taking into account the own weight of the soil [11], the influence of which is often non negligible in geotechnical applications.

The equilibrium finite element adopted in the calculations, within which the stress components σ_x, σ_y and τ_{xy} have linear variation, has triangular shape and its nodal variables are represented by the three stress components for each node. Note that while in the standard displacement elements all the nodal variables (i.e. the nodal displacements)

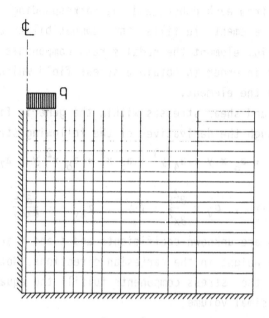

Fig. 2 Finite element mesh for the elasto-plastic analysis of a strip footing on a purely cohesive, weightless soil.

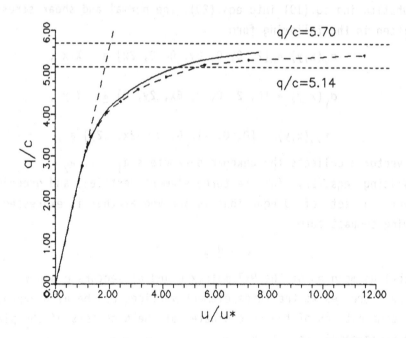

Fig. 3 Non dimensional load q vs. settlement u curve for the strip load on a purely cohesive, weightless soil. QP approach (solid line) Newton-Raphson method (dashed line); c = cohesion; u* = settlement at the elastic limit.

are independent from each other, and the corresponding strain distribution within the element fulfills the compatibility equations, in the
adopted equilibrium element the nodal stress components have to be related to each other in order to obtain a stress field which is equilibrated
everywhere within the element.

The normal and shear stresses within the generic finite element can
be expressed through the derivatives of the following stress function F,

$$F = a_1 \cdot y^2 + a_2 \cdot x^2 + a_3 \cdot x \cdot y + a_4 \cdot y^3 + a_5 \cdot x^3 + a_6 \cdot x^2 \cdot y + a_7 \cdot x \cdot y^2 , \tag{19}$$

$$\sigma_x = \frac{\partial^2 F}{\partial y^2} - X \cdot x \quad , \quad \sigma_y = \frac{\partial^2 F}{\partial x^2} - Y \cdot y \quad , \quad \tau_{xy} = - \frac{\partial^2 F}{\partial x \partial y} \quad , \tag{20a,b,c}$$

where $a_1, \ldots a_7$ are unknown coefficients and X and Y are the components
of the soil own weight in the cartesian directions. Note that the above
definitions for the stress components fulfill the equations of equilibrium on infinitesimal volume

$$\frac{\partial \sigma_x}{\partial x} + \frac{\partial \tau_{yx}}{\partial y} + X = 0 \quad , \quad \frac{\partial \sigma_y}{\partial y} + \frac{\partial \tau_{xy}}{\partial x} + Y = 0 \quad . \tag{21a,b}$$

By substituting eq.(19) into eqs.(20), the normal and shear stresses can
be written in the following form

$$\sigma_x(x,y) = [2, 0, 0, 6y, 0, 0, 2x] \underline{a} - X \cdot x , \tag{22a}$$

$$\sigma_y(x,y) = [0, 2, 0, 0, 6x, 2y, 0] \underline{a} - Y \cdot y , \tag{22b}$$

$$\tau_{xy}(x,y) = [0, 0, -1, 0, 0, -2x, -2y] \underline{a} , \tag{22c}$$

where vector $\underline{a}$ collects the unknown parameters $a_1, \ldots a_7$.

Writing eqs.(22) for the three element vertices, and grouping them
together, a set of 9 equations is arrived at that is expressed in the
following compact form

$$\underline{\sigma} = \underline{R} \, \underline{a} - \underline{W} , \tag{23}$$

with obvious meaning of the 9x7 matrix $\underline{R}$ and of vectors $\underline{\sigma}$ and $\underline{W}$.

The stresses at the three element vertices can be also expressed as
linear combinations of the co-ordinates of the N corners of the piecewise
linear yield condition

$$\underline{\sigma} = \underline{S} \, \underline{\alpha} \quad . \tag{24}$$

Here, $\underline{\alpha}$ is a 3N vector of unknown coefficients and $\underline{S}$ is the 9x3N matrix of the corner co-ordinates.

Now eq.(23) and (24) can be partitioned as follows

$$\underline{\sigma}_1 = \underline{R}_1 \, \underline{a} - \underline{W}_1 \quad , \quad \underline{\sigma}_2 = \underline{R}_2 \, \underline{a} - \underline{W}_2 \quad , \tag{25a,b}$$

$$\underline{\sigma}_1 = \underline{S}_1 \underline{\alpha} \quad , \quad \underline{\sigma}_2 = \underline{S}_2 \underline{\alpha} \quad , \tag{26a,b}$$

where $\underline{\sigma}_1$ collects the first 7 stress components at the element vertices and $\underline{\sigma}_2$ collects the remaining 2 components.

Combining eqs.(25,26), and taking into account that the 7x7 matrix $\underline{R}_1$ is non singular, the following relationship is arrived at

$$\underline{E} \, \underline{\alpha} = \underline{X} \quad , \tag{27}$$

where

$$\underline{E} = [\underline{R}_2 \, \underline{R}_1^{-1} \, \underline{S}_1 - \underline{S}_2] \quad , \quad \underline{X} = \underline{W}_2 - \underline{R}_2 \, \underline{R}_1^{-1} \, \underline{W}_1 \; . \tag{28a,b}$$

Eq.(27) represents the pair of scalar relationships which are necessary to enforce the equilibrium between the stress components $\underline{\sigma}_1$ and $\underline{\sigma}_2$ when an arbitrary vector $\underline{\alpha}$ is given. In fact, the most general equilibrated linear stress field is defined by seven independent coefficients or by 7 independent stress components. Since vector $\underline{\alpha}$ consists of the 9 stress components at the element vertices, two constraints are necessary in order to reduce to 7 the stresses that can be arbitrarily assigned.

The interelement equilibrium is fulfilled if the stress components normal and tangent to each element side are equal to the corresponding components of the adjacent element, or to the external linear load distribution along that side. Such stress components, ordered in vector $\underline{Q}$ having 12 components (cf.fig.4), are related to the nodal stresses $\underline{\sigma}$ through a matrix $\underline{C}$ depending on the direction cosines of the outward unit vectors normal to the element sides

$$\underline{Q} = \underline{C} \, \underline{\sigma} = \underline{C} \, \underline{S} \, \underline{\alpha} \; . \tag{29}$$

Now all vectors and matrices of the finite elements are assembled into super-vectors and matrices pertaining to the whole structure. For sake of simplicity these quantities will be denoted in the following with the same symbols previously used for the corresponding quantities at the element level. Then, taking into account eq.(29), the overall equilibrium of the finite element grid can be expressed as

$$\underline{B} \, \underline{Q} = \underline{B} \, \underline{C} \, \underline{S} \, \underline{\alpha} = \psi \, \underline{F} \, , \qquad (30)$$

where $\underline{B}$ is a matrix, having entries 0 or ± 1, which is used to impose the equilibrium of the boundary stresses corresponding to each element side by suitably summing the contribution of the adjacent elements, $\underline{F}$ is the vector of the normal and tangential tractions applied at the end points of the external sides of the mesh and ψ is the load factor.

The respect of the yield condition is ensured if the sum of the $\underline{\alpha}$ coefficients corresponding to each node of the finite element grid is not greater than 1 and if all α coefficients are non negative. These conditions can be expressed as

$$\underline{D} \, \underline{\alpha} \le \underline{U} \quad , \quad \underline{\alpha} \ge \underline{0} \, , \qquad (31a,b)$$

where $\underline{D}$ is a matrix having entries equal to 0 and 1 and $\underline{U}$ is the unit vector.

Finally, on the basis of the static theorem of limit analysis the collapse load is evaluated by solving a Linear Programming problem which consists in finding the maximum values of ψ and $\underline{\alpha}$ that fulfill the following set of equations

$$\underline{B} \, \underline{C} \, \underline{S} \, \underline{\alpha} - \psi \, \underline{F} = \underline{0} \quad , \quad \underline{E} \, \underline{\alpha} = \underline{X} \quad , \quad \underline{D} \, \underline{\alpha} \le \underline{U} \quad , \quad \underline{\alpha} \ge \underline{0} \quad , \quad \psi \ge 0 \, . \qquad (32)$$

A test problem solved with the described finite element technique, concerning the determination of the limit vertical load acting on a shallow strip footing, is shown in fig.5. The soil is assumed to be purely cohesive, its own weight is neglected and no lateral surface load are present. The numerical analysis leads to a value of the limit load equal to 4.96c. This can be considered as an acceptable approximation of the coefficient of bearing capacity N_c the value of which, depending on the solution adopted, ranges between 5.14 and 5.7. The difference between these values and the numerical solution is partially due to the limited number of elements adopted in the mesh and to the fact that static approaches always underestimate the actual value of the collapse load.

4. BACK ANALYSIS OF MATERIAL PARAMETERS

Back analysis or calibration procedures are often used in soil and rock engineering to refine the available information on the mechanical

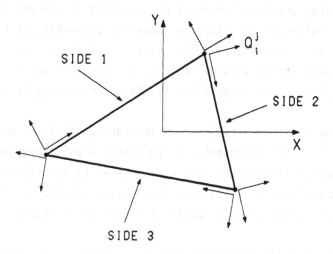

Fig. 4 Traction components on the sides of a triangular "equilibrium" element (i-th side, i=1,2,3; j-th traction component, j=1,2).

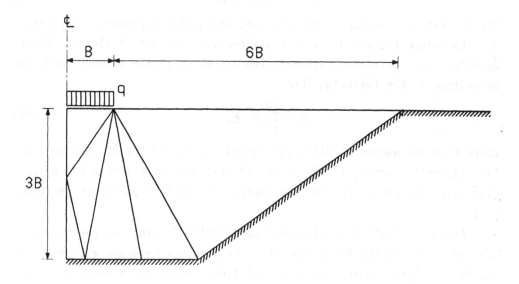

Fig. 5 Mesh of equilibrium elements for evaluating the bearing capacity of a strip footing on a purely cohesive, weightless soil.

characteristics of soil/rock masses to be adopted in the design of complex works. They are based on function minimization procedures and aim at
finding the values of the "average" material properties of the soil/rock
mass that, when introduced in the stress analysis of a construction or
excavation work or of a large scale in situ test, lead to results (e.g.
displacements) as close as possible to the corresponding in situ measurements.

   An important characteristic of the back analysis of material parameters is that it represents a non linear problem even when a linear
elastic behaviour is assumed. This can be shown by describing a technique
for the back analysis of elastic constants [12,13] based on a finite element approach originally proposed for structural engineering problems
[14].

   In order to apply this procedure it is necessary to establish a linear relationship between the stiffness matrix of each finite element $\underline{K}^e$
and the unknown material parameters. Such a relationship is obtained, in
the case of isotropic material, by describing the elastic behaviour in
terms of bulk B and shear G moduli

$$\underline{K}^e = B \ \underline{K}_B^e + G \ \underline{K}_G^e \ . \tag{33}$$

The two matrices on the right hand side of eq.(33) represent, respectively, the volumetric and deviatoric stiffnesses of the e-th element. Consequently, the stiffness matrix of the assembled finite element model can
be written in the following form

$$\underline{K} = \sum_{1i}^{2n} p_i \ \underline{K}_i \ , \tag{34}$$

where n is the number of different materials, p_i are the 2n unknown elastic parameters and $\underline{K}_i$ is the assembled stiffness matrix obtained by setting all the parameters to zero except for the i-th parameter set equal
to 1.

   Assuming that m displacement components of points of the rock mass
are measured in the field, and that these points coincide with nodes of
the finite element mesh, the system of linear equations governing the behaviour of the finite element discretization can be partitioned as follows

$$\begin{bmatrix} \underline{K}_{11} & \underline{K}_{12} \\ \underline{K}_{21} & \underline{K}_{22} \end{bmatrix} \cdot \begin{Bmatrix} \underline{u}_1^* \\ \underline{u}_2 \end{Bmatrix} = \begin{Bmatrix} \underline{f}_1 \\ \underline{f}_2 \end{Bmatrix} , \tag{35}$$

where vector $\underline{u}_1^*$ collects the m measured displacement components, and $\underline{f}_1$ and $\underline{f}_2$ are known nodal force vectors.

A static condensation of eq.(35) leads to

$$(\underline{K}_{11} - \underline{Q} \ \underline{K}_{21}) \ \underline{u}_1^* = \underline{f}_1 - \underline{Q} \ \underline{f}_2 , \tag{36}$$

where

$$\underline{Q} = \underline{K}_{12} \ \underline{K}_{22}^{-1} . \tag{37}$$

Taking into account eq.(34), eq.(36) can be written in the following form

$$\sum_1^{2n} p_i \ \underline{r}_i = \underline{f}_i - \underline{Q} \ \underline{f}_2 , \tag{38}$$

where

$$\underline{r}_i = (\underline{K}_{11,i} - \underline{Q} \ \underline{K}_{21,i}) \ \underline{u}_1^* . \tag{39}$$

The stiffness matrices in eq.(39) are obtained by partitioning matrix $\underline{K}_i$ with the same criteria used in eq.(35).

Grouping the unknown elastic parameters in the 2n vector $\underline{p}$, and grouping vectors $\underline{r}_i$ in the mx2n matrix $\underline{R}$

$$\underline{R} = [\underline{r}_1 \ | \ \underline{r}_2 \ | \ \dots \ | \ \underline{r}_{2n}] , \tag{40}$$

eq.(38) yields the following relationship

$$\underline{R} \ \underline{p} = \underline{f}_1 - \underline{Q} \ \underline{f}_2 \tag{41}$$

that governs the calibration problem.

Assuming that the number of data (m measured displacements) exceeds the number of unknowns (2n elastic constants), a standard least square minimization can be applied to eq.(41) leading to the following non linear equation system (note in fact that the matrix of coefficients $\underline{R}$ depends, through matrix $\underline{Q}$, on the unknown vector $\underline{p}$)

$$\underline{R}^T \ \underline{R} \ \underline{p} = \underline{R}^T \ (\underline{f}_1 - \underline{Q} \ \underline{f}_2) . \tag{42}$$

The solution of the above system is reached with a simple iterative procedure which requires at every step the inversion of a part ($\underline{K}_{22}$) of the assembled stiffness matrix, cf. eq.(37), calculated on the basis of the parameters determined at the end of the preceding iteration.

An alternative procedure, with respect to the previous one, can be based on the direct minimization of a non linear function representing

the discrepancy between the field measurements and the corresponding numerically evaluated quantities. The following error function ξ is adopted
to this purpose which depends on the measured displacements (denoted by a
star) and on those deriving from a numerical stress analysis in which a
given set of material parameters $\underline{p}$ is adopted

$$\xi = \sum_{1}^{m} [u_i^* - u_i(\underline{p})]^2 . \tag{43}$$

Clearly, other definition are possible considering, for instance, only
the maximum absolute value of the differences in eq.(43), or dividing the
terms of the summation by the corresponding measured value in order to
obtain a non dimensional definition of the error.

Since the error function depends through the numerical results on
the parameters to be back calculated (which in this context have a rather
general meaning and may correspond to elasticity or shear strength properties, viscosity coefficients, etc.), the back analysis reduces to determining the set of parameters that minimizes the function ξ, i.e. that
leads to the best approximation of the field observation through the
chosen numerical model. The error defined by eq.(43) is in general a complicated non linear function of the unknown quantities and in most cases,
like for non linear or elasto-plastic problems, the analytical expression
of its gradient cannot be determined. Therefore, the adopted minimization
algorithm must handle general non linear functions and it should not require the analytical evaluation of the function gradient.

A comprehensive discussion of the algorithms fulfilling the above
requirements, known in mathematical programming as Direct Search Methods,
can be found e.g. in [15]. These are iterative procedures which perform
the minimization process only by successive evaluations of the error function. In the present context, each evaluation requires a stress analysis
of the geotechnical problem on the basis of the trial vector $\underline{p}$ chosen for
that iteration.

In most practical cases some limiting values exist for the unknown
parameters. For instance, the modulus of elasticity cannot reach negative
values. These limits, expressed by inequality constraints, can be easily
introduced into a direct search algorithm by means of a penalization pro-

cedure. When a point in the space of the free variables is reached outside the feasible domain, the error function is assigned a large value so
that the minimization algorithm automatically drives back the optimization path into the feasible region. This penalty approach turns out to be
general and simple to implement. In fact, no assumptions are required on
the characteristics of the constraints (e.g. about their convexity) and
the computer program for constrained minimization can be easily obtained
with few modifications of the corresponding unconstrained code.

From the computational view point the back analysis approach requiring the minimization of the error function expressed by eq.(43) presents
non negligible differences with respect to that based on the least square
method. It turns out, in fact, that the least square technique, specifically developed for the calibration of elasticity constants, converges
towards the optimal values of the parameters faster and "smoother" than
the direct search procedure. As a consequence, the computer cost required
by the first solution method is in general smaller than that of the
second one.

The different performance of the two approaches can be easily seen,
for instance, by applying both of them to the back analysis of the elastic constants (bulk modulus B and shear modulus G) of the rock underlying
the embankment shown in fig.6. The marked points denote the locations
where vertical or horizontal displacements are known. The results of such
analyses are summarized by the diagrams in fig.7 showing the values of B
and G obtained at each iteration of the solution process. The direct
search algorithm adopted for this example is the so called "Simplex"
method presented in [16].

It has to be recognized, however, that back analysis procedures
based on direct search algorithms present also a non negligible advantage
with respect to the least square approach. In fact, while the least
square procedure requires the implementation of an ad hoc computer program, the direct search approaches can be developed on the basis of standard computer codes for non-linear function minimization in which the finite element program for stress analysis is introduced as a subroutine.
This requires some simple changes of the original finite element code and
a limited programming effort. In addition, the same stress analysis and
error minimization programs can be used for various characterization

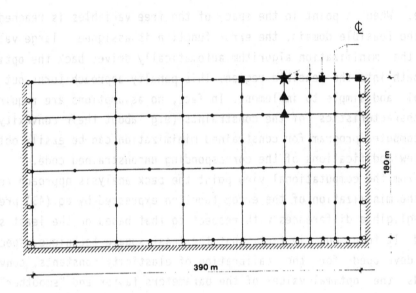

Fig. 6 Finite element mesh for the back analysis problem. The points
where the displacements are known are denoted by a black symbol.

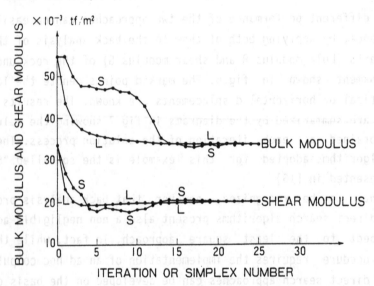

Fig. 7 Variation of the bulk and shear moduli during the error minimization process (L refers to the least square algorithm; S refers to
the Simplex algorithm).

problems, merely by considering the calculated quantities as functions of the unknown parameters, regardless their nature.

For instance, direct search algorithms have been applied to the calibration of elasto-plastic and visco-plastic material models for in situ rock masses [17] and to back analyses related to slope stability problems [18].

Another technique for the back analysis of elasticity parameters has been proposed in [19], where the conjugate gradient method [20] is adopted for minimizing the error function in eq.(43). Also in this case the final solution is reached by means of an iterative process, but while the previously mentioned direct search algorithms require at every iteration only the evaluation of the error function, the gradient type techniques require also the determination of the function derivatives with respect to the free variables. Under the assumption of linear material behaviour, the analytical expression of these derivatives can be worked out and directly programmed into the computer code developed for the solution of the calibration problem.

5. UNCONFINED SEEPAGE ANALYSIS

In unconfined seepage problems the fluid flow takes place through a domain part of whose boundary (the so called free surface) is a priori unknown and has to be determined, together with the distribution of hydraulic head, as a part of the solution. Due to the difficulties introduced by their geometrical non linearity, these problems can be solved analytically only in particularly simple cases and the use of numerical techniques is often necessary for more complex situations.

Among the numerical procedures, those based on the finite element method are perhaps the most frequently used [21,22,23]. They can be subdivided into two main groups: the so called fixed and variable mesh techniques.

The approaches of the first group operate on meshes of constant geometry and allow the free surface to pass through the elements by means of procedures that are conceptually similar to those adopted for non linear, or elasto-plastic, stress analyses. The variable mesh approaches modify

the grid geometry during the solution process so that at the end of iterations part of the mesh boundary represents the "correct" shape of the
free surface. They are in general characterized by high accuracy of the
final results, but are also affected by some drawbacks related, in particular, to the stability problems that might show up when the point of
intersection between the free surface and a pervious boundary exposed to
the atmosphere has to be determined.

In the following, after recalling the governing equation and boundary conditions for unconfined problems, a variable mesh technique for
steady state analyses is illustrated, based on a mathematical programming
algorithm for function minimization, which avoids the mentioned stability
problems. The extension of this technique to transient problems has been
discussed in [24].

5.1 Governing equations and boundary conditions

Consider a two-dimensional seepage flow through a saturated porous
medium assuming that both pore fluid and soil grains are incompressible
and that the deformability of the soil skeleton can be neglected. Under
these hypotheses the flow continuity equation is written in the following
form, where v_x and v_y are the components of the fluid discharge velocity,
and q is the assigned flux per unit volume

$$\frac{\partial v_x}{\partial x} + \frac{\partial v_y}{\partial y} = q .$$

(44)

The majority of seepage phenomena in geomechanics involve laminar flow
conditions. Consequently, Darcy's law can be adopted as a linear "constitutive" relationship between the components of the discharge velocity and
those of the gradient of the hydraulic head h. If x and y coincide with
the principal directions of permeability, and if y is the upward vertical
coordinate, Darcy's law is expressed as

$$v_x = - k_x \, i_x = - k_x \frac{\partial h}{\partial x} \quad , \quad v_y = - k_y \, i_y = - k_y \frac{\partial h}{\partial y}$$

(45a,b)

where

$$h = y + \frac{p}{\gamma} .$$

(46)

In the above equations k_x and k_y are the coefficients of permeability; p
is the pressure of the pore fluid and γ is its unit weight.

Substitution of eqs.(45) into eq.(44) leads to the following final relationship governing the problem at hand

$$\frac{\partial}{\partial x} (k_x \frac{\partial h}{\partial x}) + \frac{\partial}{\partial y} (k_y \frac{\partial h}{\partial y}) = q \ . \tag{47}$$

Note that eq.(44) is derived without introducing any assumption on the time dependency of flow. As a consequence eq.(47) holds for both steady and transient flows, neglecting for the later case the inertial effects.

For the typical problem shown in fig.8 the boundary conditions to be associated to eq.(47) are:

- pervious (constrained) boundaries

$$h = h_1 \text{ on } 1\text{-}2 \quad , \quad h = h_2 \text{ on } 4\text{-}5 \ , \tag{48a,b}$$

- pervious (wet) boundary, or seepage face (3-4)

$$h = y(x) \ , \tag{49}$$

- impervious boundary normal to the y axis (5-1)

$$\frac{\partial h}{\partial y} = 0 \quad (\text{or } v_y = 0) \ , \tag{50}$$

- free surface (2-3)

$$h = y = f(x) \quad , \quad v_n = 0 \ . \tag{51a,b}$$

Eqs.(51) represent the two conditions to be imposed on the free surface in steady state regime. They require that the hydraulic head be equal to the elevation of the free surface (eq.51a) and that the velocity of the fluid normal to it, v_n, is zero (eq.51b). Note that the function y(x) in eq.(49) is known, since it describes the shape of the downstream slope of the dam. On the contrary, the function f(x) in eq.(51a), representing the shape of the free surface, is unknown.

When the finite element method is adopted in the solution of seepage problems, the flow domain is subdivided into elements within which the hydraulic head distribution h(x,y) depends on the nodal hydraulic heads $\underline{h}$ through a vector $\underline{b}$ of interpolating functions

$$h(x,y) = \underline{b}(x,y)^T \underline{h} \ . \tag{52}$$

By writing the governing equation (47) in weak form, and taking into account eq.(52), a linear relationship is arrived at between nodal hydraulic heads $\underline{h}$ and assigned nodal fluxes $\underline{q}$

$$\underline{K} \, \underline{h} = \underline{q} \ , \tag{53}$$

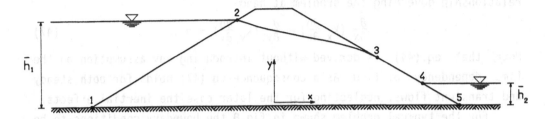

Fig. 8 Unconfined seepage problem.

where the so called "flow" matrix $\underline{K}$ is expressed by the following integral over the volume V of the element

$$\underline{K} = \int_V \begin{bmatrix} \dfrac{\partial \underline{b}}{\partial x} & \dfrac{\partial \underline{b}}{\partial y} \end{bmatrix} \begin{bmatrix} k_x & 0 \\ 0 & k_y \end{bmatrix} \begin{bmatrix} (\partial \underline{b}/\partial x)^T \\ (\partial \underline{b}/\partial y)^T \end{bmatrix} dV \ . \tag{54}$$

The system of linear equations governing the solution of the discrete problem is obtained by suitably assembling eq.(54) written for all the elements of the mesh.

The finite element solution of confined seepage problem (i.e. those problems in which all boundaries of the flow domain have known geometry) is straightforward and does not deserves particular comments. On the contrary, unconfined analyses present some computational difficulty due to the fact that, in addition to the distribution of hydraulic head, also the location of the free surface (and hence the shape of the domain within which the seepage flow takes place) is unknown.

As previously mentioned, two main groups of iterative procedures (i.e. fixed and variable mesh techniques) have been proposed for the solution of unconfined problems in the context of the finite element method. In the next Section the basis of a variable mesh technique, frequently adopted in practice, will be recalled.

5.2 A variable mesh approach

The techniques belonging to this category consist of iterative processes that modify the geometry of the finite element mesh, part of whose boundary coincide with the free surface, until an adequate approximation of the correct shape of the flow domain is reached.

A relatively simple procedure of this kind was proposed for steady state problems by Taylor and Brown [25]. At the beginning of each iteration a confined analysis is performed, on the basis of the current geometry, in which the free surface is treated as an impervious boundary (eq.51b). If the shape of the free surface is not correct this calculation leads to hydraulic heads at the free surface nodes different from their elevation. In order to fulfill the second free surface condition, eq.(51a), the nodes on the free surface are moved so that their elevation becomes equal to the corresponding hydraulic head previously computed. Then, the modified mesh is used in the subsequent iteration.

This solution method presents an ambiguity in defining the movement of the node at the intersection between the free surface and the wet boundary, or seepage face (i.e. point 3 in fig.8). In fact, this node (which is referred to in the following as seepage node) has in general a non vanishing flux, even though it belongs to the free surface. Consequently, during the confined analysis a "pervious" boundary condition (eq.51a) has to be imposed on it, and the procedure adopted for moving the other free surface nodes cannot be applied. The authors observe, however, that this difficulty is minimized by reducing the mesh size adjacent to this point.

Similar problems are shown also by other variable mesh techniques, and lead to spurious oscillations of the free surface in the vicinity of the exit point. A provision suggested to avoid them consists in describing the free surface shape by means of a reduced number of parameters, smaller than the number of free surface nodes. This implies, however, that the free surface boundary condition can be satisfied only in an average sense.

5.3 Position of the seepage point

In order to get an insight into the mentioned drawbacks affecting the variable mesh techniques, it seems worthwhile to discuss the results of a bench-mark example solved with the previously outlined algorithm.

The problem concerns the steady state seepage flow through the homogeneous and isotropic rectangular block dam. Since the standard procedure does not provide a criterion for determining the elevation of the seepage point P, a series of analyses was carried out by choosing a priori the position of this point and then applying the method proposed in [25] for evaluating the consequent geometry of the free surface. The results of these analyses are shown by dashed lines in fig.9, while a solid line represents the "optimal" shape of the free surface.

Two aspects of these solutions should be pointed out. First, no stability problems were observed during the analyses, regardless of the difference between the initial trial shape of the free surface and the final results. Second, all solutions are correct "in the model", in the sense that they exactly fulfill the boundary condition (51a) at the first and last free surface nodes (that belong also to pervious boundaries) and eqs.(51a,b) at the remaining nodes. It is obvious, however, that only one of these solutions is correct from a physical view point, i.e. the solution in which the "correct" position of the seepage point is assumed.

In other words, the finite element variable mesh technique provides an infinite number of stable solutions that fulfill the governing equation within the flow domain and all relevant boundary conditions. This indicates that the numerical analysis of the unconfined flow is affected by a non uniqueness of solution, which is not present in the continuum problem and that is apparently introduced by the discretization process.

The most straightforward way to eliminate the non uniqueness of results is to establish a criterion for choosing among them the one closest to the correct physical solution. Since this effect is not present in the continuum problem, but is introduced by the discretization process, it seems desirable to work out a criterion involving only quantities which are meaningful for the discrete model. In the finite element context these quantities are represented by the nodal variables.

It has also to be considered that the criterion, in the spirit of the finite element method, should not be of "local" nature (i.e. involving nodal variables only at one specific location), but it should rather be expressed by integrals or summations over a significant portion of the finite element mesh. Finally, from the numerical view point the influence of this criterion should become weaker and weaker with increasing refinement of the mesh, and it should virtually disappear when tending to the continuum problem.

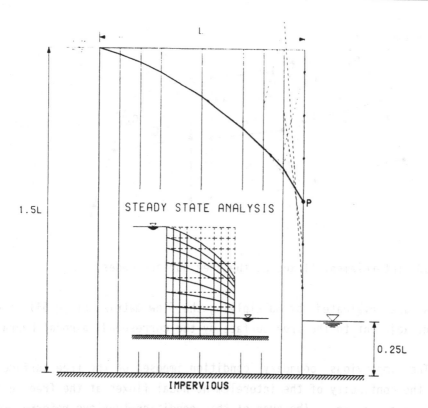

Fig. 9 Steady state flow through a rectangular block dam: influence of the position of the seepage point P on the numerically evaluated free surface.

It can be easily seen that the nodal variables of the assembled finite element mesh are not suited for developing the mentioned criterion. In fact, they should obey only the governing equation (in its weak form) and the relevant boundary conditions. Consequently, additional relationships imposed among them would interfere with the physical characteristics of the problem, leading to solutions non correct "in the model". Hence, the remaining alternative is to base the criterion on the nodal variables at the element level, in particular on the nodal fluxes leaving and entering the element facing the free surface at the free surface nodes. A pictorial representation of these quantities is shown in fig.10. Note that the nodal fluxes are scalar variables, hence they are represented in the figure by means of arrows only for graphical reasons. These

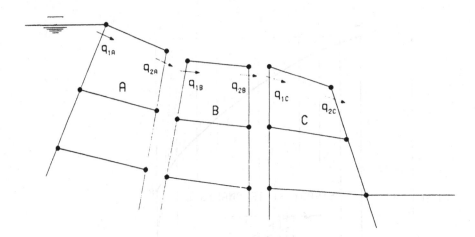

Fig. 10   Interelement fluxes at the free surface nodes.

fluxes are evaluated by multiplying the flow matrix (cf.eq.54) of each element adjacent to the free surface by the corresponding nodal hydraulic heads.

The impervious boundary condition imposed on the free surface implies the continuity of the interelement nodal fluxes at the free surface nodes, e.g. $q_{2A} = q_{1B}$ (because of this condition they are referred to in the following as fluxes "tangent" to the free surface), but no constraint is imposed between the fluxes at the nodes of the same element, e.g. q_{1B} and q_{2B}.

Now, considering that in the "correct" solution of the unconfined steady state problem the streamlines in the vicinity of the free surface are almost parallel to it, it seems reasonable to choose among the infinite possible numerical solutions the one corresponding to the most regular variation of the nodal fluxes entering and leaving the element facing the free surface. This "regularity" criterion can be translated into numerical terms by choosing the solution that minimizes the following function Q

$$\min_{\underline{y}_E} \left\{ Q(\underline{y}_E) = \Sigma_i \left[q_{1i}(\underline{y}_E) + q_{2i}(\underline{y}_E) + \ldots + q_{ni}(\underline{y}_E) \right]^2 \right\} . \qquad (55)$$

Here, vector $\underline{y}_E$ collects the elevations of the seepage points, which are the free variables of the minimization problem; the index i runs over the elements facing the free surface and n represents, for each one of these elements, the number of nodes belonging to the free surface. Note that, since inward and outward fluxes have opposite sign, function Q tends to zero if a constant flux takes place through the free surface nodes.

Many standard algorithms are applicable to the minimization of the above function. Here Rosenbrock's method (see [15]) has been adopted, which reaches the minimum through a sequence of evaluations of the function without requiring the determination of the function gradient. This characteristic leads to a simple iterative solution procedure.

At each iteration the minimization program modifies the free variable vector (i.e. the elevations of the seepage points) and an unconfined analysis is carried out obtaining (in addition to the shape of the free line corresponding to the current position of the seepage nodes) the nodal fluxes and the value of function Q. On the basis of this value the minimization program defines new elevations of the seepage points and a subsequent iteration is carried out. The process continues until a convergence criterion is fulfilled. In order to avoid oscillations or stability problems during the iterative solution process, some limiting values can be specified for the elevations of the seepage points leading to inequality constraints for the free variables. These constraints are easily enforced through the same penalty approach adopted for the back analysis problem.

The heavy solid line in fig.9 represents the shape of the steady state free surface obtained by the modified solution procedure for the rectangular block dam problem.

The above procedure, extended to the solution of transient problems as discussed in [24], has been adopted for determining the lowering of water table caused in a homogeneous aquifer by four vertical wells. The three dimensional mesh of eight node brick elements adopted for this problem is depicted in fig.11.

The three vertical planes through lines AB, BC and CD are considered as pervious boundaries with constant hydraulic head h=L, while the vertical plane through line AD is a plane of symmetry and, hence, is equivalent to an impervious surface. Due to this symmetry, only three wells (denoted by 1,2,3) are introduced in the mesh and the forth well, symme-

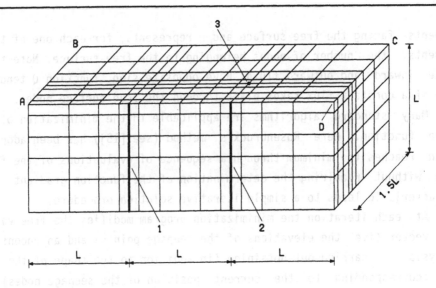

Fig. 11   Three-dimensional   drainage problem: finite element mesh and position of pumping wells 1, 2 and 3.

tric to well 3 with respect to the impervious plane, is implicitly accounted for. As a consequence, three seepage points are present, at the
intersection between the free surface and the vertical lines representing
the pumping wells.

At the beginning of calculations the free surface coincides with the
horizontal plane. Then at time t=0 the water level within the wells is
instantaneously brought to zero and the transient process is initiated.

The results of analysis are summarized in figs.12 and 13. In the
first figure the shapes of the free surface, obtained with the modified
analysis procedure, are shown for three different values of the non
dimensional time T

$$T = \frac{k \cdot t}{m \cdot L} \ . \tag{56}$$

A comparison between standard and modified techniques is shown in fig.13,
reporting the variation with time of the positions of the three seepage
points.

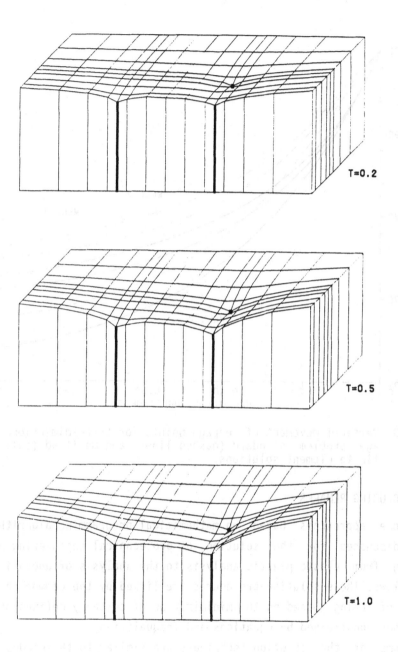

Fig. 12 Three-dimensional drainage problem: shape of the free surface for three different values of non-dimensional time T.

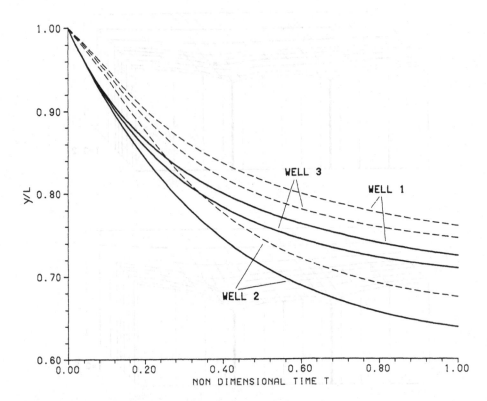

Fig. 13 Vertical movement of seepage points for three-dimensional drain-
         age problem: standard (dashed lines) and modified (solid lines)
         finite element solutions.

6. CONCLUDING REMARKS

Some approaches based on mathematical programming algorithms have
been discussed for the solution of geotechnical engineering problems
ranging from elasto-plastic analysis to the analysis of unconfined seepage flows. These solution procedures are linked by the common characteristic of being based on the minimization of suitably defined objective
functions constrained by equations and inequalities.

Some of the solution techniques are similar to those adopted in a
structural engineering contest (like those concerning the elasto-plastic
and limit analyses), while other procedures have a typical geotechnical
nature, like the one outlined for the unconfined seepage problems.

Ine choice of the minimization technique strictly depends on the characteristics of the problem under examination. For instance, a gradient type minimization procedure has been adopted for the elasto-plastic analysis since in this case the evaluation of the gradient of the objective function is possible with a limited computational effort. On the contrary, direct search minimization procedures can be adopted if the evaluation of the objective function gradient turns out to be computationally heavy, like for the back analysis problems.

Also the practical advantages of the mathematical programming approaches depend on the nature of the problem. In some cases, in fact, they can be seen as one of the possible alternatives among various solution algorithms (e.g. for elasto-plastic analyses), or as a way to refine the solution based on a different analysis technique (like for the seepage problems), while in other instances (e.g. the limit analysis) they represent probably the most convenient and "natural" solution procedure.

REFERENCES

[1] Maier G.: Mathematical programming methods in structural analysis, Proc.Int.Symp.on Variational Methods in Engineering, Southampton, 1973.

[2] Maier, G. and J. Munro: Mathematical programming applications to engineering plastic analysis, Applied Mechanics Reviews, 35 (1982), 1631-1643.

[3] Sloan, S.W.: Upper bound limit analysis using finite element and linear programming, Int.J.Numer.Anal.Methods Geomech., 13 (1989), 263-282.

[4] Maier, G.: Quadratic programming approach for certain classes of nonlinear structural problems, Meccanica, 3 (1968).

[5] De Donato, O. and A. Franchi: A modified gradient method for finite element elastoplastic analysis by quadratic programming, Comp.Methods in Appl.Mech.Eng., 2 (1973).

[6] Gioda, G. and O. De Donato: Elastic-plastic analysis of geotechnical problems by mathematical programming, Int.J.Numer.Anal.Methods Geomech., 3 (1979), 381-401.

[7] Maier, G.: A matrix structural theory of piecewise linear elastoplasticity with interacting yield planes, Meccanica, 5 (1970).

[8] Cottle, R.W.: Monotone solution of the PLCP, Mathematical programming, 3 (1972), 210-222.

[9] Kunzi, M.P. and W. Krelle: Nonlinear Programming, Blaisdell, Waltham, Mass. 1966.

[10] Zavelani-Rossi, A.: Finite element techniques in plane limit problems, Meccanica, 4 (1974).

[11] Zavelani-Rossi, A., Gatti, G. and G. Gioda: On finite element stability analysis in soil mechanics, Proc.Int.Symp.on Discrete Methods in Engineering, C.I.S.E., Milan, 1974.

[12] Gioda, G.: Indirect identification of the average elastic character-
     istics of rock masses, Proc.Int.Conf.on Structural Foundations on
     Rock, Sydney, 1980.
[13] Gioda, G. and S. Sakurai: Back analysis procedures for the interpre-
     tation of field measurements in geomechanics, Int.J.Numer.Anal.
     Methods Geomech., 11 (1987), 555-583.
[14] Kavanagh, K.T. and R.W. Clough: Finite element application in the
     characterization of elastic solids, Int.J.Solids Structures, 7
     (1971), 11-23.
[15] Himmelblau, D.M.: Applied non-linear programming, McGraw-Hill, New
     York, 1972.
[16] Nelder J.A. and R. Mead: A simplex method for function minimization,
     The Computer Journal, 7 (1965), 308-313.
[17] Cividini, A., Gioda, G. and G. Barla: Calibration of a rheological
     material model on the basis of field measurements, Proc.5th Inter-
     national Conference on Numerical Methods in Geomechanics, Nagoya,
     1985.
[18] Nguyen, V.U.: Back analysis of slope failure by the secant method,
     Geotechnique, 34 (1984), 423-427.
[19] Arai K., H. Otha and T. Yasui: Simple optimization techniques for
     evaluating deformation moduli from field observations, Soils and
     Foundations, 23 (1983).
[20] Fletcher R. and C.M. Reeves: Function minimization by conjugate gra-
     dient, The Computer Journal, 7 (1965), 149-154.
[21] S.P. Neuman and P.A.Witherspoon: Finite element method of analyzing
     steady seepage with a free surface, Water Resources Research, 6
     (1970), 889-897.
[22] K.J. Bathe and M.R.Khoshgoftaar: Finite element free surface seepage
     analysis without mesh iteration, Int.J.Numer. Anal.Methods in Geo-
     mechanics, 3 (1979), 13-22.
[23] C.S. Desai and G.C. Li: A residual flow procedure and application
     for free surface flow in porous media, Int.J. Advances in Water
     Resources, 6 (1983), 27-35.
[24] Cividini A. and G. Gioda: On the variable mesh finite element analy-
     sis of unconfined seepage problems, Geotechnique, 39 (1989), 251-
     267.
[25] R.L. Taylor and C.B. Brown: Darcy flow solution with a free
     surface, J.of Hydraulic Div., ASCE, 93 (1967), 25-33.

NUMERICAL ANALYSIS FOR THE INTERPRETATION OF FIELD MEASUREMENTS IN GEOMECHANICS

S. Sakurai
Kobe University, Kobe, Japan

1. INTRODUCTION

Numerical analyses such as the Finite Element Method (FEM) and the Boundary Element Method (BEM) are powerful tools for analyzing the mechanical behaviour of structures. In the field of geotechnical engineering, however, it is extremely difficult to quantitatively determine the mechanical properties of soils/rocks, initial stress, underground water level, permeability, etc., which are required as input data for the analyses. This difficulty is mainly due to the complex and non-homogeneous geological conditions of the ground. It is not surprising that material properties vary from place to place, although soil/rock formations seem to be identical. It is not uncommon, therefore, for the real behaviour of structures such as tunnels, underground caverns and cut slopes to differ from the predicted ones, even after a careful investigation has been made.

In order to overcome this difficulty, field measurements are conducted during the construction period, not only to monitor the stability of the structures, but also to re-evaluate the input data which have been adopted in the design analyses, such that the discrepancy between the measured and predicted behaviours of the structures is reduced to a minimum, and if necessary, the original design of the structures is modified. This is generally called an 'observational procedure' (Terzaghi and Peck, 1948).

In observing the behaviour of structures, displacement measurements which have been taken by multi-rod borehole extensometers and inclinometers, as well as surveys using, for instance, electronic distance meters, are commonly utilized because of the simplicity of their instrumentation and the reliability of the data they provide.

It must be emphasized, however, that one of the most important and difficult tasks involved with field measurements is to find a way to quantitatively interpret the results in order to assess the stability of the structures and/or to re-evaluate the input data used in the design analyses. When obtaining the measurement results, one can determine the initial stress and mechanical properties such as the modulus of elasticity and Poisson's ratio by analyzing the measured values, for instance, the measured displacements. In general, the initial stress and mechanical properties are input data in an ordinary stress analysis. Therefore, this process is just the reverse, and thus, it is referred to as 'back analysis'.

In this paper, back analysis methods for the interpretation of field measurement results are described with particular reference to both underground openings and cut slope problems. Case studies showing the applicability of the methods to practical problems are also presented.

2. MODELLING OF SOILS AND ROCKS

The ground in/on which structures are built is classified into three groups, i.e., (a) continuous, (b) discontinuous and (c) pseudo continuous types, as shown in Fig. 1 (Sakurai et al., 1988). Type (a) may be for ground consisting of intact rocks or soils and Type (b) represents jointed rock masses. Since Type (c) is for highly fractured and/or weathered rock masses, the global behaviour of this ground seems to be a continuous body. We call this the pseudo continuous type of ground.

The mechanical behaviour of the Type (a) ground can be analyzed by means of a mechanical model based on continuum mechanics, while a discontinuous model such as those proposed by Cundall (1971) and Kawai (1980) may be used for analyzing the Type (b) ground where joint elements in the finite element analysis are also useful. Concerning the Type (c) ground, one can of course adopt a discontinuous model similar to that for Type (b). In engineering practices, however, it is almost impossible to explore all the joint systems or to investigate all of their mechanical characteristics. Moreover, it seems that this type of ground behaves, in a global sense, just like a continuous body. A continuum mechanics model can therefore be used for Type (c). It should be noted, however, that the effects of discontinuities must be adequately considered in this model. Types (a) and (c), namely, the continuous and pseudo continuous types of ground are dealt with in this paper so that back analyses described here can be formulated on the basis of the theory of continuum mechanics.

3. BACK ANALYSIS PROCEDURES

The determination of the initial stress and mechanical properties of soils/rocks, and boundary conditions from the measured values of displacements, stress and strain is referred to as a back analysis or a characterization problem.

Back analysis procedures may be roughly classified into two cate-

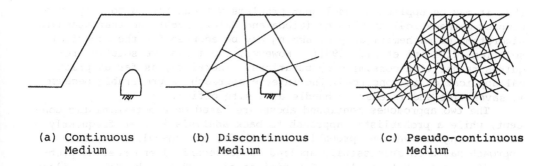

(a) Continuous   (b) Discontinuous   (c) Pseudo-continuous
    Medium   Medium   Medium

Fig. 1   Classification of the ground

gories, that is, the inverse and direct approaches (Cividini et al.,
1981). For the inverse approach, the formulation is just the reverse of
that in ordinary stress analysis, although the governing equations are
identical (see Fig. 2). However, the number of measured values should be
greater than the number of unknown parameters, so that optimization techniques can be adopted to determine the unknowns. For simple geometrical
configurations and material models, closed form solutions may be used.
For structures with arbitrary shapes under more complex geological conditions, FEM seems to be most promising. For example, Kavanagh (1973) proposed a back analysis formulation based on FEM which may make it possible
to obtain the material constants not only for isotropic materials, but
also for inhomogeneous and anisotropic materials from both measured displacements and strain. It should be noted that trouble is often encountered with the inverse approach in obtaining a stable solution for widely
scattered values of measurement data which are commonly found in geotechnical engineering problems. Difficulty is also run into when applying
this approach to nonlinear problems.

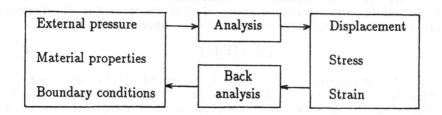

Fig. 2   Relationship between back analysis
          and ordinary stress analysis

On the other hand, the direct approach to back analysis is based on
an iterative procedure which corrects the trial values of unknown parameters by minimizing error functions. The advantages of this method are

that it may be applied to nonlinear problems without having to rely on a
complex mathematical background (Gioda and Maier, 1980) and that standard
algorithms of mathematical programming may be adopted for the numerical
solution (Cividini et al., 1981). However, the iterative solution requires rather time-consuming computations. Therefore, as far as practical application is concerned, the inverse approach is preferable because
it saves time and is easy to handle at a site office.

The two approaches mentioned above are based on a deterministic concept, while a probablistic approach to back analysis is also frequently
used. Among the various probablistic methods, the so-called Bayesian
approach has been successfully applied to geotechnical engineering problems (Asaoka and Matsuo, 1979; Cividini et al., 1983). The Kalman filter
theory is also adopted to attack back analysis problems in geomechanics
(Murakami and Hasegawa, 1985).

4. STABILITY ASSESSMENT OF UNDERGROUND OPENINGS

4.1 Direct strain evaluation technique

For assessing the stability of underground openings such as tunnels
and underground powerhouses, Sakurai (1981, 1982) proposed the Direct
Strain Evaluation Technique (DSET) which allows for a quantitative interpretation of displacement measurement results. The basic concept of the
technique is to assess the stability of tunnels by comparing the strain
occurring in the ground surrounding the tunnels with the allowable strain
of the soils/rocks. Sakurai (1981) defined critical strain as the ratio
of uniaxial compressive strength to Young's modulus, and this may be
adopted as an allowable strain. The critical strain is more or less independent of the joint systems, so that even for jointed rock masses the
critical strain can be determined from laboratory experiments carried out
on intact rocks (Sakurai, 1983).

If the amount of displacement measurement data is sufficiently large,
strain can be determined directly from the measured displacements through
use of the kinematic relationship.

Sakurai (1981) formulated a relationship between strain and measured
displacements by introducing an interpolation function for displacement
distributions and giving it in the following matrix form.

$$\{\varepsilon\} = [B]\{u\} \tag{1}$$

where $\{\varepsilon\}$ is strain, $\{u\}$ is the measured displacement and $[B]$
is a matrix, only a function of the location of the measurement points.

This approach is advantageous because no information is needed with
respect to the initial stress and material constants of the soils/rocks.
In practice, however, the number of extensometers installed around the
tunnels is generally limited and not sufficient for obtaining an overall
view of the strain distributions.

In order to overcome this difficulty, Sakurai and Takeuchi (1983)
proposed an indirect method for determining strain by means of back analysis. In this proposed method, the initial stress and material constants
are first back calculated from measured displacements and then used as

input data for an ordinary analysis by the finite element method or boundary element method to determine the strain distribution around the tunnels.
      The two different procedures for calculating the strain distribution
around the openings are shown in Fig. 3.

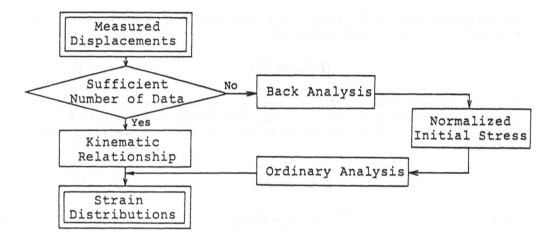

Fig. 3   Flow chart for determining strain distributions
         from measured displacements

4.2 Mathematical formulation of back analysis
      In formulating mathematical equations for the back analysis propos-
ed by Sakurai and Takeuchi (1983), the following assumptions are made.
      (1) The mechanical characteristics of soils and rocks are express-
ed by a linear homogeneous isotropic elastic model so that the material
constants are Young's modulus and Poisson's ratio. However, since Poisson's ratio has no great influence on the results of back analysis, an
adequate value can be chosen.
      (2) The elastic constants of the lining are assumed to be known.
      (3) The initial stress is uniformly distributed throughout the
ground being excavated.
      In Finite Element analysis for excavation problems of underground
openings, the genuine rock pressure can be taken into account by applying
equivalent nodal force $\{ P_0 \}$ at the excavation surface, which corresponds
to the initial state of stress $\{ \sigma_0 \}$ in the ground being excavated.
This force is determined by:

$$\{ P_0 \} = \int_V [B]^T \{ \sigma_0 \} dv + \int_V [N]^T \{ p \} dv \qquad (2)$$

where [N] and [B] are the matrices of the element shape functions
and their derivatives, respectively. $\{ p \}$ is the vector of the body

force components due to gravity and V is the volume of the excavation element. A two-dimensional formulation is presented here (Sakurai and Shinji, 1984). Hence, the initial stress components are as follows:

$$\{\sigma_0\} = \{\sigma_{x0} \quad \sigma_{y0} \quad \tau_{xy0}\}^T \tag{3}$$

The relation between nodal force { P } and nodal displacement { u } is expressed by the well-known relationship:

$$[K]\{u\} = \{P\} \tag{4}$$

where [K] denotes the stiffness matrix of the assembled finite element system. For a lined underground opening with moduli of elasticity E_R and E_L of the rock and lining, respectively, the stiffness matrix is expressed as:

$$[K] = E_R[K^*] \tag{5}$$

where

$$[K^*] = [K_R] + R[K_L]$$

$$R = E_L/E_R$$

where [K_R] represents the stiffness matrix for the finite element model of the ground where $E_R = 1$, and [K_L] for the lining where $E_L = 1$. When determining these stiffness matrices, Poisson's ratio of the rock and lining must be assumed. The finite element mesh must be chosen in such a way that the measuring points coincide with the nodes of the mesh.

Considering Eq. (5) and the equivalent nodal force given by Eq. (2) into Eq. (4), we obtain the following equation when the second term of Eq. (2) is disregarded.

$$[K^*]\{u\} = \sigma_{x0}/E_R\{P_1\} + \sigma_{y0}/E_R\{P_2\} + \tau_{xy0}/E_R\{P_3\} \tag{6}$$

where { P_i } (i = 1 ~ 3) denotes the equivalent nodal forces corresponding to the components of the unit initial stress.

Substituting $\sigma_{x0}/E_R = 1$ and $\sigma_{y0}/E_R = \tau_{xy0}/E_R = 0$ into Eq. (6) gives:

$$[K^*]\{u\} = \{P_1\} \tag{7}$$

Solving Eq. (7), we obtain displacement { u_x } at the nodal points due to the initial stress component, $\sigma_{x0}/E_R = 1$, only. By following a similar procedure, displacements { u_y } and { u_{xy} }, caused by the other components of unit initial stress $\sigma_{y0}/E_R = 1$ and $\tau_{xy0}/E_R = 1$, respectively, are obtained. Considering displacements { u_x }, { u_y } and { u_{xy} } due to each component of initial stress, the following equation is derived.

$$[A]\{\sigma_0\} = \{u\} \tag{8}$$

where

$$[A] = [\{u_x\}\{u_y\}\{u_{xy}\}]$$

$$\{\sigma_0\} = \{\sigma_{x0}/E_R \quad \sigma_{y0}/E_R \quad \tau_{xy0}/E_R\}^T$$

$\{\sigma_0\}$ is called the 'normalized initial stress'.
     Displacement $\{u\}$ may be split into two parts, namely, measured
displacments $\{u_1\}$ and unknown displacements $\{u_2\}$. Therefore:

$$[A_1]\{\sigma_0\} = \{u_1\} \tag{9}$$

Matrix $[A_1]$ is uniquely defined when Poisson's ratio for the ground and
lining materials and parameter R are given.
     It should be pointed out that in practice it is easier to measure
the relative displacement between two different measuring points than the
absolute displacement. Hence, absolute displacement $\{u_1\}$ in Eq. (9)
must be transformed to relative displacement $\{\Delta u\}$ by use of transformation matrix $[T]$ as:

$$\{\Delta u\} = [T]\{u_1\} \tag{10}$$

Hence, Eq. (9) becomes:

$$[A^*]\{\sigma_0\} = \{\Delta u\} \tag{11}$$

where

$$[A^*] = [T][A_1]$$

If the number of measured displacements is greater than three, the normalized initial stress can then be determined by an optimization procedure. If the least squares method is adopted in Eq. (11), the normalized
initial stress is uniquely determined from the measured relative displacements by:

$$\{\sigma_0\} = ([A^*]^T[A^*])^{-1}[A^*]^T\{\Delta u\} = [F]\{\Delta u\} \tag{12}$$

When determining the normalized initial stress, the displacements at all
the nodal points can be calculated by Eq. (8). The strain in each element, therefore, can be obtained by use of the following relationship
between strain and the displacements.

$$\{\varepsilon\} = [B]\{u\} \tag{13}$$

where matrix [B] is a function of the nodal point locations.

4.3 Computation procedure

The computation procedure consists of four main steps, i.e.,

(1) For underground openings with no lining, parameter R becomes unity. Hence, the overall stiffness matrix is uniquely determined by assuming Poisson's ratio and considering the location of the measuring points. Matrix [A] can then be obtained.

(2) Matrix [A_1] corresponding to the components of measured displacements is easily obtained from [A] . If the measurements are relative displacements between two different points, matrix [A_1] is transformed into [A*] . We then calculate matrix [F] . Steps 1 and 2 of the procedure can be carried out when the locations of the measuring points are known. Hence, matrix [F] can be fixed before starting the measurements.

(3) When starting the measurements, normalized initial stress $\{\sigma_0\}$ is calculated for each set of displacements measured at a certain period of construction. We then calculate the displacements at all the nodal points.

(4) Strains are calculated from the displacements and their distribution is shown on a computer graphic display. If the vertical component of initial stress is assumed, then all other components as well as the modulus of elasticity can be determined.

In the case of lined openings, we first assume an appropriate value for parameter R (say R_1) and follow the same procedure as for the unlined openings described above. Considering the modulus of elasticity obtained from the back analysis, we determine parameter R (say R_2). If R_2 is not close enough to R_1, we repeat the calculation until the following relation is satisfied.

$$\left| \frac{R_{i+1} - R_i}{R_i} \right| < \epsilon \tag{14}$$

where ε is the allowable error. It has been verified that four or five repetitions of the calculation are good enough to get sufficiently accurate results.

4.4 Case study

4.4.1 Brief description with respect to the tunnels and instrumentation

Two double-lane highways and two double-track railway tunnels were constructed at a shallow depth. They consist of two upper and two lower tunnels, parallel and adjacent to each other. The ground in which the tunnels were bored consisted of highly fractured and weathered granite. An example of the cross section of the four tunnels together with the geological formation is shown in Fig. 4.

An intensive field measurement system has been carefully planned and executed. In this measurement system, displacement measurements are highlighted, where a sliding micrometer and TRIVEC developed by Kovari and his group (Kovari et al., 1982; Koeppel et al., 1983) are extensively used. High precision inclinometers are also employed. One of the measurement sections is given in Fig. 5.

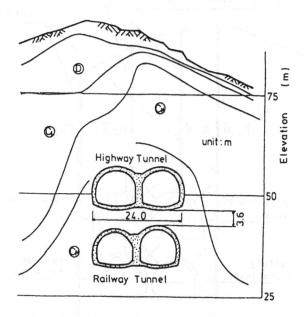

Fig. 4   Tunnel cross section with
           geological formations

The tunnel excavation was begun from the lower west railway tunnel, followed by the lower east. After completion of the two railway tunnels with a reinforced concrete lining, three pilot tunnels for the upper highway tunnels were bored in advance, and then the main portion of the upper tunnels was excavated.

4.4.2 Back analysis of measured displacements

The systematic measurements and back analyses were performed throughout the excavation. The normalized initial stress was back analyzed from measured displacements and Young's modulus was then determined if necessary. As an example, one set of results is shown in Table-1. It should be noted that the values of the normalized initial stress vary for each excavation step. This is simply because stress in the ground around tunnels is influenced by excavations, as well as by the fact that initial stress differs from place to place.

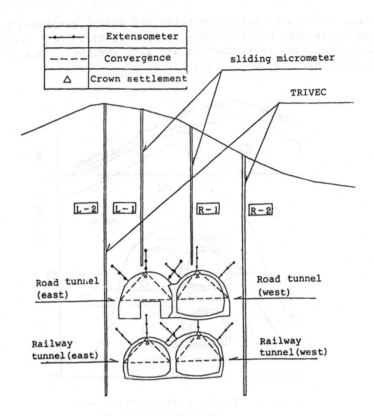

Fig. 5   One of the measurement sections

It is of interest to know that Young's modulus slightly decreases
with the progress of the excavations until the main excavation of the
upper highway tunnels is begun. This may be due to the fact that the
ground surrounding the tunnels tends to be loosened by the excavations,
and thus, large displacements occur in the tunnels excavated afterwards.
In the main excavation of the upper highway tunnels, however, a group of
horizontal pipes was installed near the tunnel crown prior to the excavation in order to stabilize the ground. This may give a large value to
Young's modulus.

It should also be noted that Young's modulus obtained by the back
analysis described here is not necessarily the mechanical constant of
the material itself, but that of the material systems with reinforcement,
such as rock bolts and other support structures.

As mentioned earlier, the normalized initial stress is then used as
input data for an ordinary FE analysis to calculate the displacement and
strain. As an example, we show the results of this calculation performed
during the main excavation of the upper highway tunnels. Some of the
displacement results are shown in Figs. 6 and 7, and are compared with
the measured displacements.

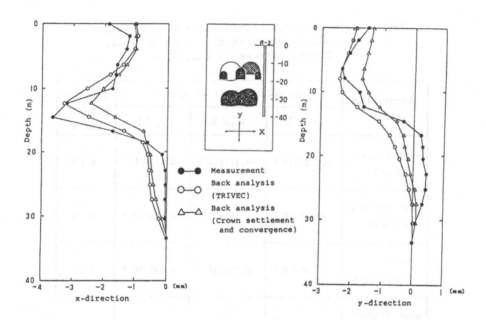

Fig. 6   Displacements obtained from measurements and back
analysis (due to the excavation of the upper half
of the west highway tunnel)

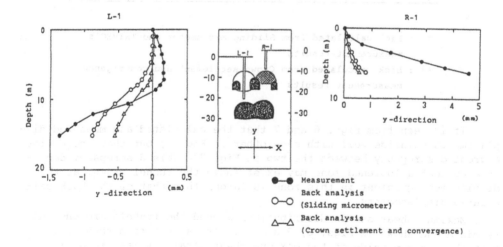

Fig. 7   Displacements obtained from measurements and back
analysis (due to the excavation of the upper half
of the west highway tunnel)

Table-1   Normalized initial stress and Young's
            modulus determined by back analysis
            (Tensile stress is positive)

Excavation steps	Normalized initial stress			Young's modulus kg/cm2
	σ_x/E	σ_y/E	τ_{xy}/E	
West railway tunnel*	-0.199×10^{-2}	-0.111×10^{-2}	0.126×10^{-3}	5,470
East railway tunnel*	-0.140×10^{-2}	-0.153×10^{-2}	-0.554×10^{-3}	3,970
Pilot tunnel**	-0.236×10^{-2}	-0.956×10^{-3}	0.899×10^{-4}	3,370
West highways tunnel**	-0.819×10^{-3}	-0.302×10^{-3}	0.700×10^{-4}	7,770
East highways tunnel**	-0.477×10^{-3}	-0.262×10^{-3}	0.171×10^{-4}	11,590

*   : Back calculated from Sliding Micrometer and TRIVEC's
      measurement results
** : Back calculated from Crown settlement and convergence
      measurement results

It is seen from Figs. 6 and 7 that the calculated and measured displacements coincide well with each other in Fig. 6, but that there lies
a great discrepancy between the two in Fig. 7. This discrepancy demonstrates that a loosened zone may exist above the tunnel crown, as the
displacement gradient in this zone is larger than that of the back calculated displacements.

Maximum shear strain distributions around the tunnels are then determined and shown in Figs. 8, 9 and 10. It is seen from these figures
that the maximum value of the maximum shear strain is less than 1%
throughout the region around the tunnels. Since the critical strain of
the highly weathered granite is approximately 1% (Sakurai, 1981), the
stability of the tunnels is verified. Special care must be given to the
loosening zone occurring above the crown, however, even though the strain
level remains small.

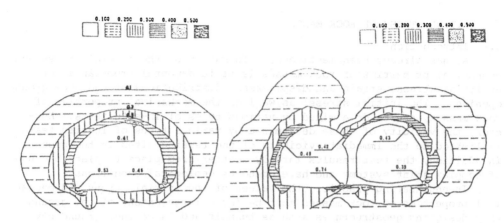

Fig. 8   Maximum shear strain (%) (after excavation of the west railway tunnel)

Fig. 9   Maximum shear strain (%) (after excavation of the east railway tunnel)

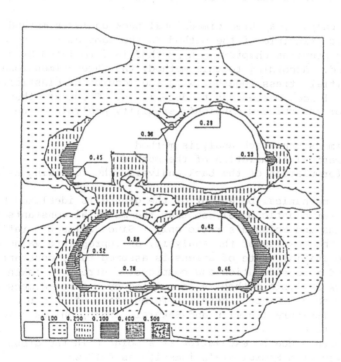

Fig. 10   Maximum shear strain (%) (after excavation of the upper half of the east highway tunnel)

5. INITIAL STRESS OF ROCK MASSES

5.1 Introduction

As has already been mentioned, information on the initial stress and mechanical properties of rock masses is of fundamental importance for analyzing stress, strain and displacement distributions around underground openings. The initial stress, as well as the mechanical properties of rock masses, can be determined by various types of in-situ tests. However, the validity of those determined by various types of in-situ tests is limited to the immediate vicinity of the testing sites in the ground. In addition, the test results largely scatter from place to place because of complex joint systems and heterogeneous geological conditions.

In engineering practice, the values of the initial stress and mechanical properties of rock masses must be for a large extent of the ground when designing geostructures such as tunnels and large underground caverns. In addition, it is also emphasized that the values should be evaluated in relationship to the size of the structures.

One of the most promising ways to determine the initial stress and mechanical properties of a large extent of rock masses, in relation to the size of the structures, is a back analysis of field measurement results obtained during the excavation of pilot tunnels and/or trial excavations.

In this chapter, a three-dimensional back analysis method is described. The basic principle of the method is the same as the one already presented in the previous chapter. The method is formulated by a boundary element method. According to this method, the three-dimensional stress tensor of initial stress together with the modulus of elasticity for a large extent of rock masses can be determined. These are then used as input data for the three-dimensional stability analysis of underground openings.

5.2 Three-dimensional back analysis method
5.2.1 Mathematical formulation of the method

In the formulation of the back analysis, the following assumptions are made:

(1) The mechanical behaviour of the ground is idealized by an isotropic linear-elastic model, so that the material constants reduce to Young's modulus and Poisson's ratio only. Since Poisson's ratio has less influence on the results of the analysis, an appropriate value can be used.

(2) The initial state of stress is assumed to be constant all over the region under consideration and compressive stress is taken as positive.

This back analysis method is formulated by a three-dimensional boundary element method. Since a detailed formulation of this back analysis is presented elsewhere (Shimizu and Sakurai, 1983), only a brief description is given here.

Displacements due to excavation at a point p in the ground (see Fig. 11) are derived from Somigliala's identity as follows:

$$u_i(\mathbf{p}) = \int_S U_{ki}(\mathbf{q}, \mathbf{p}) t_k(\mathbf{q}) dS_{\mathbf{q}} - \int_S T_{ki}(\mathbf{q}, \mathbf{p}) u_k(\mathbf{q}) dS_{\mathbf{q}} \qquad (15)$$

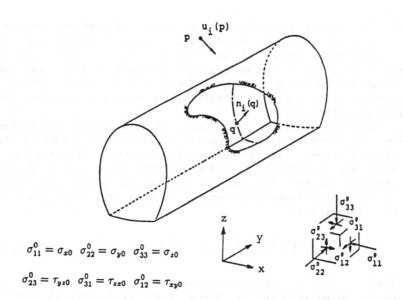

$$\sigma_{11}^0 = \sigma_{x0} \quad \sigma_{22}^0 = \sigma_{y0} \quad \sigma_{33}^0 = \sigma_{z0}$$

$$\sigma_{23}^0 = \tau_{yz0} \quad \sigma_{31}^0 = \tau_{zx0} \quad \sigma_{12}^0 = \tau_{xy0}$$

Fig. 11   Displacements due to excavation of an underground opening

where $U_{ki}(q, p)$ is the well-known Kelvin solution, corresponding to a concentrated force acting at the point p in the infinite elastic space.

$$U_{ij}(q, p) = \frac{1 + \nu}{8 \pi E (1 - \nu) r} \{(3 - 4\nu)\delta_{ij} + r_i r_j\} \tag{16}$$

where δ_{ij} is Kronecker's delta, r is the distance between points p and q, and E and ν denote Young's modulus and Poisson's ratio, respectively. $T_{ki}(q, p)$ is the fundamental solution of traction corresponding to the Kelvin solution. Traction vector $t_i(q)$ is given as follows:

$$t_i(q) = -n_j(q)\sigma_{ji}^0 \tag{17}$$

where σ_{ji}^0 represents the initial stress in the ground, and $n_i(q)$ denotes the normal unit vector at point q on the surface of the underground opening.

The following linear relationship between displacement $\{u\}$ and the initial stress can be derived from Eqs. (15) - (17).

$$\{u\} = [A]\{\sigma_0\} \tag{18}$$

where $\{\sigma_0\}$ is defined as:

$$\{\sigma_{x0}/E_R \quad \sigma_{y0}/E_R \quad \sigma_{z0}/E_R \quad \tau_{yz0}/E_R \quad \tau_{zx0}/E_R \quad \tau_{xy0}/E_R\}^T$$

is called the normalized initial stress, as already defined earlier in a
two-dimensional state. E_R denotes Young's modulus of the ground. It
should be noted that matrix [A] is a function of Poisson's ratio of
rock masses, the shape of the underground openings, and the location at
and the direction in which the displacements are measured. Knowing the
location of the extensometers, therefore, the matrix is uniquely defined
for a given Poisson's ratio.

Eq. (18) consists of the same number of equations as the number of
displacement measurement data and contains six unknown values of the normalized initial stress. If the number of measurements is six, Eq. (18)
gives a simultaneous equation to solve the normalized initial stress.
If the number is greater than six, the normalized initial stress can be
determined by an optimization procedure. When adopting the least squares
method, Eq. (18) yields:

$$\{\sigma_0\} = ([A]^T[A])^{-1}[A]^T\{u_m\} \qquad (19)$$

where $\{u_m\}$ is a vector of the measured displacements. This equation gives the normalized initial stress uniquely determined from the
measured displacements for a given Poisson's ratio. It is noted that in
Eq. (18), the measurements of the relative displacement between two measurement points as well as absolute displacements can be used as input
data (Gioda and Jurina, 1981; Sakurai and Takeuchi, 1983). It follows
that even simple convergence measurements alone carried out at the surface of the underground openings are sufficient for determining a three-
dimensional initial stress state. In addition, the three-dimensional
sequence of steps for excavation and the period for installation of measuring instruments can easily be taken into account in this back analysis.

The components of initial stress and Young's modulus can be separat-
ed from the normalized initial stress by assuming that the vertical
stress is equal to the overburden pressure, i.e.:

$$\sigma_{z0} = \gamma H \qquad (20)$$

where H and γ denote the overburden at the datum point and the
specific weight of the ground, respectively.

5.2.2 Computational stability

Measured displacements always show some scattering for a variety of
reasons, e.g., complex characteristics of rock masses, heterogeneous geological conditions, measurement errors, etc. This means that, in general, a back analysis method in geomechanics should derive a numerically
stable solution for any scattered measurement data. In order to assure
the computational stability of the method described in the previous section, a numerical simulation has been conducted, and it has been verified
that the method is stable enough to provide sufficiently accurate results

even for scattered input data. Since a detailed description for the
numerical simulation has been given elsewhere (Sakurai and Shimizu, 1987),
it is disregarded here.

5.3 Case study
      In order to demonstrate the applicability of the above-mentioned
back analysis method to engineering practice, a case study on an underground powerhouse is shown in the following (Sakurai and Shimizu, 1986).
      The ground in which the powerhouse was built consists of fresh and
solid coarse-grained biotite granite. The powerhouse is located with an
overburden of about 200 m. The cavern is 28.6 m high, 16.5 m wide and
25.1 m long, as shown in Fig. 12. The arrangement of the measurements
is also given in this figure.
      The back analysis was conducted by application of the measured dis-
placements taken at the completion of the 4th excavation stage, Lift 4
(see Fig. 13). The assumption of ν = 0.3 leads immediately to the
normalized initial stress.

$$\sigma_{x0}/E_R = 0.72 \times 10^{-3}$$
$$\sigma_{y0}/E_R = 0.90 \times 10^{-3}$$
$$\sigma_{z0}/E_R = 0.40 \times 10^{-3}$$
$$\tau_{yz0}/E_R = -0.65 \times 10^{-3}$$
$$\tau_{zx0}/E_R = -0.15 \times 10^{-3}$$
$$\tau_{xy0}/E_R = -0.69 \times 10^{-3}$$

The normalized initial stress obtained may now be split into the initial
stress and Young's modulus by assuming the vertical component of initial
stress.

$$\sigma_{z0} = \quad 23.5(KN/m^3) \times 200.0(m) = \quad 4.70MPa$$

and

$$E = 11.75GPa$$
$$\sigma_{x0} = 8.46MPa$$
$$\sigma_{y0} = 10.58MPa$$
$$\tau_{yz0} = -7.64MPa$$
$$\tau_{zx0} = -1.76MPa$$
$$\tau_{xy0} = -8.11MPa$$

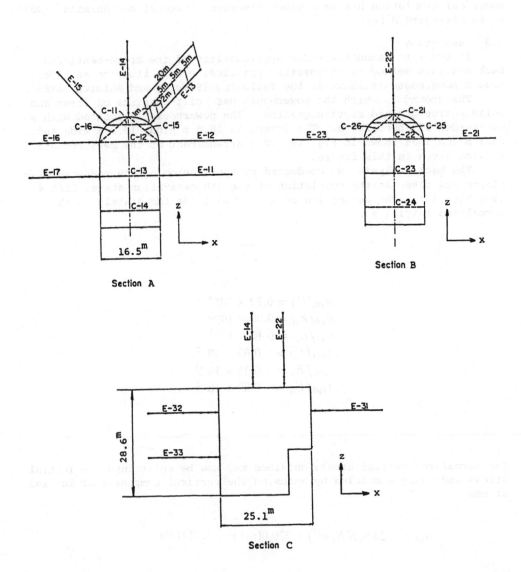

Fig. 12   Shape of cavern and arrangement of instruments

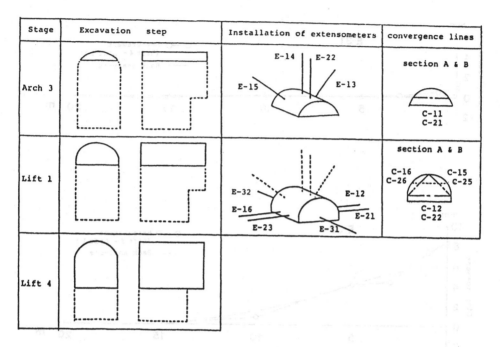

Fig. 13   Installation of measuring instruments and sequence of
           excavation steps

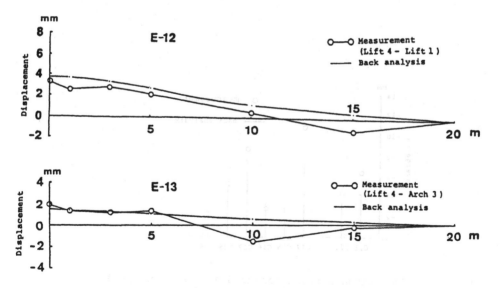

Fig. 14   Comparison between measured and back analyzed
           displacements (continued on next page)

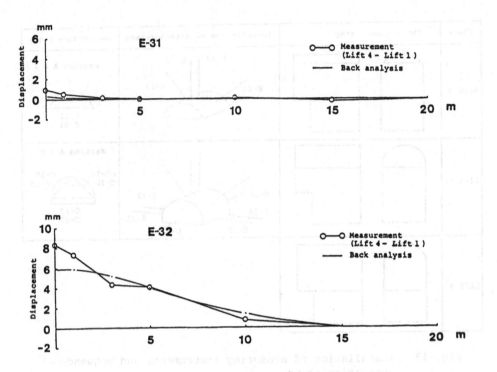

Fig. 14   Comparison between measured and back analyzed
           displacements (continued from previous page)

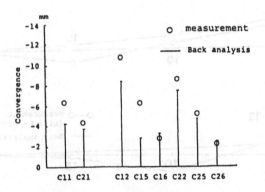

Fig. 15   Comparison between measured and back analyzed
           convergences

The displacement distribution in the surrounding ground is now computed by means of an ordinary boundary element analysis utilizing the back calculated initial stress and Young's modulus as input data. Some of the results are illustrated in Figs. 14 and 15. In comparing the measured displacements with the computed ones, one can see that they agree with each other fairly well.

6. BACK ANALYSIS FOR THE PLASTIC ZONE OCCURRING AROUND UNDERGROUND OPENINGS

6.1 Introduction

In order to assess the stability of an underground opening, it is of primary importance to evaluate the extent of the plastic zone occurring around the opening. In principle, if all the material constants for elasto-plastic characteristics of soils/rocks, as well as the initial stress of the ground are known, the plastic zone can be calculated by an ordinary elasto-plastic stress analysis. However, an ordinary stress analysis may not be an effective means to evaluate the plastic zone, because it is extremely difficult to know all such information about the ground. Up to the present, there has been no reliable method available to determine the plastic zone.

In this chapter, a back analysis method which can determine the plastic zone appearing around underground openings is presented. If the material constants and initial stress are back analyzed, the plastic zone can of course then be calculated by an ordinary elasto-plastic analysis. In general, however, a back analysis for determining the material constants of cohesion and the internal friction angle is not easily performed, because of the nonlinearity of the problems.

Sakurai et al. (1985) proposed a simple back analysis method for determining the plastic zone, which is based on an interpretation of the maximum shearing strain obtained by the measured displacements. The procedure for the method is as follows: Firstly, the normalized initial stress (initial stress divided by Young's modulus) is back analyzed from measured displacements. In this back analysis, only the linear elastic finite element analysis is necessary, because the ground is assumed to be an equivalent linear elastic material. Secondly, an ordinary linear elastic finite element analysis is carried out using the normalized initial stress as input data, and the maximum shear strain distribution is obtained. Finally, the maximum shear strain occurring around the underground openings is compared with the critical shear strain of soils and rocks. One of the contour lines of the maximum shear strain then gives the boundary between the elastic and plastic zones. In this method, however, the evaluation of the critical shear strain is in question and is a most difficult task.

A method for evaluating the critical shear strain is described in the following, and its applicability is demonstrated by means of a computer simulation.

6.2 Assumptions

The mechanical properties and initial stress of the ground are assumed as follows:

(1) The ground consists of homogeneous isotropic elastic and perfectly plastic material which conforms with Mohr-Coulomb's yield criterion (see Fig. 16).
(2) Hooke's law is assumed for the stress-strain relationship in
the elastic zone, and it is supposed that no volume change occurs in the
plastic zone.
(3) The initial stress existing in the ground prior to excavation
is constant in the excavation area (see Fig. 17).
(4) A plane strain condition is assumed in the computer simulation.
The compressive stress is taken as positive in this chapter.

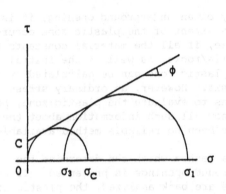

Fig. 16   Mohr-Coulomb's yielding
          criterion

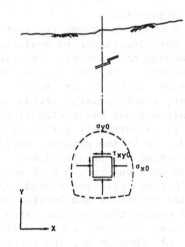

Fig. 17   Initial state of stress

6.3 Fundamental equations

6.3.1 Maximum shear strain on the elasto-plastic boundary
        Mohr-Coulomb's criterion is expressed as:

$$\sigma_1 - \sigma_3 = (\sigma_1 + \sigma_3)\sin\phi + 2c\cos\phi \tag{21}$$

where σ_1 and $\sigma_3 (\sigma_1 > \sigma_3)$ are principal stresses, and C and ϕ are
cohesion and the internal friction angle, respectively. The principal
stress on the elasto-plastic boundary must satisfy Hooke's law. Hence,
the following equation can be derived on the boundary.

$$\sigma_1 - \sigma_3 = \frac{E}{1+\nu}(\varepsilon_1 - \varepsilon_3) \tag{22}$$

where ε_1 and $\varepsilon_3 (\varepsilon_1 > \varepsilon_3)$ are principal strains, and E and ν denote Young's modulus and Poisson's ratio, respectively.

From Eqs. (21) and (22), the maximum shear strain on the elasto-plastic boundary is expressed as:

$$\varepsilon_1 - \varepsilon_3 = 2\frac{1+\nu}{E}(\overline{p}\sin\phi + c\cos\phi) \tag{23}$$

Eq. (23) is rewritten as follows by introducing the uniaxial critical strain ε_0, $\varepsilon_0 \overset{\text{def}}{=} \sigma_c/E = 2c\cos\phi/E(1-\sin\phi)$ (σ_c : uniaxial compressive strength) proposed by Sakurai (1981).

$$\gamma_c = (1+\nu)\left[2\frac{\overline{p}}{E}\sin\phi + (1-\sin\phi)\varepsilon_0\right] \tag{24}$$

where $\overline{p} = (\sigma_1 + \sigma_3)/2$ is the average value of the maximum and minimum principal stresses on the elasto-plastic boundary. γ_c defined by Eq. (24) is named the 'critical shear strain'.

6.3.2 Relationship between real and equivalent Young's modulus

In order to determine the critical shear strain, it is seen from Eq. (24) that Young's modulus E, Poisson's ratio ν , internal friction angle ϕ and critical strain ε_0 must be known in advance. The values of ν, ϕ and ε_0 can easily be determined from laboratory test results. The most difficult task in evaluating γ_c in Eq. (24) is to determine Young's modulus E of soils/rocks, particularly of the jointed rock masses. One of the possibilities for determining E is to back analyze it from the deformational behaviour of the tunnels observed during excavation. If a linear elastic model is used in the back analysis, Young's modulus can easily be obtained. This elastic back analysis does not give the real value of Young's modulus, however, if a plastic zone occurs around the openings. The value determined from the elastic back analysis is called an equivalent Young's modulus and usually differs from the real value needed for Eq. (24). Therefore, in the case where a plastic zone exists, the real Young's modulus must be determined. For this purpose, we should know the relationship between real Young's modulus E and equivalent Young's modulus E*, so that E can be derived from E*. The relationship between E and E* can be obtained as explained in the following.

The radial displacement around an unlined circular tunnel excavated in an elastic ground which is under a hydrostatic initial state of stress is expressed as,

$$u^e = \frac{1+\nu}{E^*}p_0\frac{a^2}{r} \tag{25}$$

where p_0 is the initial stress (see Fig. 18).

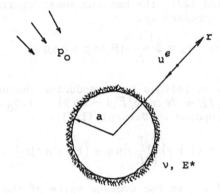

Fig. 18   Circular tunnel excavated in elastic
ground under hydrostatic initial stress

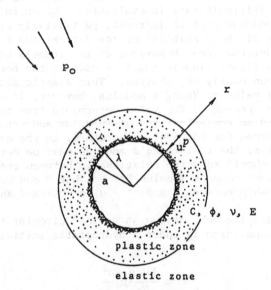

plastic zone

elastic zone

Fig. 19   Circular tunnel excavated in elasto-plastic
ground under hydrostatic initial stress

On the other hand, the radial displacement around an unlined circular tunnel excavated in the elasto-plastic ground under hydrostatic initial stress is expressed as (see Fig. 19):

$$u^p = \frac{1+\nu}{E}(p_0 \sin \phi + c \cos \phi)\left(\frac{\lambda}{a}\right)^2 \frac{a^2}{r} \tag{26}$$

$$\lambda/a = \{(1 - \sin \phi)(\frac{p_0}{c}\tan \phi + 1)\}^{(1-\sin \phi)/2\sin \phi} \quad , \quad c = \varepsilon_0 E(1 - \sin \phi)/2\cos \phi$$

where λ is the radius of the elasto-plastic boundary.

Equating Eqs. (25) and (26) gives the relationship between E and E* as follows:

$$E = \frac{E^* \sin \phi}{\{\left(\frac{2p_0}{\varepsilon_0 E} - 1\right)\sin \phi + 1\}^{\frac{1-\sin \phi}{\sin \phi}} - \frac{\varepsilon_0 E^*}{2p_0}(1 - \sin \phi)} \tag{27}$$

The real value of Young's modulus is then calculated by Eq. (27) and substituted into Eq. (24) to determine the critical shear strain.

6.4   The method for determining the elasto-plastic boundary

The procedure for the method which determines the elasto-plastic boundary from the measured displacements is as follows:

(1) Displacements around openings due to excavation are measured by convergence and borehole extensometers.

(2) The normalized initial stress is back calculated from the measured displacements, and Poisson's ratio is assumed. The method of back analysis described in Chapter 4 can be adopted for this purpose.

(3) Ordinary elastic finite element analysis is conducted through use of the normalized initial stress to determine the maximum shear strain occurring due to the excavation.

(4) The initial strain $\{\varepsilon_0\}$ is calculated by the following equation.

$$\{\varepsilon_0\} = \left\{ \begin{array}{c} \varepsilon_{x0} \\ \varepsilon_{y0} \\ \gamma_{xy0} \end{array} \right\} = \left(\begin{array}{ccc} 1-\nu & -\nu & 0 \\ -2\nu & 1 & 0 \\ 0 & 0 & 2(1+\nu) \end{array} \right)\{\sigma_0\} \tag{28}$$

where $\{\sigma_0\} = \{\sigma_{x0}/E^* \quad \sigma_{y0}/E^* \quad \tau_{xy0}/E^*\}^T$ are the normalized initial stresses. The total maximum shear strain is then calculated as,

$$\gamma_{max} = \sqrt{\{(\varepsilon_{x0} + \Delta\varepsilon_x) - (\varepsilon_{y0} + \Delta\varepsilon_y)\}^2 + (\gamma_{xy} + \Delta\gamma_{xy})^2} \tag{29}$$

where $\Delta\varepsilon_x, \Delta\varepsilon_y$ and $\Delta\gamma_{xy}$ are components of strain due to excavation.

(5) The critical shear strain of soils/rocks is calculated by Eq. (24) in which $\bar{p}$ is the maximum principal stress component of the initial stress. The real Young's modulus which is necessary for this calculation is obtained by Eq. (27) in which P_0 is the maximum principal stress

component of the initial stress. The initial stress components are determined from the normalized initial stress by assuming that the vertical
stress component of initial stress is the same as the overburden pressure. The internal friction angle and uniaxial critical strain are
assumed by considering laboratory test results.

(6) The boundary between the elastic and plastic zones can be obtained as a contour line of the total maximum shear strain being the
same value as the critical shear strain. If the critical shear strain
is greater than any of the total maximum shear strains occurring around
the openings, no plastic zone exists.

6.5 Computer simulation

6.5.1 Procedure

In order to demonstrate the applicability of the method described
above, a computer simulation is conducted. In the simulation, the displacements around an opening due to excavation are firstly calculated by
an ordinary elasto-plastic finite element method, and they are used as
'measured displacements' in the computer simulation.

In order to verify the accuracy of the method, the elasto-plastic
boundary obtained by the back analysis is compared with the one calculated by the ordinary elasto-plastic finite element analysis.

6.5.2 An example problem and simulation results

A horseshoe-shaped unlined tunnel with a radius of 5 m is considered here as an example (see Fig. 20). The initial state of stress is non-
hydrostatic and the magnitudes of principal stress are $P_1 = 3.92\text{MPa}$
and $P_2 = 1.96\text{MPa}$. Their directions are shown in Fig. 20. The mechanical constants are listed in Table-2.

Table-2   Mechanical constants used in
the simulation

Young's modulus	E	(MPa)	196
Poisson's ratio	ν		0.3
cohesion	c	(MPa)	0.49
internal friction angle	ϕ	(deg.)	30
critical strain	ε_0	(%)	0.87

The elasto-plastic finite element analysis is firstly conducted to
obtain the measured displacements. The extent of the plastic zone obtained by the analysis is shown in Fig. 21 as a shaded area.

The back analysis of the measured displacements provides the following normalized initial stress.

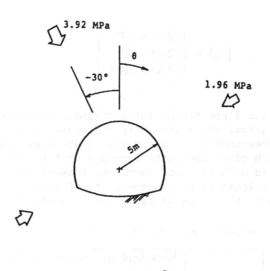

Fig. 20   A horseshoe-shaped unlined tunnel under
           non-hydrostatic initial stress

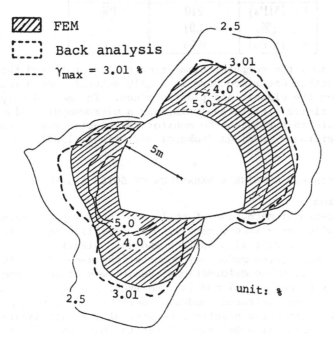

Fig. 21   Plastic zone evaluated by the back analysis method

$$\{\sigma_0\} = \left\{ \begin{array}{c} 2.17 \times 10^{-2} \\ 2.69 \times 10^{-2} \\ 0.62 \times 10^{-2} \end{array} \right\}$$

Some of the contour lines of the total maximum shear strain are shown in Fig. 21. The critical shear strain in this case is $\gamma_c = 3.01$ % . The contour line corresponding to this value is shown by a dotted line in this figure, which gives the outer boundary of the plastic zone. It is seen that there is a fairly good agreement between the back analyzed results and those calculated by elasto-plastic finite element analysis. The results of this back analysis are summarized in Table-3.

Table-3   Results of the back analysis

	Back analysis results	Exact value
σ_1^0 (MPa)	3.96	3.92
σ_3^0 (MPa)	2.25	1.96
θ (deg.)	-33.37	-30
E^* (MPa)	128	-
E (MPa)	210	196
γ_c (%)	3.01	-
c (MPa)	0.53	0.49

In conclusion, the computer simulation has demonstrated that the back analysis method described here is highly accurate for determining the boundary between the elastic and plastic zones. The method only requires linear elastic finite element analysis, and a microcomputer placed at the construction site is sufficient to monitor the plastic zone occurring around the underground openings (Sakurai and Shinji, 1984).

7.   BACK ANALYSIS FOR NONLINEAR BEHAVIOUR OF SOILS AND ROCKS

7.1 Introduction

The deformational mechanism of soils and rocks may be classified into the following three modes, namely, (1) spalling of joints, (2) sliding along a particular slip surface and (3) plastic flow. And, any one or more of these three modes may occur at the same time. It is possible to analyze all these deformational behaviours of soils and rocks by use of different mechanical models. However, a single model capable of analyzing all these different modes of deformation is recommended, particularly in engineering practice, because of its easy handling. Therefore, the mechanical model adopted for the back analysis must include all three modes of deformation, depending on the initial state of stress, material properties, geological conditions, and so on. In addition, the model should allow for usage of a microcomputer placed at the construction site to monitor the stability of the ground during the construction of the structures.

In this chapter, a back analysis method which can deal with the three different modes of deformation is presented (Sakurai and Ine, 1986). A single linear mechanical model is introduced in the method by which all three modes can be analyzed. A computer simulation is shown to indicate the validity of the method. Case studies are also presented here to demonstrate the applicability of the method to engineering practice.

7.2 Constitutive equation

In order to represent all three modes of deformation, i.e., spalling, sliding and the plastic flow, the following constitutive equation is adopted (Sakurai and Ine, 1986).

The equation is expressed in terms of the x'-y' local coordinates as (see Fig. 22):

$$\{\sigma'\} = [D']\{\varepsilon'\} \tag{30}$$

where

$$[D'] = \frac{E_2}{(1+\nu_1)(1-\nu_1-2n\nu_2^2)} \begin{bmatrix} n(1-n\nu_2^2) & n\nu_2(1+\nu_1) & 0 \\ n\nu_2(1+\nu_1) & 1-\nu_1^2 & 0 \\ 0 & 0 & m(1+\nu_1)(1-\nu_1-2n\nu_2^2) \end{bmatrix} \tag{31}$$

hence, it is transformed into the x-y global coordinates as follows:

$$\{\sigma\} = [D]\{\varepsilon\} \tag{32}$$

where

$$[D] = [T][D'][T]^T \tag{33}$$

[T] is a transformation matrix expressed as:

$$[T] = \begin{bmatrix} \cos^2\alpha & \sin^2\alpha & -2\sin\alpha\cos\alpha \\ \sin^2\alpha & \cos^2\alpha & 2\sin\alpha\cos\alpha \\ \sin\alpha\cos\alpha & -\sin\alpha\cos\alpha & \cos^2\alpha - \sin^2\alpha \end{bmatrix} \tag{34}$$

where α is the angle between the x'- and x- coordinate systems.

It should be noted that Eq. (32) can represent all three modes of deformation by changing the material constants, particularly n and m, which are called anisotropic parameters.

(a) Spalling of joints

The spalling of joints shown in Fig. 22 can be represented by increasing anisotropic parameter n, i.e., by reducing the value of E_2 against E_1 . Poisson's ratio ν_2 is taken to be zero because spalling in the direction of the y'- axis makes no movement in the direction of the x'- axis. In this case, the other anisotropic parameter m must be taken as m = 1/2(1+ν_1).

(b) Sliding along joints or a slip plane

When sliding occurs along the joints parallel to the x'- axis, anisotropic parameter m is reduced to a small value (m<1/2(1+ν_1)), while n = 1.0 and $\nu_1 = \nu_2$.

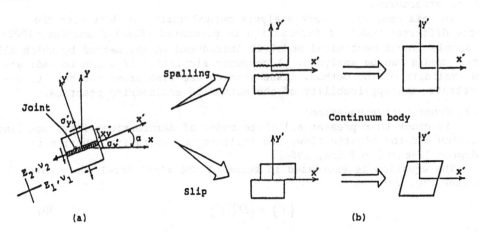

(a)   (b)

Fig. 22   Modelling for discontinuous deformation
           in continuum mechanics

(c)   Plastic flow
      Material under a plastic state tends to slide along the two
families of potential sliding planes with an angle of ± (45° + ϕ /2)
from the maximum principal stress direction, as shown in Fig. 23. ϕ
denotes the internal friction angle.

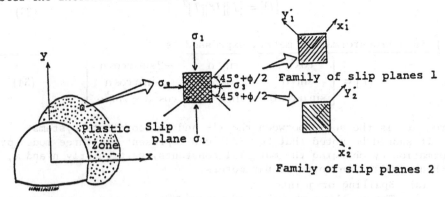

Fig. 23   Families of slip planes in the plastic zone

Let us take two different coordinate systems in order to consider the
two conjugate slip planes, as shown in Fig. 23.   The stress-strain relationship for each family of slip planes is given in the same form as
Eqs. (30) and (32) for the local and global coordinate systems, respec-

tively. The total strain is assumed to be expressed as:

$$\{\varepsilon\} = \frac{1}{2}[\{\varepsilon_1\} + \{\varepsilon_2\}]$$ (35)

where $\{\varepsilon_1\}$ and $\{\varepsilon_2\}$ are strains due to the families of the two slip planes, respectively. Considering Eq. (32), Eq. (35) becomes:

$$\{\varepsilon\} = \frac{1}{2}[[D_1]^{-1} + [D_2]^{-1}]\{\sigma\}$$ (36)

This is a stress-strain relationship for representing the plastic behaviour of soils and rocks. Eq. (36) is expressed in the following common form.

$$\{\sigma\} = [D]\{\varepsilon\}$$ (37)

where

$$[D] = [\frac{1}{2}\{[D_1]^{-1} + [D_2]^{-1}\}]^{-1}$$ (38)

7.3 Computer simulations
     In order to assure the validity of the constitutive equation described above, computer simulations have been conducted for analyzing tunnelling problems.

.3.1 Spalling of joints
     The tunnel under consideration is shown in Fig. 24. A main discontinuous plane is located 4 m above the crown of the tunnel. A joint element (Goodman et al., 1968) is used to represent the mechanical behaviour of the joints. The input data employed in this computer simulation is shown in Table-4.
     The displacement distribution is firstly calculated by an ordinary FE analysis. The displacement along the discontinuous plane is shown in Fig. 25. It clearly shows the spalling of joints due to the tunnel excavation. The calculated displacements along the extensometers shown in Fig. 24 are assumed as measured displacements.
     Back analysis is then carried out to determine the initial stress and material properties under consideration of the measured displacements so as to minimize the following error function.

$$\delta = \sum_{i=1}^{N}(u_i^c - u_i^m)^2$$ (39)

where u_i^c and u_i^m are the calculated and measured displacements, respectively. In the back analysis, two different constitutive equations are used. One is Eq. (32) which takes into account the anisotropic parameter and the other is that for the isotropic elastic material.
     The back analysis results are shown in Fig. 26. It is seen that the back calculated displacements and the measured ones coincide well, while

the isotropic elastic constitutive equation naturally cannot simulate
the spalling of joints.

Table-4   Input data and back analysis results
           in the case of spalling of joints
           (Tensile stress is positive)

| | FE ANALYSIS WITH JOINTS | BACK ANALYSIS | |
		Anisotropic	Isotropic
σ_x kgf/cm²	-25.0	-25.7	-7.7
σ_y kgf/cm²	-50.0	-50.0	-50.0
τ_{xy} kgf/cm²	0.0	0.0	0.0
E kgf/cm²	10000.0	10546.3	6835.0
Poisson's Ratio ν	0.3	0.3 (assumed)	
n	—	10.0	1.0
Data for joints			
Wall rock compressive strength (kgf/cm²)	-70.0		
Ratio of tensile to compressive strength	0.1		
Shear stiffness (kgf/cm³)	3850.0		
Ratio of residual to peak shear strength	0.33		
Maximum normal closure (cm)	10.0		
Seating load (kgf/cm²)	-50.0		
Friction angle of a smooth joint (degree)	30.0		
Dilatancy angle (degree)	0.0		

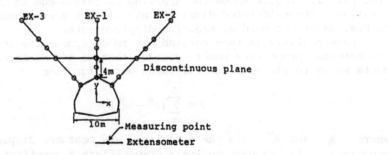

Fig. 24   Tunnel excavated near a discontinuous
           plane and location of the measuring
           points

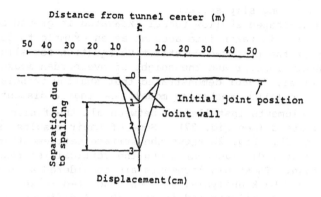

Fig. 25   Spalling of joints

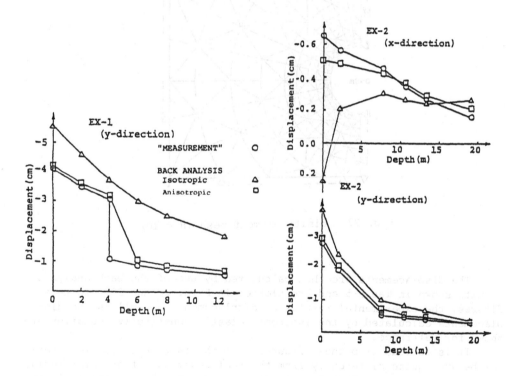

Fig. 26   Comparison between measured and back analyzed displacements

7.3.2 Sliding along slip surfaces
A circular-shaped shallow tunnel excavated in ground consisting of
granular material is taken into account as an example to demonstrate
the validity of the proposed constitutive equation for sliding problems.
Both the tunnel diameter and the height of overburden measure 8 m.
Adachi et al. (1986) demonstrated that shallow tunnels cannot be
analyzed by a continuum mechanics model, and that a discontinuous model
such as joint elements installed in between all the finite elements is
recommended instead (see Fig. 27). Some of their results are indicated
in Figs. 28 and 29. Fig. 28 shows the surface settlement and Fig. 29
is for the vertical displacement along the vertical reference line above
the tunnel crown. These displacements are considered as measured displacements, and a back analysis has been conducted in which the slip surfaces are assumed as illustrated in Fig. 30. A different anisotropic
parameter m is given for each slip zone. The results of the back analysis are shown in Table-5.

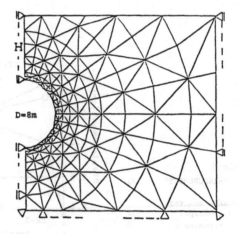

Fig. 27   Finite element mesh (H = 1D)

The displacements are then calculated by use of the back analyzed
results shown in Table-5 and are plotted in Figs. 28 and 29. In these
figures, the experimental results of displacements as well as the displacements calculated by the isotropic elastic constitutive equation are
shown for reference.
It is obvious from these figures that the isotropic elastic material behaves quite differently from the real behaviour of granular media,
while the proposed constitutive equation is acceptable for the analysis
of shallow tunnels. The joint element model proposed by Adachi et al.
seems to be reasonable. However, the physical meaning of the elements
is ambiguous for representing the behaviour of granular materials.

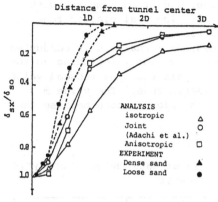

δ_{sx}:surface settlement
δ_{so}:surface settlement at tunnel center

Fig. 28   Surface settlement due
           to tunnel excavation

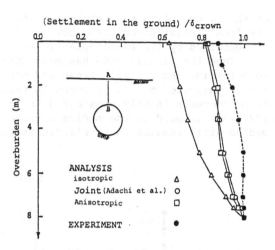

Fig. 29   Distribution of vertical
           displacement along line
           AB above tunnel crown

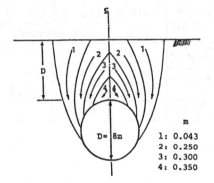

Fig. 30   Shallow tunnel excavated
           in sandy ground with
           slip zones

Table-5   Back analysis results

Unit weight(tf/m³)	2.00
Young's modulus (tf/m²)	2,000
Poisson's ratio	1/3
Coefficient of earth pressure at rest K_0	0.5
m_1	0.043
m_2	0.250
m_3	0.300
m_4	0.350

7.3.3 Plastic flow

It is assumed that a horseshoe-shaped tunnel, 5 m in radius, is
bored in an elasto-plastic medium (see Fig. 31). Elasto-plastic FE
analysis is firstly conducted by assuming the Drucker-Prager yielding
function and von Mises type of plastic potential function. The initial
stress and material properties employed here are listed in Table-6. The
calculated displacements along three reference lines (see Fig. 31) are
taken as measured displacements which are used as input data for back
analysis.

A plastic zone appearing around the tunnel is then back calculated

by the method already described in Chapter 6, and it is compared with
the 'real' plastic zone, as shown in Fig. 31. The real plastic zone is
one obtained by the elasto-plastic FE analysis.

Once the plastic zone has been obtained, back analysis is conducted
to determine the initial stress and material constants, including anisotropic parameter m. It is noted in this analysis that the reduced value
of parameter m is only considered in the plastic zone. The value m =
1/2(1+ν) is used in the medium outside of the plastic zone because the
medium still remains in an elastic state.

Table-6   Input data and back analysis
           results

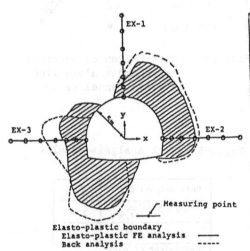

Elasto-plastic boundary
Elasto-plastic FE analysis   ———
Back analysis   -----

	ELASTO-PLASTIC FE ANALYSIS	BACK ANALYSIS	
		Anisotropic	Isotropic
σ_x kgf/cm^2	−25.0	−24.4	−28.4
σ_y kgf/cm^2	−35.0	−35.0	−35.0
τ_{xy} kgf/cm^2	8.66	11.26	10.02
E kgf/cm^2	2000.0	1961.0	1257.4
Poisson's Ratio ν	0.3	0.3 (assumed)	
Cohesive C Strength (kgf/cm^2)	5.0	———	———
Internal Friction Angle φ (degree)	30.0	30.0	———
m	———	0.0962	1/2(1+ν) =0.3846
Critical Strain ε_0 %	———	0.866	———

Fig. 31   Horseshoe-shaped tunnel excavated
          in elasto-plastic ground

The back analyzed displacements along the three reference lines are
indicated in Fig. 32 together with the measured displacements. In this
figure, the back analysis results obtained on the assumption of an isotropic elastic material are also shown for reference. It seems from this
figure that the displacements obtained by the back analysis which adopts
anisotropic parameters coincide well with the measured displacements.

7.4 Loosened and plastic zones occurring around underground openings
When excavating underground openings, like tunnels, a loosened zone
and/or plastic zone may occur in the vicinity of the excavated free surface depending on geological conditions, the joint system, the mechanical
characteristics of the soils/rocks, the initial stress and even on the
excavation method. A loosened zone may be defined as a zone in which
discontinuities tend to open or slide along certain slip planes due to
stress relief caused by excavation. On the other hand, a plastic zone

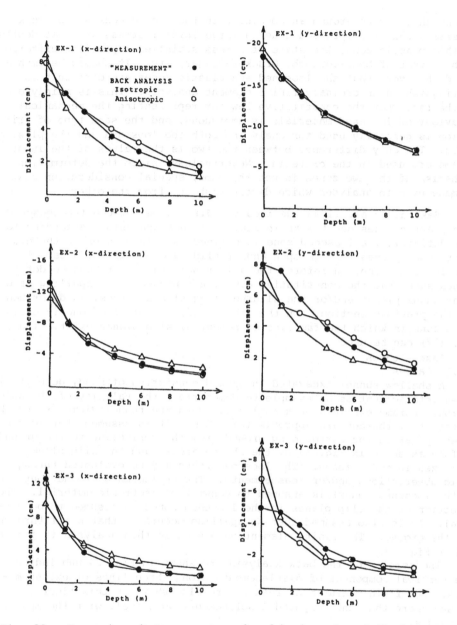

Fig. 32   Comparison between measured and back analyzed displacements

is one that occurs around an opening when the ground consists of weak
materials and is compressed under a large initial stress at great depths.
In this plastic zone, the state of stress satisfies a yielding criterion
such as that of Mohr-Coulomb, Drucker-Prager, etc. It should be emphasized, however, that the loosened and plastic zones are difficult to
distinguish in an ordinary finite element analysis. This is mainly due
to the fact that the constitutive law for representing the mechanical
behaviour of loosened materials is ambiguous, and the same type of criterion is generally used for analyzing both the loosened and the plastic
zones. The only difference between the two is the values of the input
parameters used in the criteria. Nevertheless, since the deformational
mechanism of the two zones is not the same, special consideration must
be made even in analyses which distinguish one from the other.

Sakurai et al. (1988) defined the difference between the loosened
and plastic zones dealt with in Finite Element analysis. According to
the definition, a loosened zone is defined as a zone in which the discontinuous planes tend to open and/or slide along slip surfaces. In
the loosened zone, therefore, the anisotropic parameters can be determined such that the constitutive equation represents the spalling of discontinuous planes and/or the sliding along slip surfaces, as described
in the previous section. On the other hand, the plastic zone is defined
as a zone in which two families of potential slip planes exist. Thus,
Eq. (37) can be used.

7.5 Case studies

7.5.1 Case-A

A shallow tunnel excavated in ground consisting of sandy deposits is
taken into account as an example to demonstrate the validity of the anisotropic parameters in a back analysis. Both the tunnel diameter and the
height of overburden are approximately 10 m. It is assumed that loosened
zones appear in the surrounding ground from the springline to the ground
surface as shown in Fig. 33 by the shaded areas, and the slip planes are
mobilized in these zones. The loosened zones may be evaluated through
field observations and/or measurements. The un-loosened ground (outside
of the loosened zones) is assumed to consist of isotropic material. The
direction of the slip planes in the loosened zones is assumed to be vertical. It is also assumed that no spalling occurs so that n = 1 throughout the ground. The finite element mesh used in this analysis is indicated in Fig. 33.

The results of the back analysis are shown in Fig. 34 and indicate
the vertical component of displacements along three lines of extensometers. The field measurement results are also shown in this figure. It
is seen here that the computed displacements agree well with the measured values.

The maximum shear strain distribution is given in Fig. 35. It should
be noted that the strain distribution is largely different from the one
derived by the ordinary isotropic elastic analysis. One of the direct
field measurement results for maximum shear strain distribution occurring
around a circular tunnel is shown in Fig. 36 for reference (Hansmire and
Cording, 1985). It is interesting that a similar strain distribution
appears in both Figs. 35 and 36.

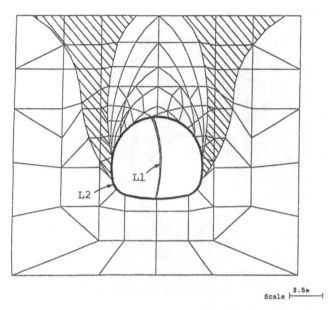

Scale |—— 3.5m ——|

Fig. 33   Finite element mesh and loosened zones

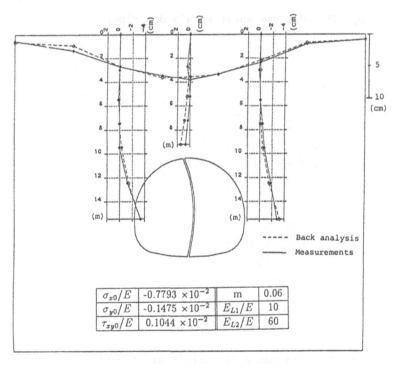

| ----- | Back analysis |
| —— | Measurements |

σ_{x0}/E	-0.7793×10^{-2}	m	0.06
σ_{y0}/E	-0.1475×10^{-2}	E_{L1}/E	10
τ_{xy0}/E	0.1044×10^{-2}	E_{L2}/E	60

Fig. 34   Back analysis results and comparison between
           computed and measured displacements
           (Tensile stress is positive)

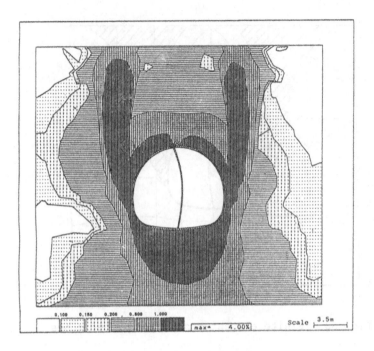

Fig. 35   Maximum shear strain distribution

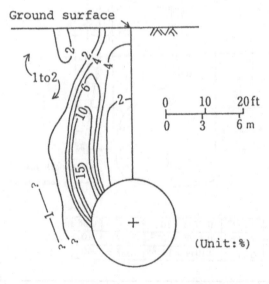

Fig. 36   Maximum shear strain distribution
           (after Hansmire and Cording)

7.5.2 Case-B

     For the second example of a demonstration showing the validity of
the anisotropic parameters adopted in back analysis, a large underground
powerhouse cavern is shown. The ground in which the cavern was excavated
consists of mainly tuff breccia and andesite. The dimentions of the cavern are given in Fig. 37. Displacement measurements were conducted during the excavation by use of multi-rod extensometers installed from the
inner surface of the cavern. The locations of the extensometers are shown
in Fig. 38 along with the measurement data (Kamemura et al., 1986).

     The finite element mesh used in the back analysis is shown in Fig.
39. The back analysis is conducted with anisotropic parameters. Both
sliding along potential slip planes and spalling of joints are taken into
account. Therefore, both the anisotropic parameters m and n are back
calculated (Sakurai and Tanigawa, 1989).

     Fig. 40 indicates the loosened zones occurring around the cavern
which are assumed by considering the results of displacement measurements. The results of the back analysis are shown in Fig. 41. The anisotropic parameters m = 0.038 and n = 20 are obtained. In this figure,
the displacements calculated by use of the results obtained from the
back analysis are compared with the measured values. It is seen from
the figure that the values coincide well with each other. The maximum
shear strain distributions in this case become as shown in Fig. 42.

     For reference, the results of the back analysis conducted by assum-
ing the isotropic constitutive equation are shown in Fig. 43. The measured values are also indicated in this figure for reference.

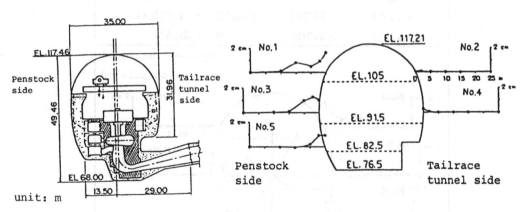

Fig. 37   Cross section of the cavern   Fig. 38   Displacements measured by
                                                         multi-rod extensometers

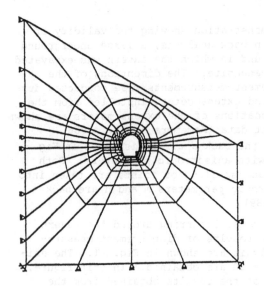

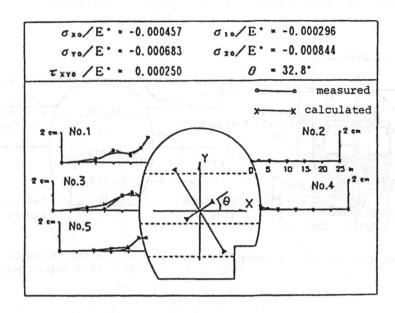

Fig. 39   Finite element mesh

Fig. 40   Loosened zones occurring
           around the cavern

Fig. 41   Results of back analysis
           (anisotropic model)
           (tensile stress is positive)

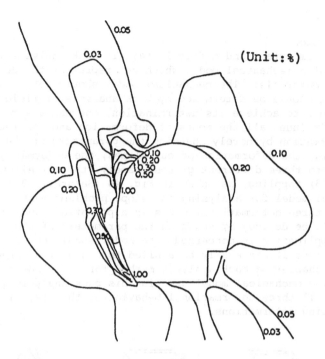

Fig. 42   Maximum shear strain distribution

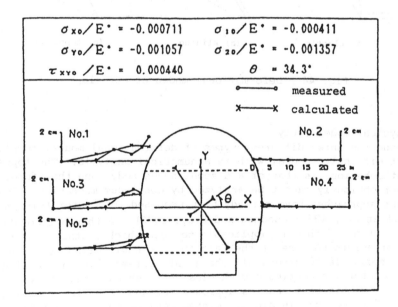

Fig. 43   Results of back analysis
          (isotropic model)

8. CUT SLOPES

8.1 Introduction
     The most important and difficult task in back analysis is the de-
termination of a mechanical model which can represent the deformational
behaviour of materials. The mechanical model should not be assumed beforehand, but should be determined by back analysis of field measurement
data. In order to achieve its determination, one can assume a mechanical
model which includes all the modes of deformation and can represent any
type of deformation by merely changing the parameters of the model.
     Concerning the deformation of cut slopes, the deformational mode is
classified into three different groups, namely, the (1) elastic, (2)
sliding and (3) toppling, as shown in Fig. 44 (Sakurai, 1987). Therefore,
the mechanical model for analyzing cut slope problems must be one which
includes all three deformational modes as a potential, and one or more of
the modes will be derived by changing the parameters of the model. As
far as the application is concerned, the model should be as simple as
possible, so that it can easily be applied to engineering practice.
     In this chapter, a back analysis method for cut slope problems is
described. The mechanical model used in this back analysis is capable of
dealing with all three deformational behaviours, that is, elastic, sliding and toppling deformations.

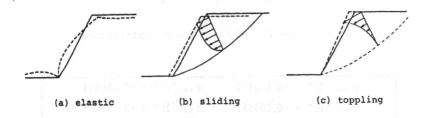

(a) elastic   (b) sliding   (c) toppling

Fig. 44   Deformational modes of cut slopes

8.2 Physical model study
     Among the three different types of deformational modes, toppling is
the most difficult to deal with in a numerical analysis. The toppling's
inherent behaviour is that of a discontinuous body, and therefore, its
behaviour obviously cannot be analyzed by continuum mechanics. Several
different approaches have already been proposed to counter this problem
(Cundall et al., 1975; Goodman and Bray, 1976). Although theoretically
possible, none of them is applicable to rock which contains an infinite
number of joints because of the fact that in engineering practice it is
almost impossible to detect all the joint systems. In engineering practice, therefore, a continuum mechanics approach may still be preferable.
To adopt a continuum mechanics approach for analyzing the toppling behaviour of slopes cut in highly jointed rock masses, the deformational
mechanism must first be thoroughly investigated.
     In an effort to tackle toppling, a physical model study has been

performed. For this study, a block model consisting of·a large number of aluminum bars has been used. The experiments are performed by firstly stacking the aluminum bars on the platform and then removing some blocks of bars step by step to simulate cut slopes. The test results reveal that when a small movement of the toppling occurs, continuous movement behaviour of the model still exists (Deeswasmongkol and Sakurai, 1985, 1986). This means that even for such a discontinuous body as the model consisting of a large number of aluminum bars, a continuum mechanics approach may be applied to analyze its deformational behaviour. However, when applying this approach, a constitutive equation for the materials must be carefully investigated.

8.3 Constitutive equation

From careful investigations of the deformational behaviour of the discontinuous model, it has been found that the shear deformation of the material above the base line takes place easily, and the largest shear deformation occurs almost parallel to the direction of the cross joints. The base line is defined as a line under which no displacement occurs at ι, and may be slightly steeper than the angle of the cross joints.

Considering these deformational behaviours of the jointed material, an anisotropic constitutive equation is assumed, which has the smallest shear rigidity in the direction parallel to the cross joints (Sakurai, Deeswasmongkol and Shinji, 1986). Letting the local coordinate system x'-y' be taken as shown in Fig. 45, where the x'- axis is parallel to the direction of the cross joints, the anisotropic constitutive equation can then be described as follows:

$$\left\{ \begin{array}{c} \sigma_{x'} \\ \sigma_{y'} \\ \tau_{x'y'} \end{array} \right\} = [D'] \left\{ \begin{array}{c} \varepsilon_{x'} \\ \varepsilon_{y'} \\ \gamma_{x'y'} \end{array} \right\} \tag{40}$$

where

$$[D'] = \frac{E}{1 - \nu - 2\nu^2} \begin{bmatrix} 1 - \nu & \nu & 0 \\ \nu & 1 - \nu & 0 \\ 0 & 0 & m(1 - \nu - 2\nu^2) \end{bmatrix} \tag{41}$$

$\{\sigma_{x'} \; \sigma_{y'} \; \tau_{x'y'}\}^T$ and $\{\varepsilon_{x'} \; \varepsilon_{y'} \; \gamma_{x'y'}\}^T$ are stress and strain in the x'-y' coordinate system, respectively. E is Young's modulus, ν is Poisson's ratio and m is a parameter representing the anisotropy of the material characteristics. When m = $1/2(1+\nu)$, the equation turns to the one for an isotropic material. It should be noted that the constitutive equation, Eq. (40), can be derived from Eq. (30) by substituting the anisotropic parameter n = 1.

When the constitutive equation for the local coordinate system is known, it is easy to extend it to the x-y global coordinate system, that is,

$$\left\{ \begin{array}{c} \sigma_x \\ \sigma_y \\ \tau_{xy} \end{array} \right\} = [D] \left\{ \begin{array}{c} \varepsilon_x \\ \varepsilon_y \\ \gamma_{xy} \end{array} \right\} \tag{42}$$

where

$$[D] = [T][D'][T]^T \qquad (43)$$

where [T] is a transformation matrix given as a function of the angle α between the x- and x'- axies, as shown in Eq. (34).

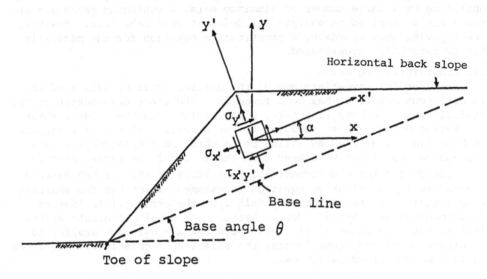

Fig. 45   Coordinate system

8.4 Mechanical model of slopes

In the analysis of cut slope problems, the material above the base line is divided into N layered elements of zones parallel to the base line, as shown in Fig. 46. The model may be extended to more general slope problems with a curved failure plane using curved layers. It is assumed that each layer has a different value for m and α , but the same value for E and ν . It is also assumed that the material below the base line behaves as a homogeneous, isotropic material, so that only two material constants (E_0 and ν_0) exist.

8.5 Determination of mechanical constants and initial stress

The location and inclination of the base line can be evaluated through the careful investigation of observations/measurements taken during the cutting of slopes. The number of layers above the base line can also be estimated by considering the results of measurements, depending on the displacement distributions along the vertical axis. The material constants (E, ν , m_1 , ..., m_N), as well as the initial stress existing in the ground prior to slope excavation, are obtained so as to minimize the following error function:

$$\delta = \sum_{i=1}^{M} (u_i^c - u_i^m)^2 \rightarrow min. \tag{44}$$

where,

u_i^c :   calculated displacement at measurement point i

u_i^m :   measured displacement at measurement point i

M   :   number of measurement points

For this minimization analysis, computer programs such as Simplex, Rosenbrock, etc., supplied in the program library, can easily be used.

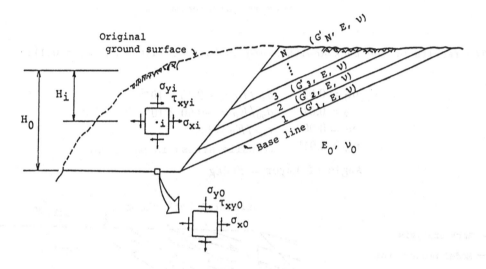

Fig. 46   Layered elements of zones
parallel to the base line

8.6 Applicability of the back analysis method to toppling problems

In order to verify the applicability of the back analysis method to
the toppling of cut slopes, the method is applied to the analysis of the
displacements measured at the physical model study described earlier
(Sakurai, Deeswasmongkol and Shinji, 1986). The medium above the base
line is divided into four layers, as shown in Fig. 47, so that the anisotropic parameters m_1, m_2, m_3 and m_4 and the normalized initial stresses
σ_{x0}/ E , σ_{y0}/ E , τ_{xy0} / E must be back analyzed. The base angle is assumed to be 30° considering the results of the physical model
study.

The displacements along line A-B of the model shown in the figure
are taken to be the measured displacements which are used as input data
for the back analysis.

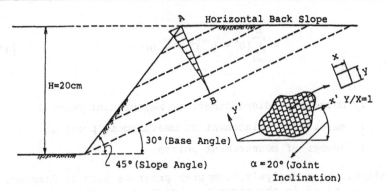

Fig. 47   Mechanical model and
           measured displacements

The results of the back analysis are as follows (assuming $\nu = 0.1$):

$$m_1 = 0.01 \qquad \sigma_{x0}/E_R = -2.15 \times 10^{-3}$$
$$m_2 = 0.01 \qquad \sigma_{y0}/E_R = -0.60 \times 10^{-3}$$
$$m_3 = 0.008 \qquad \tau_{xy0}/E_R = 0.0$$
$$m_4 = 0.015 \qquad \text{(tensile stress is positive)}$$

Angle of Layer = 20 deg.

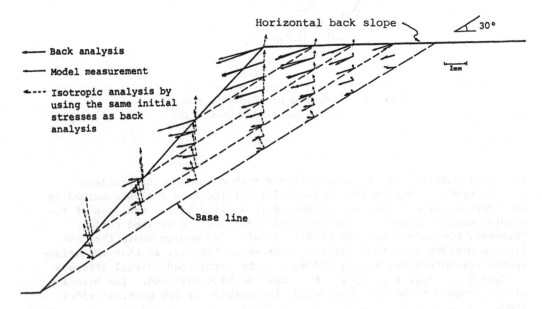

Fig. 48   Comparison between calculated and measured displacements

    The values of the initial stress and material constants determined
by back analysis are used as input data for ordinary FEM, and the overall
displacement distribution of the model is obtained. The computed displacements are plotted in Fig. 48 and are compared with the measured
values. From this figure one can see that the computed and measured displacements agree well. In the same figure, the displacement distribution
of a slope consisting of homogeneous, isotropic elastic materials is also
shown. It is interesting to note that an extremely large difference appears between the displacements obtained from the anisotropic model and
those obtained from the homogeneous, isotropic model. It is surprising
to know that the continuum mechanics approach described here is quite
applicable to the analysis of the toppling behaviour of such a highly
jointed material.

8.7 Practical application

8.7.1 Case study (A)
    A cut slope appeared adjacent to the portal of a highway tunnel.
The stability of the slope became a serious problem, and therefore, field
measurements were performed to monitor the slope during excavation.
    The casing tube for a borehole inclinometer was installed prior to
excavation 2 m apart from the slope surface and 9 m below the floor of
the excavation, as shown in Fig. 49. The geological formation of the
ground consists of horizontal layers of sand and gravel.

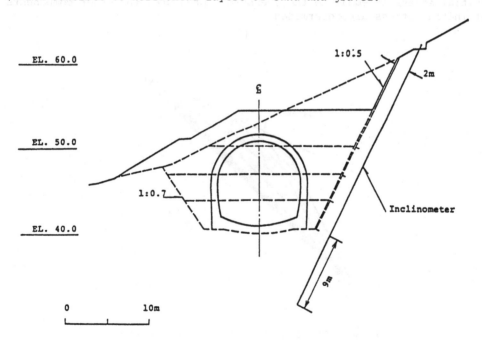

Fig. 49   Configuration of slope and location of inclinometer

The displacements due to the cutting of the slopes were measured by the
inclinometer, and the results were used for a back analysis to determine
the initial stress and mechanical constants. As mentioned earlier, the
back analysis requires that firstly the configuration of the potential
sliding surface be assumed. This can be done by a careful investigation
of geological conditions, as well as by field observational measurements.
In this case study, a flat plane is assumed, and the ground above this
potential sliding plane is divided into three layers, as shown in Fig.
50. Each layer may have a different value for the anisotropic parameters.
All the material constants, including the anisotropic parameters as well
as the initial stress existing before excavation, are then back calculated from measured displacements. The results are as follows (Kondoh and
Shinji, 1986):

$$m_1 = 0.385 (isotropic) \qquad \sigma_{x0}/E_R = -0.274 \times 10^{-2}$$
$$m_2 = 0.385 (isotropic) \qquad \sigma_{y0}/E_R = -0.478 \times 10^{-2}$$
$$m_3 = 0.025 \qquad \qquad \tau_{xy0}/E_R = -0.113 \times 10^{-2}$$
$$\nu = 0.3 \ (\mathbf{assumed})$$

where σ_{x0} , σ_{y0} , τ_{xy0} are the components of initial stress acting at the toe of the slope. Assuming that the vertical component of
initial stress is equal to the overburden pressure, the other components
of initial stress are determined.

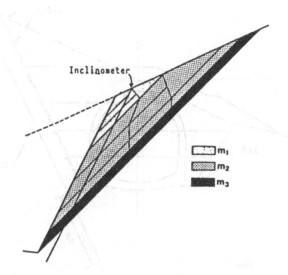

Fig. 50   Three layers of block above the
           potential sliding surface

Once all these values are known, one can calculate the displacements caused by the cutting of slopes by means of an ordinary finite element analysis. Then they can be compared with the measured values to verify the accuracy of the back analysis. Fig. 51 illustrates this comparison and shows that a good agreement exists between the measured and calculated displacements. In this figure, the results for the isotropic elastic material are also shown for reference. It is seen from this figure that only a small discrepancy appears between the results obtained by the isotropic model and those obtained by the anisotropic model. This means that the behaviour of this cut slope is similar to that of isotropic elastic materials. Thus, the slope is classified as being of an elastic type. However, the maximum shear strain distribution shown in Fig. 52 demonstrates that the potential sliding surface seemingly starts to occur, although it is not too serious.

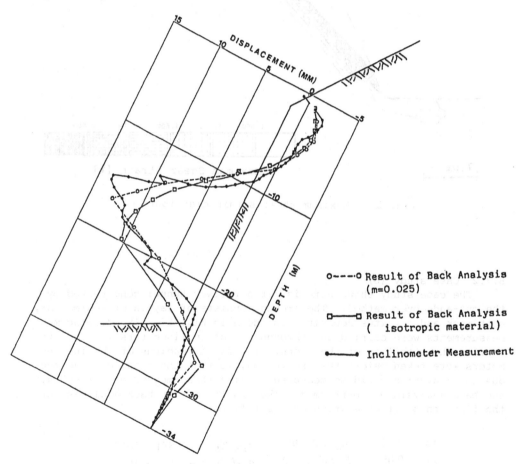

Fig. 51   Comparison between measured calculated displacements

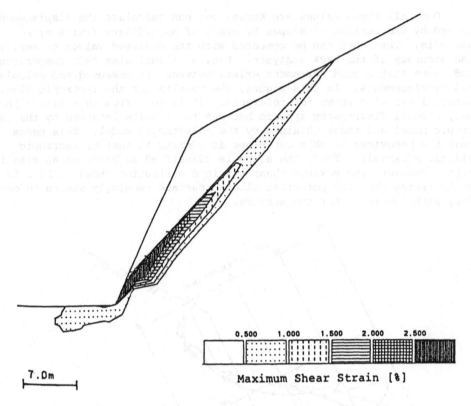

Fig. 52   Maximum shear strain distribution

8.7.2 Case study (B)

The case study shown here is for a railway tunnel constructed by the cut-and-cover method. The ground consists of layers of sedimentary soft rock and detritus deposits, as shown in Fig. 53. The displacement measurements were carried out through use of inclinometers in order to monitor the stability of the slope. The first readings of the inclinometers were taken before the excavation, so that the total displacements due to excavation could be measured. Using these measurement results, the back analysis was performed. The results of the back analysis for the final phase of excavation are as follows:

$$m_1 = 0.01 \quad\quad m_5 = 0.006 \quad\quad \sigma_{x0}/E_R = -0.871 \times 10^{-2}$$
$$m_2 = 0.05 \quad\quad E_L/E = 400 \quad\quad \sigma_{y0}/E_R = -0.148 \times 10^{-2}$$
$$m_3 = 0.1 \quad\quad\; E_B/E = 0.01 \quad\quad \tau_{xy0}/E_R = 0.473 \times 10^{-3}$$
$$m_4 = 0.004 \quad\quad\quad\quad\quad\quad\quad\quad\;\; \nu = 0.3 \;\textbf{(assumed)}$$

These results are used as input data for determining the displacements by an ordinary finite element method.

The back analyzed displacements are compared with the measured values, as shown in Fig. 54, and indicate a good agreement between the two. This proves that the anisotropic model proposed for analyzing the toppling of cut slopes is also applicable for analyzing the sliding mode of cut slope deformations.

The maximum shear strain distribution is given in Fig. 55, and it is obvious that there is a clear potential sliding surface occurring along one of the layers of sedimentary rock. Fig. 55 also reveals the fact that a large strain occurs at the floor of the excavation. Thus, it has been recommended that floor concrete be placed immediately and that rock anchors be installed in a downward direction so as to prevent a heave of the ground. These support measures will increase the stability of the slope tremendously.

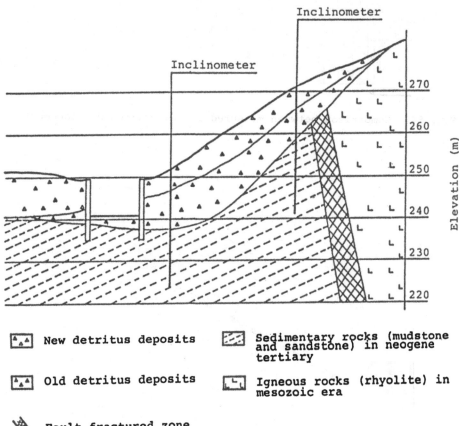

New detritus deposits   Sedimentary rocks (mudstone and sandstone) in neogene tertiary

Old detritus deposits   Igneous rocks (rhyolite) in mesozoic era

Fault fractured zone

Fig. 53   Geological conditions

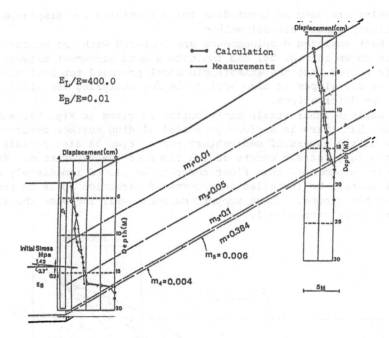

Fig. 54   Comparison between measured and calculated displacements

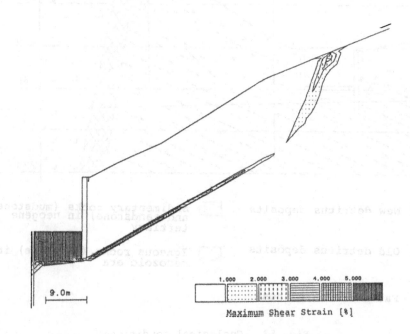

Fig. 55   Maximum shear strain distribution

9. CONCLUDING REMARKS

It should be emphasized here that field measurements performed during the construction of geo-structures such as tunnels, underground caverns, cut slopes, etc., are of extreme importance. The purpose of these measurements is to monitor the stability of the structures, as well as to re-evaluate the various mechanical parameters such as Young's modulus, Poisson's ratio, cohesion and the friction angle, which have been used in the design works. It is obvious that the field measurement results must be properly interpreted in order to achieve reliable monitoring and to provide useful information for re-evaluating the design parameters. To accomplish this, back analysis must be a powerful tool and should be investigated carefully.

In this paper, various methods of back analysis mainly proposed by the author and his co-workers have been presented. The back analysis methods described here are formulated on the basis of continuum mechanics, such that they can be applied to a continuous and/or pseudo continuous type of ground which represents soils and/or highly jointed rock masses.

The most important yet difficult task in back analysis is the determination of a mechanical model to represent the behaviour of soils and rocks. The mechanical model should not be assumed, but should be determined by a back analysis. In this paper, a method of determination has been described with particular reference to cut slopes.

ACKNOWLEDGEMENTS

The author wishes to thank Mr. N. Shimizu for his assistance in the preparation of this paper. Special thanks are also due to Ms. H. Griswold for proofreading and typing this manuscript.

REFERENCES

1. Terzaghi, K. and R.B. Peck: Soil Mechanics in Engineering Practice, John Wiley & Sons, 1948, 627-632.
2. Sakurai, S., T. Ine and M. Shinji: Finite element analysis of discontinuous geological materials in association with field observations, Proc. 6th Int. Conf. Numerical Methods in Geomech., Innsbruck, 3 (1988), 2029-2034.
3. Cundall, P.A.: A Computer model for simulating progressive large-scale movements in blocky rock systems, Proc. Sympo. Int. Soc. Rock Mech., Nancy, II (1971), Art 8.
4. Kawai, T.: Some considerations on the finite element method, Int. J. Numerical Methods in Engineering, 16 (1980), 81-120.
5. Cividini, A., L. Jurina and G. Gioda: Some aspects of 'characterization' problems in geomechanics, Int. J. Rock Mech. Min. Sci. & Geomech. Abstr., 18 (1981), 487-503.
6. Kavanagh, K.: Experiment versus analysis: Computational techniques for the description of static material response, Int. J. Numerical Methods in Engineering, 5 (1973), 503-515.
7. Gioda, G. and G. Maier: Direct search solution of an inverse problem in elastoplasticity: Identification of cohesion, friction angle and in-situ stress by pressure tunnel tests, Int. J. Numerical Methods in

Methods in Engineering, 15 (1980), 1823-1848.

8. Asaoka, A. and M. Matsuo: Bayesian approach to inverse problem in consolidation and its application to settlement prediction, Proc. 3rd Int. Conf. Numerical Methods in Geomechanics, Aachen, 1 (1979), 115-123.

9. Cividini, A., G. Maier and A. Nappi: Parameter estimation of a static geotechnical model using a Bayes' approach, Int. J. Rock Mech. and Mining Sciences, 20 (1983), 215-226.

10. Murakami, A. and T. Hasegawa: Observational prediction of settlement using Kalman filter theory, Proc. 5th Int. Conf. Numerical Methods in Geomech., Nagoya, 3 (1985), 1637-1643.

11. Sakurai, S.: Direct strain evaluation technique in construction of underground openings, Proc. 22nd U.S. Sympo. Rock Mech., MIT, 1981, 278-282.

12. Sakurai, S.: Monitoring of caverns during construction period, Proc. ISRM Sympo. Rock Mech.: Caverns and Pressure Shafts, Aachen, 1982, 433-441.

13. Sakurai, S.: Displacement measurements associated with the design of underground openings, Proc. Int. Sympo. Field Measurements in Geomechanics, Zurich, 2 (1983), 1163-1178.

14. Sakurai, S. and K. Takeuchi: Back analysis of measured displacements of tunnels, Rock Mech. and Rock Engineering, 16 (1983), 173-180.

15. Sakurai, S. and M. Shinji: A monitoring system for the excavation of underground openings based on microcomputers, Proc. ISRM Sympo. Design and Performance of Underground Excavations, Cambridge, 1984, 471-476.

16. Sakurai, S.: Strain distribution around tunnels determined by displacement measurements, Proc. 5th Int. Sympo. Deformation Measurement and 5th Canadian Sympo. Mining Surveying and Rock Deformation Measurements, Fredericton, New Brunswick, 1988, 451-461.

17. Kovari, K., Ch. Amstad and J. Koeppel: New developments in the instrumentation of underground openings, Proc. 4th Rapid Excavation and Tunnelling Conference, Atlanta, 1979.

18. Koeppel, J., Ch. Amstad and K. Kovari: The measurement of displacement vectors with the 'Trivec' borehold probe, Proc. Int. Sympo. Field Measurements in Geomechanics, Zurich, 1 (1983), 209-218.

19. Shimizu, N. and S. Sakurai: Application of boundary element method for back analysis associated with tunnelling problems, Proc. 5th Int. Conf. Boundary Elements, Hiroshima, 1983, 645-654.

20. Gioda, G. and L. Jurina: Numerical identification of soil-structure interaction pressures, Int. J. Numer. Anal. Methods Geomech., 5 (1981), no. 1, 33-56.

21. Sakurai, S. and N. Shimizu: Back analysis method for measured displacements of underground openings by using 3-D boundary element method, Proc. Int. Conf. Numerical Methods in Geomechanics, Vysoke Tatry, 2 (1987), 153-159.

22. Sakurai, S. and N. Shimizu: Initial stress back analyzed from displacements due to underground excavations, Proc. Int. Sympo. Rock Stress and Rock Stress Measurements, Stockholm, 1986, 679-686.

23. Sakurai, S., N. Shimizu and K. Matsumuro: Evaluation of plastic zone around underground openings by means of displacement measurements, Proc. 5th Int. Conf. Numerical Methods in Geomechanics, Nagoya, 1

(1985), 111-118.

24. Sakurai, S. and T. Ine: Strain analysis of jointed rock masses for monitoring the stability of underground openings, Sympo. Computer Aided Design and Monitoring in Geotechnical Engineering, Bangkok, 1986, 599-608.

25. Goodman, R.E., R.L. Taylor and T.L. Brekke: A model for the mechanics of jointed rock, J. Soil Mech. Found. Div., ASCE, 94 (1968), no. 5143, 637-659.

26. Adachi, T., T. Tamura, A. Yashima and H. Ueno: Surface subsidence above shallow sandy ground tunnel, Proc. Japan Society of Civil Engineers, No. 370/III-5 (1986), 85-94 (in Japanese).

27. Hansmire, W.H. and E.J. Cording: Soil tunnel test section: Case history summary, J. Geotech. Eng., ASCE, 111 (1985), 1301-1320.

28. Kamemura, K., N. Homma, K. Shibata, T. Harada and S.W. Soetomo: Observational method on large rock cavern excavation, Proc. Int. Sympo. Large Rock Caverns, Helsinki, 2 (1986), 1503-1512.

29. Sakurai, S. and M. Tanigawa: Back analysis of deformation measurements in a large underground cavern considering the influence of discontinuity of rocks, Proc. Japan Society of Civil Engineers, 403/VI-10 (1989), 75-84 (in Japanese).

30. Sakurai, S.: Interpretation of the results of displacement measurements in cut slopes, Proc. 2nd Int. Sympo. Field Measurements in Geomechanics, Kobe, 2 (1988), 1155-1166.

31. Cundall, P., M. Voegele and C. Fairhurst: Computerized Design of Rock Slopes Using Interactive Graphics, Proc. 16th Sympo. Rock Mechanics, Design Methods in Rock Mechanics, Minneapolis, 1975, 5-14.

32. Goodman, R.E. and J.W. Bray: Toppling of Rock Slopes, Proc. Specialism Conference Rock Engineering for Foundations and Slopes, Boulder, Colorado, ASCE, 2 (1976), 201-233.

33. Deeswasmongkol, N. and S. Sakurai: Study on rock slope protection of toppling failure by physical modelings, Proc. 26th US Sympo. Rock Mech., Rapid City, SD, 1 (1985), 11-18.

34. Sakurai, S., N. Deeswasmongkol and M. Shinji: Back analysis for determining material characteristics in cut slopes, Proc. Int. Sympo. Engineering in Complex Rock Formations, Beijing, 1986, 770-776.

35. Kondoh, T. and M. Shinji: Back analysis of assessing for slope stability based on displacement measurements, Proc. Int. Sympo. Engineering in Complex Rock Formations, Beijing, 1986, 809-815.

Printed in the United States
By Bookmasters